The
Uncertain
Sciences

The Uncertain Sciences

Bruce Mazlish

With a new introduction by the author

Routledge
Taylor & Francis Group

LONDON AND NEW YORK

Originally published in 1998 by Yale University Press

Published 2007 by Transaction Publishers

Published 2017 by Routledge
2 Park Square, Milton Park, Abingdon, Oxon OX14 4RN
711 Third Avenue, New York, NY 10017, USA

Routledge is an imprint of the Taylor & Francis Group, an informa business

New material this edition copyright © 2007 by Taylor & Francis.

Notice:
Product or corporate names may be trademarks or registered trademarks, and are used only for identification and explanation without intent to infringe.

Library of Congress Catalog Number: 2006053017

Library of Congress Cataloging-in-Publication Data

Mazlish, Bruce, 1923-
    The uncertain sciences / Bruce Mazlish.
        p. cm.
    Includes bibliographical references and index.
    "With a new introduction by the author."
    ISBN 978-1-4128-0630-5 (alk. paper)
        1. Social sciences. I. Title.

H85.M394 2007
300- dc22

                                                                    2006053017
ISBN 13: 978-1-4128-0630-5 (pbk)

Contents

Introduction to the Transaction Edition

I

It is almost a decade ago that this book was first published. The intervening years have been troubled ones. Many political conflicts have flared up across the world, and have taken many lives. We have experienced the threat and reality of diseases—AIDS, SARS, avian flu—flowing across borders. Globalization, while bringing benefits, has also brought increased hardship and divisions in different parts of the world, as well as arousing forces of anti-globalization. A resurgence of fundamentalist religions, challenging Enlightenment values of reason and peace, has often merged with terrorist movements, adding an increased sense of anxiety to contemporary societies. One could go on in this vein, but the examples given are sufficient to convey the conclusion that ours is a deeply troubled and insecure world.

Do the "uncertain sciences" have anything to contribute in this situation? Those that sit on the natural science side of the spectrum keep achieving, but often their results seem to produce new problems along with any gains. A prime example would be molecular biology's new cures on one side and the potential dangers of genetically modified food on the other. As for the human sciences, our major concern in the book, their instrumental advantages appear slim, and they frequently seem to play an ideological rather than an ameliorative role. Such is the short-term and low-level picture of the happenings of the last number of years.

The short-term and instrumental is not, however, the ground on which the original book was written, and it is not the frame in which I am writing the present introduction. My subject from the beginning has concerned the kind of knowledge that could be expected from the human sciences. In pursuing my inquiry, I asked how "scientific" such sciences might be, and what could be meant by "science" in this case. That we need better knowledge in regard to human problems, if it can be obtained, hardly needs saying. That we have made major advances in our understanding of the human species also seems to me a "fact." The core problem, it appears, is how to implement the scientific insights we have achieved in respect to our everyday lives.

I must repeat, emphatically, that "science" is not the only form of knowledge humans can seek. There are many other "knowledges": artistic, religious, intuitive, to cite a few. Further, there are many directions in which humanity is headed, not just one. Some of these may be served by scientific knowledge, others less so. My concern, however, is with knowledge gained through the application of what I call the "scientific method," whose nature I attempted to spell out in the book. My framework is an evolutionary one, and I was, and am, focused on the way we humans develop science as a way of dealing with reality, that is, the environmental and social conditions surrounding our lives. I am most concerned with the knowledge we need in regard to human behavior, as brought to us by the human sciences, which must, however, be viewed as on a spectrum with knowledge gained in regard to nature, the natural sciences. The two, human and natural sciences, are, in fact, intertwined though we necessarily in most cases must pursue them separately.

In this introduction I will attempt to say something more about where I think we are now in this effort, and where I am myself in thinking anew about some of the issues involved. *The Uncertain Sciences* put out a call for certain institutional steps to advance the cause. One was the creation of departments of the history and philosophy of human sciences to match those devoted to the history and philosophy of the natural sciences. This step would bring those working in particular disciplines, such as anthropology, history, philosophy, sociology, etc., into closer contact with their counterparts in the other disciplines. Transcending their own disciplines, they would attempt to forge a new one, advancing the effort to create the human sciences in a new way.

This institutional suggestion has not been put into place anywhere I know. This non-development is hardly a surprise. It would have required a leap into unknown territory, over disciplinary boundaries, and the rewards that attend those who stay within them. A conceptual obstacle also exists. In the minds of many there is no such thing as "human science." As one reviewer stated, it is doubtful that the human sciences constitute a meaningful grouping at all. He recommends abandoning what he called Hegelian universalism, and embracing more fruitful particularism. This is a legitimate challenge. In my view, however, to go that way simply reiterates the defects of particularism and does not advance us toward what I invoked as a scientific community. That particular reviewer and I are operating in different "thought universes."

Indeed, I have treated this question at great length in the original text, and forebear to repeat much of that argument here. In this regard, however, it is well to remember that the natural sciences faced the same problem. Only in 1833 did William Whewell recommend the term "science" for what practitioners in geology, chemistry, biology, and so forth were engaged upon (thereby substituting for the phrase "natural philosophy"). One could then inquire into the history and philosophy of science, stressing the overlaps and common interests between particular fields. Of course, one did not practice "science" as such, but did chemistry or geology. It is a question as to the level and intent of inquiry. The same holds for the human sciences.

One can also have a scientific community. Al least this is the ideal. And in regard to this idea I have gone on to try to explore the forces compacted in the notion of globalization as fostering such a community, going well beyond my handling of the subject in my original text. In the intervening decade I have thought hard about what is involved in the globalization process, and have researched and written about it under the heading, New Global History (NGH).[1] Here, in the present-day process of globalization, I have discerned a move toward a global scientific community. The NGH initiative itself delves into the various factors making for increased human interdependence and interconnection, transcending existing boundaries in the process: for example, multinational corporations, non-governmental organizations, an emerging global civil society (think of the UN, the International Criminal Court, and other such entities), and an expanding global culture.[2]

While there are many cultures, surrounding and modifying something called global culture, we nevertheless recognize the movement afoot. Needless to say, it is a contingent and uncertain movement, with reactions succeeding immediately upon actions. One piece of that movement, however, is the spread of a common global "civilization," a part of which is a shared devotion and adherence to natural science and a scientific method.[3] While there are loud voices proclaiming an Islamic or an Asian science, these are ideological statements and not legitimate claims about science itself (any more than earlier ones concerning Jewish science). Eloquent testimony to this fact is the recent denunciation in South Korea and China of claims made by some of their ambitious molecular biologists who have violated the tenets of "universal" science. The scientific community in these cases has stood firm.

The human sciences stand on a different footing from the natural sciences, exactly in regard to the problems and for the reasons elucidated in my original book. The need for greater knowledge, and then action in accord with it, becomes more and more acute as the natural sciences continue their "progress." One road to an increased scientific community with respect to the human sciences, I am now arguing, is that of globalization, whereby local, parochial differences are overcome, as humanity faces common problems of survival and development. With globalizing forces at work, existing social boundaries are being breeched; so, too, I am arguing, must the boundaries between disciplines be transcended, for it is only our present way of knowing reality that is cut up into economic, political, social and cultural compartments.

I am, of course, taking the long view. And the broad one, too. In this introduction I will not be going into the details of globalization as seen from an historical perspective. Instead, I want to stress the way the present-day process is creating a community prepared to adhere to the scientific method in both the natural and human realm. Divided by subject matter, the two realms are united in the way they seek to link theory and observation. Both are devoted to the exercise of "rationality" in a community open to the individual's free and critical examination—testing—of evidence and coherence. For the reasons given in this book, the path to this community is far more difficult and uncertain in the human than the natural sciences. Nevertheless, it is this path on which a globalized humanity must set foot and proceed however slowly and tortuously, in an effort to gain greater understanding and knowledge that may help it to survive in an increasingly risky and uncertain world.

With my eyes on the planetary space into which we increasingly penetrate with our explorations, I fully recognize the pitfalls at my feet into which I may fall; the example of Thales is there for all to see. Evolutionary theory should be our guide, but as a framework, not a set of specific notions imported into the human from the natural sciences.[4] Consilience between the two is a false siren, beckoning us onto the rocks of a presumed unity other than that of a scientific method appropriate to its particular subject. Within an evolutionary framework we can speculate about *Homo sapiens* as a "species," which as part of a thought experiment we can think of as changing into another species. The species does this by science and technology, so that "modern" man is different from ancient man, not in body but in mind. And that "mind," that is, cultural achievement, can be put on or taken off, which obscures for us the way the species is changing. This fact becomes clearer if we think of the human as putting on wings—an airplane or spaceship—which in any other specie change we would recognize as becoming a different species. It is because we can also take off such wings that we remain human at the same time as we enter upon a new species state.

II

These are unsettling and problematic thoughts. Yet, even if eventually shown to be unrewarding or wrong, they are heuristic ways of furthering our inquiry. Now I want to consider the hoary assertion made from all sides—including many practicing scientists—that science has nothing to do with morality. This seems to me egregiously wrong. It is the scientific method, a way of thinking, not any particular finding that provides us with a moral position. That method, adjusted, to the phenomena it is studying, teaches us to be truthful about our findings, including a willingness to subject them to refutation. The scientific way of thinking is prepared to accept uncertainty, and then by means of inquiry into the data, constantly inter-larded with theorizing about them, to elaborate hypothesis that are then further tested against the phenomena.

This "moral" attitude is widespread in natural science. It is equally applicable to society, though encountering much greater obstacles due to the greater complexity of the materials. Science is based on "true witnessing," as recommended by Robert Boyle and others. It also serves as a prototype for a legal system dedicated to fair and informed verdicts. It is the basis of our belief, for example, in the modern jury system, in accordance with our dedication to rule by law rather than by a despotic ruler. Such a method is frequently correlated with a democratic political system. In sum, the same scientific method, though under greatly different conditions and con-victions, permeates at least in principle both the natural and the human sciences.

Needless to say, there are other claimants to moral authority. Political and re-ligious leaders have their say, as rival voices. They often claim a different kind of "truth." I cannot refrain in this connection from referring to a quote by a leading figure in the Moslem Brotherhood, who says "Our people look like slaves to dictator-ships, and Islam came to release people from slavery to be only slaves of God."[5] In the scientific way of thinking, one is not a slave to either political or religious authority, but is free to pursue the truth according to the scientific method outlined above.

A decade ago, in *The Uncertain Sciences,* I tried to treat the problem of religion and science in as neutral a fashion as possible. The events of the last ten years, how-ever, have made me more uneasy about religion and its increasing involvement in public life. It seems ever more difficult to see religion as a private belief, helpful to the individual and innocuous in terms of a larger society. More and more the claims of religious truth are proclaimed in louder and louder tones. And more and more I am pushed by such religious assertiveness to take a firmer stance on the matter.

A few random provocations. The first concerns the issue of blasphemy. Any ques-tioning of received dogma is conceived as an attack on religion itself, to be punished in whatever ways available. The fatwa against Rushdie was an early example. The recent outcry against the novel, *The Da Vinci Code,* is a more recent one. Thus, a Nigerian Cardinal of the Catholic Church, Francis Arinze, declared that "Those who blaspheme Christ and get away with it are exploiting the Christian readiness to forgive.... There are other religions which, if you insult their founder, they will not be just talking."[6] In this hardly veiled threat, the Cardinal is prohibiting any other view than his own, on the grounds that those who hold such views are showing "disrespect." It never occurs to him to ask whether he is disrespecting those who hold alternate views to his own.

Another of his countrymen, Peter Akinola, the Anglican Archbishop of Nigerian, incensed over the matter of homosexuality, sees his church as betraying its faith in ordaining a gay bishop in the far distant diocese of New Hampshire, USA. "What is written of God is for all time, for all people," Akinola declared, undeterred or unaware of the controversy over accurate texts and the entire hermeneutic movement dealt with in the body of *The Uncertain Sciences.*[7] Much less would Akinola be interested in the findings of some scientists that homosexuality has strong genetic roots.

Of course, if such controversies remained solely within the churches, one might shake one's head but than agree that it is their business. The problem is that it appears that the issue cannot be detached from the political activism of religious bodies. The. same can be said in regard to stem cell research and similar scientific enterprises. Much of religion seems dedicated to preventing the exercise of scientific method not only in regard to itself but in all other "public" realms. These are to be converted into private, religious kingdoms. One writer, Philip Pullman, commenting on theocracies, says that they demonstrate "the tendency of human beings to gather power to themselves in the name of something that may not be questioned."[8] Not to question or to contest power is, indeed, to be a slave to someone else hiding behind an inerrant God.

As a matter of fact, historically the natural sciences developed in the seventeenth and eighteenth centuries in the context of religions, undeterred by the latter's claim to inerrancy. In this light, religion, along with myth, has served as a great creative expression of humanity's needs, helping its survival. Alas, the same cannot be said in the case of the human sciences and their future. Their further development is significantly, if not totally, inhibited by any authority that claims to hold the keys to knowledge and stands aside and opposed to the exercise of the scientific method. The world today demonstrates more and more that the human species is threatened by self-extinction. Admittedly, much of this threat is tied to the ever-increasing "success" of science and technology. One reaction to the anxiety provoked by this development is to turn "back" to unquestioned authority and truth. Instead, I am arguing, it is by forging ahead with scientific method in *regard to the human sciences* that *Homo sapiens* has the best chance of surviving. And not only of surviving, but doing so in a world of increased peace and justice.

In the course of the book, I made a number of observations concerning anthropomorphism, that is, the human tendency to inject its personal feelings and desires into nature. For example, the Greeks portrayed their gods as if they were mortal in their lives and lusts, even if in fact being "immortal." Thunder and lightening were not "natural" phenomena, but the gods displaying their anger. The Jewish religion projected its people's sentiments onto Moses, the Christian religion onto Jesus, and the Islamic religion onto Mohammed, all of whom entered into communion and communication with God. This general fact of anthropomorphism is jarringly true, for example, in the specific stickers proclaiming "Jesus loves me."

Only by depersonalizing nature could progress in scientific understanding take place. This was Bacon's message, and it was Comte's especially. Only thus could the universalism desired in science be reached. To be objective meant to try to eliminate the human factor. Mathematics was a superb tool by which to achieve this end. The heavens had to be emptied of the gods before astronomy could succeed to astrology. The result has often been announced in terms of the "disenchantment of the world." A heavy price—the feeling of alienation from the creation, of living on the edge of an abyss—has hovered over the achievements of natural science.

It is to the human sciences that we must turn to deal with this apparent void. Here, as I have argued, the "person" must return, but in a controlled fashion. Religions, with their heavy investment with anthropomorphism, and in their effort to offer consolation and "salvation" in another world, are a major obstacle to the possibilities of the human sciences I sought to describe. Held as a personal belief, and as a source of social bonding, religions can and should be tolerated. Injected into political life, and put forth as a public "truth" they are a threat to peace and justice, and to human survival dependent on scientific understanding. To put ourselves in the hands of the gods, or God, is to resign ourselves to humans, unaccountable to reason or any other restraint in their quest for unquestionable power.

III

As stated earlier, there are many different ways to many different "truths." On a personal level, each person can choose his or her own "way." Clearly, I value the scientific way highly. In my view, it transcends particular, more parochial, ways and offers great survival advantage to the human "species." It promises and delivers a "truth" that can be believed in by all, on a public level, and not just by true believers. True, it requires a scientific community, prepared to accept certain rules comprising what I have called the scientific method. How could it be otherwise? In this, for example, it resembles communities of religious and political believers, but with the difference that its truths are not held dogmatically and are constantly open to revision and new truths.

In *The Uncertain Sciences* I have devoted a number of pages to discussing the reasons why religions survive, in spite of (perhaps because of?) their sacred texts offering little or nothing in the way of public truths. Without repeating those arguments, I want here simply to remind ourselves of the Durkheimian argument for religion as a survival mechanism. A common belief system, though based on magic, illusion, and mythical thinking—existing long before the development of scientific method, and offering deeply needed assurance—functions as a way of drawing the primary group together. It is the glue, though others are possible, that binds a community. No matter to the believers that it is not a scientific community; it is a community.

In our uncertain and increasingly anxious and alienated world, this is no mean feat. Alas, it also justifies hostility, frequently unto death, to those not of the given religious community. Many Enlightenment thinkers thought that the removal of religion from public life would entail greater peace and justice. They and many subsequent thinkers believed that the way of the future meant increased secularity. For a while, they seemed to be right. By the end of the twentieth century, however, with the resurgence of religious faith in many parts of the world, often of a fundamentalist nature, this hope has been called into question, if not destroyed. The possibility of the kind of scientific community about which I had been speculating appears more and more remote.

Appearances can be deceiving. Indeed, I would argue that the resurgence of religion is largely fueled by the continued success of secularism. It has seeped or surged into all aspects of life, even in the most religious quarters. The images of the secular life are everywhere in our globalized world, to be found on the television and movie screens so ubiquitous in daily life. Secularism is carried by the modernity that uproots and dislocates so many people today, and is now doubly so carried in the form of globalization. No wonder that there has been a massive reaction to the march of secularity. True believers rightly feel that they are constantly besieged by the forces of reason and science. The priests and preachers of the religious word have had their positions as authority figures usurped by other voices. Education has slipped from their grasp.

In the ebb and flow of history, we constantly witness action and reaction. Much of the world is now in the throes of reaction. It takes an effort to realize that this is in reaction to the success of non-religious social forces. My observation is that secularity is not on the wane but becoming more effulgent. Further, the long view suggests that while the move to secularity can be reversed, so can the move back toward religion. A scientific community of the kind that I have speculated about in regard to the human sciences is still on the horizon. It is necessary to remember that in the natural sciences more scientists are alive today than existed in the whole of past history. A similar pattern may well be found in the human sciences.

It is important, therefore, to note that while most attention presently is devoted to the return of religion as the center of attention (mainly, I would argue, because of its connection to politics) the long-term trend to increased secularization may be being neglected. Against Christian and Islamic fundamentalism, one must also place in the balance the increasing turn to science and secularity of the European Union. Other signs of the times are the appearance of such books as Sam Harris,

The End of Faith: Religion, Terror, and the Future of Reason, and Daniel Dennett, *Breaking the Spell: Religion as a Natural Phenomenon.*

Especially worth noting is the tentative rise of questioning in Islamic religion. It is well to recall that Christianity in Europe went through similar turmoil in the nineteenth century. Embraced by reactionary monarchs, it was gradually attacked in the theological schools. Both Hegel and Feuerbach began their studies there. The Young Hegelians were originally believers who began to question their faith, on the grounds of reason. It was from inside as much as from the secularizing forces of the modern state that Christianity as a public matter began to be hollowed out. Something similar may be happening in the madrassas and theological academies of the Islamic world. Even in the theocracies of Iran and Saudi Arabia, public life, with ordinary "citizens" pushing the bounds of what is permissible, is escaping bit by bit from the control of the mullahs.[9]

Am I an incurable optimist (who should be burned at the stake by true believers)? Or an historian who has seen much of it before, and has learned from past episodes? Or some of each? What I see in the present situation are many peoples, in many places, being dragged kicking and screaming toward the increasing universality of the scientific way of thinking. In denial themselves, the exigencies of modern life drive them in unintended directions. There is no reason why the historian must join them in a denial of much of present-day reality in this regard. Is the move toward secularity and an expanding scientific community a sure thing? Hardly. The historian keeps two things simultaneously in mind: actions lead to reactions, in a sort of dialectic (though with no certain aim), and contingency reigns.

IV

Perspective is all. For example, in the subset of the human sciences called social sciences the picture today seems gloomy. Economics has become more and more powerful as an ideological defense of and justification for the expanding free market without counterbalancing restraints.[10] Sociology departments have tilted increasingly to the pursuit of the ephemeral data collecting and statistical analyses so beloved of funding agencies that want "hard" results. Anthropologists are enamored of cultural studies with major political implications. Historians place the highest value on monographic research, based on the archives.

It appears that the sort of development in the social sciences that I favor in my book is in recession. However, this is not an all or nothing situation. The sort of work described in the preceding paragraph, positivistic as its tendency may be, has its own value. While it also tends to shut down the more hermeneutic type of work, it is not wholly successful. Thus, work in economics outside the neo-classical paradigm goes forward, treating of economic activity in terms of context, and seeking to include in its theory sustainable development and the environment. Sociologists have been in the forefront of theorizing about globalization. There they are joined by some anthropologists. And historians seek to break out of the national and monographic mode by moving in the direction of transnational and world/global history.

Such moves in the social sciences are in affinity with the thesis that the human sciences mainly accumulate increased consciousness, self-consciousness and historical consciousness. They are necessarily heavily involved in interpretation. Such interpretation, in accord with the dictates of scientific method, appropriately applied to their subject, allow for the human being as agent, who then becomes part of the science. Such agents increasingly form part of an amorphous scientific community, no less real for all that, which adheres to the requirement to submit its interpretations to the constraints of empirical evidence, tied to theory, tested by further observation, and constantly subjected to the rigor of logic and the willingness to abjure dogma and allow for change in a world of continued emergence.[11]

This is the perspective that I advance in *The Uncertain Sciences*. In the greater number of the reviews accorded the book, this view was understood, and the book, often judged warmly, in its terms. A few reviewers missed the point. To my surprise, for example, one reviewer accused me of being a neo-positivist and of downplaying the role of psychology in the human sciences (though my treatment of Freud was applauded). In another I am condemned for not having an interest in why religion has persisted—the reviewer obviously not having read the pages devoted explicitly to that question—and brought to task for judging political science a failure and ignoring the ninety-two annual volumes of the *American Political Science Review*! (This review, alas, was in the *New York Times Sunday Book Review*.)

Reviewers cannot be held at fault for not sharing a perspective. They can, however, be faulted for not having indicated what the book under review is about, and thus giving the reader a fair shot at its contents. Every cloud as we all know has its silver lining. Aware that many colleagues had had similar experiences I wrote a short piece, "The Art of Reviewing," calling for training in graduate schools in the writing of competent reviews and arguing for a minimum of fair practice.[12] My mantra was that "The book reviews the reviewer as much as the reviewer reviews the book." It was a pleasant surprise when this article received numerous letters of support.

V

There is, of course, no single correct perspective. It is essential, however, to make clear what one's perspective is, and even perhaps some of the counter-arguments that can be brought against it. Such an approach is in accord with the scientific method that I am extolling as lying at the heart of the human, as of the natural, sciences.

In the present introduction, I am trying to trying to underline as forcefully as I can the perspective that animates my book. Thus, I hope the reader will give it a fair reading in its own terms, before judging it in terms of his or her own predilections, a legitimate next step. In this way, we embark on a mutual discourse of the sort that has the potential to contribute to the greater understanding as well as the development of the human sciences.

I have also taken the opportunity to assess a little more closely where we are now, and where I am in my thinking further about the issues. The most significant step in this direction is the contribution to an increasing scientific community by what is happening under the heading of globalization. This topic, controversial and in the making, is given a fuller hearing in my new book, *The New Global History.*[13]

Ours is a time of troubles, which may call forth profound changes in the nature of our attempts to understand "scientifically" some of the causes of our distress. And such understanding may bring about amelioration of some of our troubling problems. As the reader will realize, both in this introduction and the volume into which it hopes to attract him or her, I seek to retain my optimism about this strange species, *Homo sapiens,* a frustrating combination of the rational and the irrational, the wise as the nomenclature implies, and the mad and the destructive. Without optimism, there is little hope that we will try to push on to the scientific knowledge which can help determine the shapes of our future.

Notes

1. My initial foray was in the introduction to the volume, *Conceptualizing Global History,* ed. Bruce Mazlish and Ralph Buultjens (Boulder: Westview Press, 1993; reprinted NGH Press, 2004), and my most recent *The New Global History* (New York and London: Routledge Press, 2006). In between have been a number of articles and *The Global History Reader,* ed. Bruce Mazlish and Akira Iriye (New York and London: Routledge, 2005). Readers who are further interested in the NGH initiative should look at the web site www.newglobalhistory.org.

2. For further work on some of these topics see *Leviathans: Multinational Corporations and New Global History,* ed. Alfred Chandler and Bruce Mazlish (Cambridge: Cambridge University Press, 2005). The New Global History Initiative has also conceived of and promoted the mapping of the MNCs, that is, giving them visual representation, in *Global Inc. An Atlas* of *the Multinational Corporation* (New York: New Press, 2003).

3. The notion of "civilization" is a constantly invoked one, by politicians as well as scholars. In fact, it is a fairly recent concept—the first reification appears to have taken place in 1756—and has been used and misused ever since. For its history, usage, usefulness, and relation to globalization see my book, *Civilization and Its Contents* (Stanford, CA: Stanford University Press, 2004).

4. For a fuller development of this argument, see my chapter, "Evolving Toward History" in *The Return of Science: Evolution, History, and Theory,* ed. Philip Pomper et al. (Lanham, MD: Rowman & Littlefield Publishers, 2002). See further *Historically Speaking,* May/June 2006, Vol. vii, No. 5, for my article "Progress in History," with commentaries and my response to them, with further discussion.

5. Quoted in the *New York Times,* March 25, 2006, 4.
6. *New York Times,* May 8, 2006, B2.
7. *New Yorker,* April 17, 2006, 57. In an interesting article, "Behavior, Belonging, and Belief: A Theory of Ritual Practice (*Sociological Theory,* Vol. 20, No.3, November 2002), Douglas A. Marshall builds on Emile Durkheim's work to explore the way belief is tied to rituals. This suggests one reason why it is especially difficult to shake beliefs no matter how "irrational" they are. Another suggestive article is by Thomas David Dubois, "Hegemony, Imperialism, and the Construction of Religion in East and Southeast Asia," *History and Theory* Theme Issue 44 (December 2005). Dubois remarks that "the missionary discourse of transformative conversion located it in the very personal realm of sincerity and belief." (113). We can then connect such belief to ritual. Even after shaking off colonial rule, many countries still remain under the hegemony of the Western powers, especially in regard to religion. This subject is hardly a simply one, however, for rather than sheer imposition we must see religion in these situations as an exchange between colonizer and colonized. Still, many non-Western voices, especially in Africa, seem to be identifying with the former aggressor in many ways. This is a topic that needs much further exploration.
8. *New Yorker,* December 26, 2005 and January 2, 2006, 54.
9. Cf. the article in the *Economist,* May 6, 2006, 48.
10 For a classic alternative to this position, see Karl Polanyi, *The Great Transformation: The Political and Economic Origins of Our Time* (Boston: Beacon Press, 1944).
11. An example of what I have in mind is given by Dominic Sachsenmaier in his article "Global History and Critiques of Western Perspectives," when he remarks that "there are certainly profound differences between various national historiographic traditions. Yet at the same time historians can be seen as members of a professional community, which is a result of a global transformation in the categorization and institutionalization of knowledge" (ms., 12). Specifically, Sachsenmaier looks at how Chinese historians are accepting Western methods of historiography, while at the same time contributing to their change, all in the framework of a common scientific community.
12. *Perspectives,* February 2001.
13. *The New Global History, op. cit.*

Acknowledgments

My situation is like that of some developing countries whose debts are so great that they can never hope to repay them. However, if I can't repay them, I can at least make a start at acknowledging them. None of what is in this book would have been possible without the work of the great thinkers who have pondered previously the problems addressed here. Their inspiration is obvious throughout my text. Nor could I have entered upon my project without the prior work of innumerable scholars, in many fields, whose researches are referred to in both the notes and the Critical Bibliography. Countless other scholars, though uncited, figure critically in the thinking that has gone into this book.

I have benefited from exactly the sort of truth community that I have argued for in this book. I have gained as well from the comments of anonymous reviewers of the manuscript. One of them, fundamentally antagonistic to my whole enterprise, inspired me to clarify my arguments in a way that I believe has resulted overall in a more persuasive book. Others were helpful in more expected, but much appreciated, ways.

Various individuals have read all or parts of the manuscript in various of its stages of composition. Here my debts are unconscionably large. Among those who read and helped with comments on the entire manuscript, I thank Michael M. J. Fischer, Christopher Fox, Diane Greco, William McKinley Runyon, and Dmitri Shalin. Parts of the manuscript or outlines thereof were read by Gerald Holton, Alvin Kibel, and Pauline Maier. I owe especial appreciation to Daniel Jones, who, as a program officer at the National Endowment of the Humanities, encouraged me to continue with what was clearly a daunting project, and did so at a critical stage. I am grateful to the Massachusetts Institute of Technology for funds allocated by Philip Khoury, Dean of the School of Humanities and Social Science, to assist in the compilation of the index.

More than a quarter of a century ago, I made a start toward this book. At the time, the manuscript was read and commented on by Joyce Appleby, J. W. Burrow, Robert J. Bernstein, Gerald Platt, and Lester Thurow. They will long since have forgotten their kindness, but I have not. Their keen appraisals made me put the manuscript aside for a number of years and, when I took it up anew, led me to conceptualize and to compose it in almost entirely different form.

My relations with Yale University Press have been such as dreams are made of: a wonderfully understanding editor-in-chief, Charles Grench; a gracious and helpful editor, Otto Bohlmann; and a splendid manuscript editor, Mary Pasti, who exerted herself well beyond the call of duty to save me from errors of fact and expression. Would that all authors might have the happy experience that I have enjoyed with my publisher.

Lastly, for my wife, Neva Goodwin, I do not know how adequately to express my feelings of appreciation. We work together in the same study, constantly exchanging ideas, information, and inspiration. She has read the manuscript in its various incarnations. She has been a superb editorial consultant. She is a beloved friend, and so much more. It is to her that I dedicate this work, so many years in the making.

Introduction

"Man is but a reed, the most feeble thing in nature; but he is a thinking reed. . . . If the universe were to crush him, man would still be more noble than that which killed him, because he knows that he dies and the advantage which the universe has over him; the universe knows nothing of this." — Pascal

" 'My project,' he told us, 'is to learn where to go by discovering where I am by reviewing where I've been — where we've all been. There's a kind of snail in the Maryland marshes — perhaps I invented him — that makes his shell as he goes along out of whatever he comes across, cementing it with his own juices, and at the same time makes his path instinctively toward the best available material for his shell; he carries his history on his back, living in it, adding new and larger spirals to it from the present as it grows.' " — John Barth

*M*y purpose in this book is to inquire into the condition of the human sciences — accomplishments, weaknesses, and possibilities. I deal with the questions What sort of knowledge do the human sciences claim to be offering? To what extent can that knowledge be called scientific? and What do we mean by "scientific" in such a context? I also seek to contribute, however modestly, to changing the way we think about the subject.

It is hardly an original insight to say that we are humans, who wish to know not only the world around us but also ourselves. When what we seek is scientific knowledge of the "outside" world, we speak of the natural sciences. In respect to the "inside," we speak of the human or social sciences. In fact, the two sides are united, for we humans are the lens by which we view all: what is outside ourselves as well as ourselves and our construction of a social world. Quite a feat.

Our context for such knowing must be evolutionary, as one of the sciences informs us. We now know, whatever controversies swirl about the matter, that we are a species of a particular kind, *Homo sapiens.* Biology tells us that we are a part of evolutionary nature, and helps us realize what our "nature" is in that context.

Most recently the human species has evolved more culturally than physically; in fact, for the past thirty-five thousand years or so, *Homo sapiens* has been more or less physically stable. Change has occurred through societies and cultures (to use recent terms, coined by the emerging human sciences). The population increase, highlighted by a modern "social scientist," Malthus, has been phenomenal, going with relative suddenness from about ten million, the estimated population of the earth for thousands and thousands of years, to almost six billion today, with a likely doubling ahead. In this sense, certainly, the species has successfully survived.

In the course of surviving, humans have moved from hunting-gathering bands to agricultural settlements to industrial societies (though all these groups continue to coexist, even if unequally and uneasily). Tribes, kingdoms, and empires, cities and civilizations, have risen, flourished, and disappeared. Languages have come into existence, alphabets have been invented, and observations and knowledge on an incredible range of subjects have been accumulated (as well as lost). An extraordinary array of human phenomena—wars, trading relations, religious beliefs, technological innovations, modern science—make up the emanations of the human mind and behavior.

How can we bring order out of this chaos? On what sources might we draw, and from what fields? At the heart of the problem is the fact that the species that has constructed the scientific theories giving order to the biological world is itself a product of evolution—the subject of one of those theories. Hence, members of the species have had to come to know themselves as knowing beings. They have had to step outside their own shadow and observe themselves as they would any other animal. But *Homo sapiens* is not any other animal; it is the animal that symbolizes and that is challenged to know itself scientifically.

Whether the challenge can be met is not clear. Perhaps the species can achieve scientific knowledge of the natural world but not scientific knowledge of itself, an "unnatural" being. Or such knowledge may turn out to be scientific, but not in the terms of the natural sciences as we know them now.

To answer our questions we shall be ranging widely, crossing many disciplinary lines and struggling with difficult forms of thought. My aim is a synthesis, based on the work of innumerable others. In this connection, I take solace in Joseph Schumpeter's remark about the incomparable *Wealth of Nations*: that it did not contain "a single *analytic* idea, principle, or method that was entirely new."[1] The component parts of *The Uncertain Sciences* are not particularly original (although I have tried to achieve some freshness of either expression or perspective). The originality lies, I hope, in the whole.

I will be raising topics, developing them from a particular angle, then dropping them, only to return to them later, perhaps to repeat the process with additions. In short, the same topic will be addressed from differing perspectives and in varied contexts to illuminate its meaning more fully. In fact, perspective is itself a key topic in this study, as is context. By circling and recircling around our subjects I hope to achieve the whole for which the claim to originality has been made.

Within that whole, two theses are of special importance. The first (to put it bluntly and before the historical demonstration that can make it persuasive) is that the development of the human sciences is ultimately dependent on the development of consciousness. Such a development must take specific form: it must be embodied in a scientific community.

The situation for the human sciences differs from that for the natural sciences, where the community at issue can be, and generally is, a small number of professionals. The community that is willing to accept the knowledge acquired in the pursuit of the human sciences and that is prepared to act on the basis of such acquisition ideally has to be humanity at large; only then can the sciences whose subject matter is evolving humanity go forward. Exactly how this objective is to be realized remains a central problem, involving the notion of historical consciousness. For now, we can know only the human sciences as they exist *in potentia*.

This thesis about historical consciousness illustrates, for better or worse, the attempt at original synthesis. Here, the idea of historical consciousness, the subject of much thought by many others, is combined with recent work in the history of science on "witnesses," on how natural science has been "verified" in the recent past, and to this is added a notion about a Habermasian truth community. The synthesis is augmented by way of a brief look at the globalization so prominent today. Without embracing either the inevitability or the desirability of globalization, the concept, in the context established here, needs to be studied as a tendency with certain possible consequences for the extension of a scientific truth community in relation to the human sciences.[2] This is all done in an effort to rescue some form of validation for the human sciences that does not resurrect outdated positivism but that tries to combine aspirations to positivism with hermeneutic realities in a novel manner.

My second thesis is closely related to the first. It centers on the notion of emergence. Different versions of the human sciences have emerged in response to different circumstances; for example, economics and sociology, as modern social sciences, appeared only when societies became capitalistic and industrialized. As circumstances change in ways we cannot predict, new and different forms of human science can be expected to make their appearance. But we in our time must rely on what already exists while stretching toward the possibilities inherent in the human sciences.

There is a price to be paid in the choice I have made to construct a synthesis. In many areas, depth has had to be sacrificed to breadth. Some readers can be expected to be critical; others may be grateful. Let me briefly mention some of the important issues that would belong in some ideal, expanded version of this book.

An attempt at a scholarly exposition of the full position of important thinkers has had to be abandoned. In using Habermas, for example, I am concerned mostly with my version of a particular idea of his, extracted from the corpus of his work. Although the treatment of Bacon is more thorough, the same comment might be made even in his regard.

In other of my writings, I have stressed the historicity of a work or a thinker. Here, it is not possible, though the *notion* of historicity is of central importance for the human sciences.[3] Having said this, it is important to add that in considering a work it is also legitimate to look at it from a present perspective, asking how some of its ideas, even if lifted out of context, can serve in regard to contemporary ways of thinking. It is possible, for instance, to examine Bacon for hints as to the construction of modern sociology without having to fall into the trap of Whiggish history. For a different example, we might look at bourgeois ideas and actions at the time of the French Revolution, recognizing their historicity and their function as ideology and yet also recognizing the way the bourgeois enunciation of human rights carried with it a universal possibility that could affect later developments. Thus, actual behaviors in 1789 betrayed the principles in whose name they were undertaken, but left the principles standing as an inspiration for the actions of future others.

The subjects of women, gender, and race should, ideally, be given extended attention. Women had an important role in the eighteenth-century construction of the public sphere. They played a dominant role in the holding of salons, and one would wish to look further at their involvement (and, perhaps more important, lack of involvement) in science. Their inclusion in any future scientific community of the sort envisaged here — that I take for granted. There must also be consideration of race in a history of the public sphere. An awareness of these issues has, I would like to think, informed what is said in *The Uncertain Sciences;* but the inquiry pursued here is basically of another kind. Fortunately, an abundance of literature has been building in these areas, to fill the gap left in this book.

The same must be said for a host of other concerns. Problems of war, poverty, pollution, the negative effects of globalization and capitalism, the commercialization of culture, the destructive tendencies of science and technology—in short, the darker side of much of contemporary life—would need to take their proper place in an ideal study. In large part I have attempted to understand the human sciences, and have elaborated the notion of a historical consciousness that embodies a form of science, in an effort to see past these aspects of the human condition and to help to ameliorate them—while recognizing that in one form or another they will probably persist. To give up the attempt at reason, no matter how fragile (and no matter how horrible some of its results when pursued as dogma rather than practical method), is the ultimate in human bankruptcy. Advocates of such a solution seem to be uttering a cry of despair and nihilism, which requires our sympathy but not our acquiescence.

I should also make clear that I am not trying to write a history of the social sciences (which I treat as a subset of the human sciences). Neither do I emphasize ways of rethinking or reordering the social science disciplines, such as sociology or anthropology (though I do have thoughts thereon).

Choices have had to be made because of both practical considerations (the mundane issues of time and space) and intellectual intent. What is being primarily sought in this book is the very possibility of knowledge in the human sciences, especially scientific knowledge.[4]

Perhaps a sign should have been posted on the title page: Warning! This Book Is Not a Traditional Account! As is well known, oranges should not be compared with apples. This book is like an orange, to be slowly peeled. It is an extended essay, not a monograph. Of course, one could have written a very different sort of book, invoking, for example, such philosophers as Richard Rorty, Nelson Goodman, Hilary Putnam, Charles Taylor, and Stephen Toulmin or citing eminent historians who have written in relation to the social sciences: Thomas Bender, Thomas Haskell, David Hollinger, Dorothy Ross, James Kloppenberg. But that would not have suited the purpose of this book. (And as it happens, many other scholars have undertaken the tasks described above, and carried them out well.)

The Uncertain Sciences does not fit easily into existing disciplines or genres. It is resolutely interdisciplinary in its ambitions; and although lip service is often given to that procedure, in practice there is much resistance.

Put as simply as possible, what I hope to do in this book, drawing on a wide range of materials and others' work, is help to create or redefine the field of the human sciences. Again, what I am not doing is writing a book on, say, the pragmatists' or the idealists' view of science. Instead, I will range widely across time, disciplines, and conceptual orientations. Starting from the remote evolutionary past, I shall move through the Age of Discovery to the twentieth century. I shall employ the insights of anthropology and history, economics and sociology, psychology and philosophy, political science and the history of the natural sciences. In addition, I shall incorporate analyses from a variety of positivistic and hermeneutic researches.

I shall also be seeking to link the human and the natural sciences. More tentatively, I shall explore the possibilities of drawing the humanities within the orbit of the human sciences and, reciprocally, bringing the human sciences more securely into the circle of the more philosophical humanities.

In pursuing my overall aim, I will be drawing upon work on such topics as globalization, evolutionary theory (embracing machines as well as humans), witnesses and their role in scientific attestation, footnotes in historical think-ing—subjects that rarely figure in traditional accounts of the human sciences. I do so in the belief that such topics not only throw a revealing light on the human sciences but establish a fresh context in which to probe their nature and meaning.

Given my thesis about the necessity of a scientific community composed of much or most of humankind, it is only fitting that I write both for the general public and for specialists prepared to think outside their own field. Accordingly, my language will be as straightforward as I can make it. While I have tried to benefit from work that has gone forward under the labels of deconstruction and postmodernism, to internalize some of their insights, I will avoid the style often favored in such work.

Although postmodernism, especially in the figure of Foucault, is significantly concerned with human sciences, I will be dealing only in passing with the content of postmodernist observations. There are two reasons for this. The first is that many, if not most, postmodernist thinkers are hostile to the very idea of science, whether human or natural. We are engaged, therefore, in different enterprises: mine is to construct, theirs is to deconstruct. The second reason is that to engage here in a major discussion of postmodernism would be to settle for its terms of discourse.[5]

Nietzsche, the inspiration for so much of postmodernist thought, entitled one of his writings *Untimely Meditations* (*Unzeitgemässe Betrachtungen*); in that work he attacks the pseudo-culture of many of his contemporaries. Ironically, my form and content of discourse may appear equally "untimely" in regard to some of what is fashionable in the present Zeitgeist. For those who immediately declare a work outmoded if it does not use the current "in" or "privileged" words and make constant reference to certain hermetic writings, *The Uncertain Sciences* will be a hopeless case. For Nietzsche himself, I like to think, this book might appeal as a kind of errant "joyous science."[6]

The time has arisen—or is it still untimely?—to attempt a transcendence of both "old," pre-postmodern ideas and postmodernism. The word *attempt* is used advisedly. This book aspires to be just one contribution to the effort to think forward, as well as backward, in respect to the human sciences.

Until now I have been using the phrase *human sciences* as if its meaning and content were self-evident. Indeed, it is frequently used in a catchall, unreflective manner. Quentin Skinner, for one, in his otherwise excellent collection *The Return of Grand Theory in the Human Sciences,* typically employs it in an unexamined fashion. [7] In fact, the content and definition of the human sciences is a problematique. One encounters the phrase *human sciences* in an array of circumstances. In almost all cases, its use elsewhere is far different from its use in this book. The Institute for Human Sciences (Institut für die Wissenschaften vom Menschen) in Vienna, for instance, turns out to be an institute devoted to the promotion of democratic institutions in central and eastern Europe.[8] Cheiron is described as an international society for the study of the human sciences; its focus turns out to be psychology.

Closer to my use of the term is that set forth in the journal *History of the Human Sciences.* In a recent issue, the editors declared their interest in "traditional humanistic studies as well as the social sciences." They added that they published "wide-ranging material in the fields of sociology, psychology, political science, the history of science, anthropology, classical studies and literary theory."[9]

Another way of trying to understand what is meant by the human sciences is to study their development. The book *Inventing Human Science,* edited by Christopher Fox, Roy Porter, and Robert Wokler, roots their emergence firmly in the eighteenth century, arguing (an oft-stated notion, but now supported in detail) that the human sciences were the result of the desire to emulate the successes of the seventeenth-century scientific revolution in the natural sciences. Alternatively labeled the "science of human nature" or the "sciences of man," the human sciences then embraced what later came under more specialized attention as psychology, anthropology, and social theory, or, to put it another way, sociology, political economy, and political science.[10]

Another investigator, Carlo Ginzburg, argues that historically, something called the human sciences seems to have emerged sometime between the eighteenth and the nineteenth centuries. He offers the intriguing suggestion that these sciences were linked by "a common epistemological model," which was "summoned up from way back, out of various contexts—hunting, divining, conjectural, or semiotic." In fact, they also seem to share a common inspiration in "Zadig's method" (modeled after Voltaire's story), which, in more modern nomenclature, can be called the abductive method. It involves the making of retrospective predictions and the linking of diverse phenomena by creative insight, then checking the predictions against further evidence.[11]

In principle, all the materials discussed in the *History of the Human Sciences, Inventing Human Science,* and Ginzburg's work fall under the rubric "human sciences." I shall be dealing with only some of them, and those selectively. Thus, the merits, the achievements, the aspirations of such disciplines as philology, linguistics, semiotics, aesthetics, art, music, and literary criticism will not be discussed here. Although there are discussions of narrative and music in the book, they form a small part of the whole.

One major reason for my neglect of the humanities—in contrast, I devote an entire chapter (5) to examining the achievements of the social sciences— is that to tackle them as part of the human sciences (though I do make stabs in this direction in regard to literature) would be to take on a task where few hands have been at work. The relevance of scientific method to humanistic subjects has been little explored by other scholars. I hope that in the future this omission will be remedied.[12]

Within the limitations of my treatment of the human sciences there is an important distinction in my usage between the singular and the plural (though at times I use *human science* to mean both). The singular, *human science,* is specifically the study of human evolution as it moves increasingly from the physical to the cultural, and is primarily an effort to define the human species as such, that is, as the object of study. The plural, *human sciences,* although it also includes the first, embraces all disciplined efforts to understand specialized aspects of human behavior—political, economic, sociological, and so on—not excluding the humanistic approaches mentioned earlier.

One approach to advancing the human sciences is to institutionalize their historical and philosophical study in more formal ways than at present. There are departments of the history and philosophy of natural science in many major universities. These have their own faculty and graduate students, who ensure continued research in the field. No less is necessary if the study of the human sciences is to advance. It must take its proper institutionalized place in the academic world. One of my aspirations is that this book will help the human sciences achieve such status. In line with this hope, the notes and bibliography may serve as an initial archive, for no obvious collection of documents exists in the history and philosophy of the human sciences, as they are being defined here, as it does in other fields.[13]

It might be argued that the study of the human sciences is already included in existing departments of the history and philosophy of science. But if so, it is generally relegated to the sidelines; it is considered a soft feature of what is already a soft—meaning "historical"—study. There is generally no sufficient collegial effort in the established departments to tackle this largest and perhaps most amorphous of human inquiries.

Furthermore, the historical and philosophical study of the human sciences occupies a space different from that occupied by the comparable study of the natural sciences (to which, however, it must always be connected). The historical and philosophical study of which we are speaking is itself a constitutive part of the human sciences in ways denied to the study of the history and philosophy of the natural sciences in relation to the natural sciences. For the latter, historical and philosophical findings do not particularly affect the practice of science itself, whether in the laboratory or in theoretical constructions. In contrast, in the human sciences, the study of their history and philosophy can help shape human consciousness in a major way and thus enters into its subject matter directly and forcefully.

In writing about the nature and meaning of the human sciences, or what I refer to as the uncertain sciences, I am arguing that their historical study is also their future. I am attempting to help create an audience, to bring into being a scientific community that is composed of most, or at least a large part, of humanity itself. I have been animated through the long process of researching and writing this book by the hope that it might do just that: help bring into reality what is only potential in the present moment.

1 / The Problem
of the Human Sciences

*T*he natural sciences allow us to attain relatively certain understanding of, and predictive control and manipulative power over, natural processes — or so it is widely assumed. This view holds, assuredly among scientists, even in the face of postmodernism and the social constructionism of science. Many believe that the natural sciences may, and perhaps should, serve as a model for any attempt at comparable human sciences. Indeed, at the time of the Enlightenment it was widely held that nothing seriously stood in the way of the extension of scientific method to human matters, although there was disagreement over how to carry out the assignment.

As we shall see, this view went forward forcefully as positivism. The arguments in favor of it, and the difficulties with it (for many opposed, and oppose, it, root and branch), will be discussed in the next chapter. Here, in this chapter, I want to suggest the range of hopes and problems associated with any attempt at the human sciences, whether on the model of the natural sciences or not, making only a partial effort at this point — more in the way of hints, if anything — to judge or to offer alternative possibilities.

In carrying out this task, I will be employing a historical perspective. My approach might be called philosophical history, that is, history with a philosophical intent, in contrast to the more widely accepted philosophical anthropology.[1] In taking this approach, however, it is imperative to cross existing boundary lines, including those around the discipline of history. Only a broad, synoptic treatment of the subject can offer the possibility of a real change in our thinking.

In this chapter, where the general problem is posed, I shall arrange what I have to say under two major headings: the aims of a human science and its preconditions. We shall begin, however, by further examining (following on the Introduction) the definitions attached to the words *human science(s)* and by next imagining a Martian observer on Earth. Then we shall consider such topics as secularization, symbolization, emergent phenomena, interpretation, and accumulation as vital parts of our initial inquiry. Last, I shall introduce the perspectives brought to bear on the problem of the nature and meaning of the human sciences by presenting three prerequisite events: the Age of Discovery, the Enlightenment, and the Darwinian revolution.

The Human Sciences: Further Definitions

In the Introduction, we explored various definitions of the human sciences. Now we must recognize that numerous other names for what we are in general talking about have been, and still are, used: *the science of man, social sciences, behavioral sciences, moral sciences,* and *Geisteswissenschaften.* There are perhaps others, including *Kulturwissenschaften* and even *cultural studies* (although this last does not have explicit scientific aims). Often the social sciences are viewed as equivalent to the human sciences. In fact, I would argue, they are simply a subset, usually containing economics, anthropology, sociology, and political science. History, and psychology, in the form of social psychology, are given uneasy entrance. But even literature can at least be imagined as a legitimate form of science.[2]

We must leave open, therefore, the issues of (1) what is meant by the term *science* and (2) what falls under its dominion. In the simplest etymological sense, the word *science* means "knowing." One "knower" was the erudite English classicist Benjamin Jowett, Master of Balliol College, Oxford, who announced: "I am the master of this college; / What I know not is not knowledge."[3] Such a claim, whether by an arrogant classical scholar or by contemporary social scientists, should not preclude further investigation.

Although the division between natural and human sciences is a useful one, it can be seriously misleading. Does the nomenclature imply that the natural sciences are "inhuman"? Such a conclusion might be a remnant of the view of some humanists that the natural sciences are cold and calculating, instead of being part of human creativity. Are the human sciences, in turn, not "natural"? This view would also seem to be a remnant, in this case of the belief that humans are separate from nature, somehow a special creation. We now know that humans are natural creatures, formed by evolution, and that the natural sciences are socially constructed. In this light, then, the division between the sciences must be viewed as one of convenience (which sometimes becomes inconvenient) and emphasis.

The modern use of the term *science* itself is just that — modern. As many scholars know, what today we place under the heading "natural science" was formerly thought of as natural philosophy. Although a recognizable scientific method was being practiced by the seventeenth century, the nomenclature did not change appropriately until the nineteenth century. At that time, in the 1830–40s, William Whewell asked why there was a generic term, *artists,* for sculptors, painters, and engravers and no such term for botanists, zoologists, geologists, and physicists. To repair this omission, he suggested *scientist* and *science* as the names under which to group the emerging specialities formerly lumped under *natural philosophy.*[4] The term caught on, and Michael Faraday was perhaps the last major inquirer to continue calling himself a natural philosopher.

At about the same time, the rubric "social sciences," presumably inspired by the natural ones, came into common usage. It was a loose term, verging on the inclusion of social work on one side and social theory on the other. *Social science* as a phrase seems to have come into existence around 1789. Subsequently, the Marquis de Condorcet, in his *Sketch for an Historical Picture of the Human Mind* (1793), spoke of a "social art."[5] In the 1820s, Adolphe Quetelet, the Belgium statistician, was writing about "social physics." By the 1840s, Auguste Comte, the founder of positivism, coined the term *sociology* to differentiate his larger and more ambitious science from Quetelet's limited statistical approach. Gradually the earlier overall term *social science* came to mean what Comte meant by *sociology* — a scientific inquiry into human phenomena — and to include economics (which had only recently emerged, in the 1770s, out of work in moral philosophy, vide Adam Smith) as well as the newly professionalized disciplines of sociology and anthropology.

John Stuart Mill thought that he was pursuing moral philosophy when he took up the task of inquiring into what he came to call the moral sciences. The German translation of this term was *Geisteswissenschaften,* where the moral or spiritual sciences were placed in contrast to the *Naturwissenschaften.* The assumption was that the two kinds of sciences dealt with radically different phenomena and required radically different methods.

By using the term *human sciences* I am trying to escape from the attitudes attached to the various designations. The subject matter of human science is a particular species, *Homo sapiens,* which can be studied by natural science, as we would study ants, and also by another set of disciplines, which have as their central feature the fact of human subjectivity and meaning.

What this statement entails will become clearer in the rest of this book. We get a hint of the problem in A. R. Radcliffe-Brown's wish that anthropology be a science of "concrete, observable facts" and his plaintive realization that relationships are not observable phenomena.[6] If we would understand such institutions as cannibalism or totemism, we are told, we must also understand the values attached to them. (How this is to be done is the subject of much thought in cultural anthropology.) One writer has reduced the notion to its simplest expression: the human sciences are "the sciences of subjectivity."[7]

As we go on, I think that we will find that this short definition is not so much wrong as too simple. The human sciences involve any disciplined *scientific* attempt to understand the human species, which, though not excluding the external (how could it?), tends to emphasize the internal experience. The scientific *method* used in any particular human science, as I shall try to show, must be judged not in terms of any a priori model but in terms of the particular human science itself, as it develops in regard to its view of the subject. The method, I am arguing, is continuous with that to be found in the natural sciences. There is only one way of thinking scientifically, whatever the field, but innumerable methodical modes by which so to think.

What I have given here is a brief historical background to our use of the rubric "human sciences." We need to keep the definition open-ended, at least provisionally, allowing for what is described by the terms *social science* and *literature, psychology* and *cultural studies, positivism* and *hermeneutics* to be seen as part of the domain of the human sciences. Any useful definition must be an evolving one, as demanded by its own subject matter.

The Martian Experiment

To offer perspective, let me introduce a scientist who has come from Mars and has landed on Earth to do a study of this alien species, humanity, the way we Earthlings study ants. In fact, our Martian is even further removed from "his" subject (for literary reasons, I will use the masculine pronoun), for he neither coexists with it nor has any inherited feelings about it. Thus, he is completely objective — that is, he can treat his subject matter as solely an object.

How good a scientist would our Martian be? He would observe dispassionately all that was external, he could record data, he might even run experiments forbidden to human scientists, for he would have no human moral scruples (of course, he might have his own, just as ours might interfere with what we do with ants). He would observe humans engaging in economic activities, practicing science, and making wars. On a more private scale, he could study their mating customs and sexual mores and their child-rearing habits.

He would also see buildings and art objects; he would observe humans reading books and would read them himself. And it is somewhere near this point that the question of meaning, inherent in all the human activities and practices cited above, would come front and center. How would our Martian know what all these activities "mean"? How could he understand what the Earthlings were saying hypocritically or ironically, with humor or with malice?

A facsimile of the problem is presented to us by an ethnographer studying an alien tribe—one like Bronislaw Malinowski, who specifically enjoins us to "imagine an ethnographer from Mars." [8] In an early, positivistic mode, Malinowski pictured the Ethnographer (he capitalized the term) as a being removed from the scene he was observing. From on high, so to speak, he could view the whole of the society in question. As Malinowski declared, the natives "have no knowledge of the *total outline* of any of their social structure. . . . Not even the most intelligent native has any clear idea of the Kula as a big, organized social construction, still less of its sociological function and implications." Only the Ethnographer knows: he "has to construct the picture of the big institution, very much as the physicist constructs his theory from the experimental data." [9]

Malinowski's confident tone slackens as his experience in the field deepens. He still believes that his report from the field is analogous to the "experimental contribution of physical or chemical science," but as he is drawn further into the question of how to describe the conditions under which his observations have been made, and his own role as observer, doubts begin to surface. He becomes increasingly aware that his observations are being made not by a mechanical device but by a self, *his* self. [10] Even Malinowski's Martian ethnographer, who might wish to treat the native as a particle, à la the physicist, will find himself encountering difficulties. His questioning of the natives—the interview technique so basic to anthropology in the field—will not be trustworthy, for example, for the informant's recital of a moral rule may not coincide with actual behavior. The ethnographer must observe what is done, and if he is to understand its meaning, he must have a knowledge of the native's language and of the meaning context in which the act is performed.

What the Martian discovers is that human phenomena—the messy details of rules and rituals, belief systems and social practices—are filled with ambiguity and, worst of all, with meanings that pose problems for an imagined objective observer. When the human ethnographer is himself part of what is observed — both in terms of his awareness of self and the other's awareness of him — his confidence erodes. Is he, therefore, left with merely subjective impressions? With sheer uncertainty?

Our Martian could be a follower of B. F. Skinner and insist that behaviorism is all. Or insist that he can observe "unintended consequences" with mathematical exactitude. I do not wish to pursue our *jeu d'esprit* with more rigor, however; I will simply let it signal the sort of problems to be encountered in thinking about the human sciences. In place of further cogitations on the deus ex machina — read, Ethnographer — from Mars, let us turn to the specific problem of aims.

Aims of the Human Sciences

The presumed aims of a human science are usually many and varied, frequently vague, often utopian. Still, we can identify a number of them and indicate the sorts of claims on which they are based. Without exploring any of these aims and claims in depth, we can gain a purchase on the problem in regard to the human sciences. The comparison with the similar problem facing the natural sciences should be implicit.

Before engaging in this task, however, I should enter a demurrer. Much of the discussion generally undertaken in this area is highly formalistic, employing abstract philosophical distinctions and committed to terminological disputes. The discussion is frequently more concerned with logical principles than with immersion in the human sciences. Often, this kind of philosophy of science offers benefits, especially clarity and precision, and I have sought such gains. My basic approach, however, while also philosophical in intent, is not of this sort.

In attempting to understand scientific method, I have approached the subject first and foremost from a historical perspective, seeking to understand its emergence from the actual practices and changing thoughts of those working in the disciplines. Yet the present work is not a history of the human sciences any more than it is a philosophy of them, although I have tried to benefit from such histories.

There is neither an ur-model of science nor a Platonic ideal, nor, in the case of the human sciences, a history that suffices. There is only the human attempt to understand the natural and social environment in ways that come to be considered scientific and that eventually adhere to acceptable criteria for collecting and weighing of evidence, validating procedures, making testable predictions where feasible, verifying results, and so forth, by an agreed-upon scientific community. Here, however, I will say no more, for much of what follows is intended to exemplify and spell out rather than merely describe the perspective being advocated.

Understanding

Almost all effort at science, human and natural, is an attempt to increase understanding. More specifically, *understanding* (which comes from the Old English for "a mental grasp" or "comprehension") is used here as a synonym for *knowledge*, for *science*, as the etymology of the word in Latin informs us, means "to know." Not all that we understand, or know, is science, however. For it to be science, it must be "systematized knowledge derived from observation, study, and experimentation carried on in order to determine the nature or principle of what is being studied" (*Webster's New World Dictionary*).

In earlier times, the key word was "systematized." Thus, in the Christian Middle Ages, theology was Queen of the Sciences, for it certainly systematized our knowledge of the supernatural and, by derivation, the natural as well. We are less sympathetic to its claims today, after the scientific revolution of the seventeenth century, for we find it short on controlled experimentation and long on intuitive assertion and unverifiable observation.

Although systemization is a critical component of what we take to be scientific understanding, it is not sufficient: the telephone book is systematized, yet we would hesitate to place it in the canons of scientific literature. Still, systematization we must have if we are to have *scientific* understanding as opposed to or distinct from other kinds of understanding, such as isolated insights or intuitions.

But is scientific understanding the only kind deserving of the name? Many positivists and logical positivists are happy to take an affirmative position, ruling out as irrelevant or nonsensical other claims to knowledge (as distinct from belief). For them, the arts, for example, would also fall into this irrelevant category. My position on this issue is implicit in what has already been said; further comments will be reserved for the chapters on positivism and hermeneutics.

Understanding, then, with all the definitional nuances and qualifications that adhere to it, is one all-encompassing aim of any human science. The question of what we are trying to understand—the content of the human sciences — remains open. Is it human nature? Is it how societies function? Is it the laws of social change? This is a major subject to which we shall return.

Prediction

If understanding is an aim of any human *science,* one test of understanding may be predictive power. If we really know, so it is often asserted, we should be able to predict the course of events. The French scientist Pierre Simon de Laplace, who lived in the late-eighteenth and early-nineteenth centuries, may have indulged in hyperbole (and obviously did not anticipate quantum theory) when he boasted that if he knew the location and motion of every particle in the universe at a particular moment, he could predict every possible future state. Although this view is no longer seen as tenable, even sober physicists still take it as a commonplace that, knowing the present position of electrons in a closed system, they can predict the force of a nuclear blast. Even probability theory bows to determinism where the boundary conditions can be known.

Is prediction possible in regard to the human sciences? This was certainly the hope and belief of early social scientists. Where religion indulged in prophecy, secular science was to substitute true prediction. What this meant was a subject of great contention. The Marquis de Condorcet, the eminent French progressivist, believed that mathematics could construct a "social art" whereby predictions could be made on the basis of probability theory, so that insurance schemes and welfare provisions could be scientifically constructed to take care of future happenings.[11] Karl Marx went much further and in his confident Hegelian moods predicted the inevitable rising of the proletariat against the bourgeoisie (in his better moments, Marx was aware of the contingent nature of his prediction).

The reductio ad absurdum of this belief in social science prediction can be found in the science fiction (a contradiction in terms?) of Isaac Asimov. In *Foundations* he posits a science of "psychohistory." That this has nothing to do with the hermeneutic-based science of Erik Erikson and others is made clear by Asimov's definition: it is "a profound statistical science."[12] It offers predictions, in some cases with a 98.4 percent probability as to the future history of the Galactic Empire. This is sheer hocus-pocus. When, however, Asimov has his scientist-hero Hari Seldon remark, "Were you to discover the ins and outs, our plan might fail," he is touching on on important point: if humans know how they are supposed to behave, they may behave otherwise.

In any event, prediction of some sort is a recurring problem for the human sciences. Practically, the question can be asked, if the human sciences cannot offer reliable predictions of some sort or another, what good are they? Even if we take prediction to mean foreseeing scientifically a course of events that will occur *unless* we take certain actions to head them off, thus falsifying the narrow prediction, ought we not still to expect any science worthy of the name to be able to predict? If not, we are forced to rethink the aims and purposes of the human sciences.

Prescription and Self-Fulfilling Prophecy

Closely aligned with prediction are prescription and the self-fulfilling prophecy, with each verging on the other. Hari Seldon in *Foundations* had to hide what was to happen so that no actor could purposefully change the outcome. In prescription, the actors' knowledge of the ins and outs can become essential to the realization of the prediction.

Let us consider some examples. The department store magnate Edward Filene observed in 1919 that "mass production demands the education of the masses; the masses must learn to behave like human beings in a mass production world." [13] Another example, this one from Charles Lindblom and David Cohen in their book *Usable Knowledge:* "Inflation appears to be an example of a problem that cannot be solved until certain groups of people learn different behavior," which is to moderate their income demands. [14] (I am not concerned here with the correctness of the prescription, only its form.) In both examples, the predicted behavior—mass consumption or the ending of inflation—is predicated (and I use the word advisedly) on people acting in a certain way laid down for them. Such behavior can also be "unconscious" or "unintended," that is, appearing as a consequence of conscious behavior whose aim is other than that which finally emerges.

Perhaps the most famous example is connected to Max Weber's Protestant Ethic. His thesis is that the life of economic man is prepared for by the inculcation of a highly controlled, calculating, and self-monitoring way of life. Derived as this character is from Calvinist religious admonitions, it has the unconscious and unintended consequence of being advantageous in capitalist society. As Thomas Haskell sums it up, Weber "recognized that the market itself, by rewarding some character types and penalizing others, encouraged the development of the utility-maximizing sort of person that its efficient operation demanded." [15]

Only if a sufficiently large number of individuals behave as prescribed, as utility-maximizers, does the market operate in a fashion that lends itself to reliable prediction. If enough humans act in an economically self-interested manner, the outcome of their behavior takes on a regular, and thus calculable, form. Such behavior is not "natural" (although Adam Smith sought to root bartering, and by extension the market, in a propensity of human nature); it is prescriptive.[16]

It is not only in economics that prescription can operate. The seventeenth-century English philosopher John Locke insisted that men contract, as rational beings, to enter civil society and to establish a political community. Though based on a fiction of natural rights, this description-become-prescription played a role in causing his fellow Englishmen to establish a political community on the Lockean model. In this case, moreover, what began as a "what should be"—a nation as a community of free-willing individuals—became a "what is," thus presumptively solving this perpetual dilemma of the human sciences.

If humans will behave as prescribed, then their behavior can be predicted. This is one formula (which we will take up again later in the book). In reality, humans are erratic, inconsistent, changeable, and even perverse. Only machines are, in principle, totally programed and thus predictable. The Industrial Revolution had in mind to create not only novel inorganic machines but new mechanical men as well.

Typically, Josiah Wedgwood, founder of the pottery manufacture that bears his name (and grandfather of Charles Darwin), described one of his tasks as "to make such *machines* of the *Men* [his factory "hands"] as cannot err." Jeremy Bentham, his contemporary and designer of the infamous Panopticon, wondered "whether the liberal spirit and energy of a free citizen would not be exchanged for the mechanical discipline of a soldier. . . . And whether the result of this high-wrought contrivance might not be constructing a set of machines under the similitude of men?"[17]

It is not an accident that Bentham also tried to construct a human science, his "felicific calculus," that could give apodictic, that is, certain, knowledge concerning such Man-like machines. If humans are indeed machines, as not only Wedgwood and Bentham hoped and as many behaviorists and even artificial intelligence researchers believe, then they may be as predictable as other mechanical devices are. In this light, prescription may be viewed as a means of turning humans into machinelike subject-objects whose behavior is then predictable.

The other, more limited form that prescription takes may be viewed as the self-fulfilling prophecy, made famous by the sociologist Robert Merton. Stated simply, if people believe something is going to happen, and act on that presupposition, then it comes about. For example, if I (or, better still, a megafinancier) predict that the German mark will go up in value, and act accordingly, enough people will follow suit so that the mark will go up. A more portentous example: if Marx predicts the proletariat will inherit the earth, and enough workers act on that prediction, they may indeed inherit the earth.

Prediction, prescription, self-fulfilling prophecy—they are all linked. But in whatever form, prediction in the human sciences is fraught with difficulties not to be found in many of the natural sciences. For many observers, the free-willing aspect of the human subject makes it different from other objects. In fact, as in all else, we are faced with a matter of degree. How predictive is the theory of evolution by natural selection in the fine grain of nature? We can predict that natural selection will operate, but in exactly what manner may escape us. More certain seems the prediction that a missile launched at the moon will reach its target. Moving to the social sciences, we can also predict with some confidence that births tend to go up after a war and that lowering the interest rate will stimulate capitalist investment—more or less. As with all sciences, however, at every point we must add "other things being equal" and "given the following conditions." We must acknowledge that the setting of conditions—boundaries—appears far more tentative and uncertain in the human than in the natural sciences. As a result, the problem of prediction, as of understanding, can be dealt with significantly only in regard to a particular human science and *its* attempt at establishing useful boundary conditions.

For our purposes, we need only note at this point the importance of prediction as an aim professed by many in the human sciences, observing at the same time, however, that, in practice, prediction of any important consequence turns out not to be the strong suit of the human sciences.

Control and Power

If prediction is chancy, what sort of utility can the human sciences claim? Some practitioners answer that the aim is to control, rather than leaving matters to chance. Thus, typically, an early sociologist, Edward Alsworth Ross, in 1898 urged that "the right persons" (presumably social scientists like himself) should undertake "the study of moral influences . . . for the *scientific control of the individual*." [18] The same idea is implicit in the motto of the Chicago Exposition of 1933: "Science Finds. Industry Applies. Man Conforms." The assumption behind such statements is that science is sure and acts upon Man, the object, in no uncertain terms. Yet, at the very beginning of the attempt at human science, in this case history, Herodotus voiced doubts about the proposition: "Of all the sorrows which afflict mankind, the bitterest is this, that one should have consciousness of much, but control over nothing." [19] The issue is joined.

Control can be envisioned by its supporters in terms of precise, calculable laws or in vaguer and more general terms. When Michel Foucault announces that knowledge is power, he means that "society"—in his case, bourgeois society—imposes a discipline upon its members, whether in asylums, penitentiaries, or the economic and social sphere, in the guise of social science. Thus, social science offers control, but, in fact, in the form of ideology rather than real knowledge.

A less extreme or "ideological" version of this position can be found with Lindblom and Cohen. They put little stock in the notion that social science is the pursuit of verifiable propositions, or "truth," or an increasingly correct representation of reality. Instead, they believe that its aim must be to raise new issues and to stimulate debate and, if possible, to cause policy makers to reconsider their political or social philosophies. In the end, they claim, "some kinds of issues have to be settled by the interactions called 'politics' rather than by analysis."[20]

Politics comes very close to ideological battle (whatever its basis as well in personalities and interests striving after power). In this battle, the claim to scientific certitude, in the form of social science, becomes a potent device. Following Lindblom and Cohen, we can argue that the "nine out of ten doctors agree" syndrome (my example), so helpful in selling products, can also sell social policies. The aim of a human science, then, would not necessarily be to offer scientific knowledge but merely to offer the claim to it.

Some argue that we need not settle for this disappointing conclusion. The human sciences may not be able to offer us apodictic knowledge or Panoptical control (which we may not even want), but it can offer us an analysis of connections that makes sense out of what otherwise is chaos. Such an analysis will have to match up to the data and theory requirements of the particular human science in question—and, to anticipate a later argument, to do so by mixing positivist and hermeneutic perspectives.

Such a claim (whatever its validity in practice) is made by Malinowski in his classic work *Argonauts of the Western Pacific.* His argument is that ethnography introduces law and order into what otherwise seems chaotic and freakish. The result, in his study of the Trobriand Islands, is the transformation of the "sensational, wild and unaccountable world of 'savages' into a number of well ordered communities."[21] We are granted, or so the claim goes, an understanding—anthropological science—denied to the natives themselves.

Thus, it can be argued (an argument to be taken up again at length in the concluding chapter) that control may take the form of consciousness, that is, mental comprehension, as an aim of the human sciences, for conscious analysis and understanding *is* a form of control over phenomena; and it is a form of science. This may sound paradoxical, for, indeed, analysis of this sort, à la Malinowski's transformation, is almost always multicausal and historically conditioned — making control along the lines of positivist or behavioral aspirations more or less impossible. In short, even when we are conscious of phenomena and their causes, achieving a kind of mental control, we may have to conclude that we are probably unable to affect them in a desired way.

Yet the human desire for control, for power, over social phenomena, as well as natural phenomena, persists. We see the paradox dramatically presented by even so astute a philosopher as Jürgen Habermas, when, as Thomas McCarthy paraphrases him, he calls for a "social theory [read, "human science"] designed with a practical intent." Yet Habermas wishes for a life "free from unnecessary domination." [22] How is one to have practical intent without domination? Is this to hanker after utopia? The implicit hope is that human science will have practical consequences — this is the famous Marxist theory and praxis argument — by providing scientific certainty, an acceptable form of control, in the context of which freedom can grow. It is a position, suitably modified, toward which I am not unsympathetic.

Accumulation and Interpretation

We have pondered some of the possible aims of the human sciences: understanding, prediction, power, and control. We need now to examine one other aim, which can be understood only when placed in the context of the others. It is the problem of accumulation. To qualify as scientific knowledge, knowledge must presumably grow and heap up. This is a modern view, and is certainly the positivist view. It seems to be an achievable aim in the natural sciences. As Newton said, he stood on the shoulders of giants and could therefore see further than they. We do not need to read Newton himself to share in his discoveries. They stand available to us in textbooks on physics. On these formulas and laws, we can build further science. We are in the presence of symbolic formulations that allow for precise transmission and scientific elaboration and accumulation.

In the light of this ideal, what is the situation in the human sciences? Faced with grand political and social philosophical systems, the positivist's reaction, as phrased by Richard J. Bernstein, is to say that the "trouble [with such systems] is their tendency to confuse fact and value, descriptive and prescriptive judgments. Whatever worth such a study might have, these traditional systems do not lend themselves to systematic, rigorous formulation by which they can be empirically tested." [23] Or, to put it another way, they offer little in the way of accumulated scientific knowledge, or even the possibility thereof.

What shall we make of this criticism? The first thing to note is that the same criticism is made today even of the natural sciences. Some historians of science, such as Georges Canguilhem and Thomas Kuhn, and such philosophers as Louis Althusser and Michel Foucault, as well as many postmodernists, speak of "ruptures," or breaks, or paradigms in science, rather than of scientific accumulation. As Alan Ryan sums up their position, "Science was not the accumulation of particular pieces of knowledge, but the imposition on experience of changing conceptions of reality, of how it operated, and what counted as knowledge of it. Intellectual history was not the history of individuals but of these structures. Galileo's heliocentric universe was not a development of the geocentric universe that preceded it, but its revolutionary overthrow. Its acceptance was not a matter of adding one new belief to an old stock but of agreeing to see the world differently." [24]

There is something of value in this view. As Anthony Pagden and others have pointed out, it is not always new facts but the new way of looking at them that can be said to mark the advance of knowledge. [25] What is overlooked by the rupture adherents is precisely that the seventeenth-century break shown in the history of science was marked most fundamentally by the elaboration and acceptance by a newly formed scientific community of the practice of scientific method. It is the very practice of this method—changed over time in accord with its own findings and in accord with the nature of the phenomena to which it is extended, but unchanged in its fundamental attributes—that constitutes modern science and is the basis of its accumulated findings.

If this view is accepted, the cumulative nature of science is at least partially rescued. What implication does this finding have for the human sciences? If they are devoid of true scientific method, are they not by definition a matter simply of constant ruptures and no cumulative value? We can see now why the effort in this book to understand and employ the scientific method of the human sciences is so important, for it stands at the heart of the cumulation problem.

In this light, let us tentatively address the question Why read Auguste Comte or any other human scientist? Why not extract from him whatever is of verifiable value, put it in a textbook, and read that, along with other, similar materials? The answer must be that in the human sciences we are faced with a different sense of accumulation than in, say, physics. We read Comte because he is both part of the past of humanity that we are trying to understand and a theorist about the emergent phenomena of industrialization and science. His views have a different kind of validity from that of a scientific experiment.

The human sciences, though they, like the natural sciences, are subject to the constraints of employing a proper, suitable scientific method, do work differently. It is important to realize that in the natural sciences there is a relatively fixed text, in the sense of the objects to be interpreted. For Newton, whereas stars and apples may move and even decay, their properties as objects susceptible to the force of gravitation remain constant. The earth's history is fixed in a way that written texts are not, so stratigraphy, studying the strata of the earth, is uniform for the entire globe. Written texts, a major part of the human sciences, are different from apples and rocks. They shift in a way that is not tectonic but emergent; consciousness changes as to what constitutes their meaning.

In the human sciences, we are often faced with the distinction between primary and secondary texts. The primary texts, such as Comte in intellectual history, have meaning mostly, if not solely, in terms of the subjective meaning that we attach to them. The secondary texts amount to an ersatz form of accumulation, a potted form of the reality studied by the human sciences (other than the behavioral). What is more, even if consensual knowledge, based on the texts of, say, the nature of feudal society, could be imagined — in other words, accumulated — later historians would be faced with the shift from feudal to commercial society, an emergent phenomenon with novel structures based on new principles of human consciousness. Such a shift would seem to erode the earlier accumulation almost beyond recognition.

Does this mean that accumulation in the human sciences is impossible? I shall argue later that this is not the case, that the question is poorly put. For much of the human sciences, accumulation will take place in a manner different from that of the natural sciences but no less real for all that. The natural sciences accumulate verifiable knowledge, on which further knowledge can be built in order to achieve greater understanding, prediction, and control. The human sciences mainly accumulate increased consciousness — understanding, but without the attributes of the natural sciences other than that of accordance with scientific method. What this last means I will detail later.

Before proceeding, one major qualification must be made to what has been said. In the human sciences, there is also the possibility of acquiring scientific understanding of processes that are lawlike and calculable. These are the unintended consequences that economists especially strive to comprehend. They can perhaps be found in other areas of human life, and it is the conjunction of our understanding of unintended consequences and the interpretation of texts wherein may lie the center of the human sciences.

Interpretation

It should be obvious by now that the heart of the human sciences is interpretation, not positivist observations, experiments, verifications, and predictions. One of the elements that I will seek to add is the reinstatement of positivism, in the form of scientific method, as an integral aspect of the human sciences.

As the Latin etymology of the word *interpret* informs us, an interpreter is an "agent between two parties, broker [does this allow for extension to economic activities?], negotiator." Such an agent makes things "understandable," supplies "meaning" (*Webster's New World Dictionary*). Interpretation clearly requires at least two parties; and in one sense, a symbol is a form of currency permitting exchange. I will use the word *hermeneutics* to encompass all of these definitions.

Merely to say that interpretation is central leaves us with the question How do we decide among competing interpretations? This fits in with the larger question, applicable to unintended consequences as well: Given the variety of different explanations that may appear to satisfy the criteria for truth, how do we know which to select? These questions will be addressed in the other chapters of this book. In the immediate chapter after this one, however, I shall focus on positivism.

One conclusion I can state now, even before a fuller examination of positivism, is that the human sciences in a purely positivist mode have, to a great and perhaps overwhelming extent, failed. This appears to me beyond contestation (though not argumentation). On the other hand, that an increase in "scientific" understanding and consciousness has occurred also seems to me an obvious fact. We know today not only much more about the external and physical world but also about the internal and social world. It is the conditions, or rather preconditions, of this knowing that I need now to make clearer, thereby preparing to answer the questions Is a human science possible? and What shape would it take if it is possible?

Preconditions

In discussing the possible aims of the human sciences I am aware of how much deeper other thinkers have gone into some of these matters. I am also aware of the innumerable other problems that bedevil any discussion of our subject. Questions about verification, experimentation, repetition of observations, the role of classification, objectivity, value-free research, and so on, are rightly addressed as much to the human as to the natural sciences. Again, I would point out that the most useful discussion can occur only in regard to the discipline involved (see Chapter 5).

There are other perennial problems. Richard J. Bernstein, mainly with the social sciences in mind, identifies some of them when he says, "Primary questions have been raised about the nature of human beings, what constitutes knowledge of society and politics, how this knowledge can affect the ways in which we shape our lives, and what is and ought to be the relation of theory and practice." Further on, he expands on the function of theory, which is supposed to be "its ability to help us distinguish appearance from reality, the false from the true, and to provide an orientation for practical activity." [26] For the moment, we will not pursue these questions in the abstract — the task primarily of philosophers and social theorists — but just acknowledge their existence.

But one postulate must be firmly established from the outset: the secular nature of the human sciences. This is a historical note sounded from the eighteenth century on. It becomes a trumpet blast by the mid-nineteenth century, after Darwin. A typical forceful statement of the position can be found in the twentieth century with Edward O. Wilson, who starts from the conclusion that "Darwinian natural selection, genetic chance and environmental necessity, not God, made the species" and ends by stating categorically that scientific naturalism "is the essential first hypothesis for any serious consideration of the human condition." [27]

Whatever the ultimate origins of the universe, our study of both natural and human phenomena must be in naturalistic terms if we hope for scientific understanding. Even William James, who had a strong religious streak, insisted that the notion of a soul was a "superfluity for scientific purposes." [28] We must take it, therefore, as given that science cannot appeal to deities for its explanations, but only to secular forces.

Still, the search for scientific truth, natural or human, has been frequently tied to religious inspirations. We must recognize that, historically, there is no sharp dividing line between religious attempts to explain and understand phenomena and what we call scientific ones. The pursuit of science is continuous with prescientific attempts to understand, predict, and control events (the difference being the critical attitude). Science also developed with strong connections to magic. To speak of the warfare between science and religion (actually, theology) is a gross simplification, as the modern history of science has shown us.

The "progress" of science, we now realize, has been highly mediated by religious concerns. It was, for example, inspirations drawn from the Bible and its account of the Deluge that encouraged much of the research that underlay the emergence of geology as a modern science. So, too, was geological theory influenced by religious belief, illustrating in this case the way science draws on nonscientific fields for many of its insights.[29] We can expect that nonscientific inspiration will be even more acute in regard to the human sciences.

Science must therefore define its method in secular terms, yet the scientific pursuit of knowledge often has religious inspirations. The paradox is only apparent. We commit the etiological fallacy if we confuse the social or personal grounds for engaging in science with its logical requirements. Secularization is a logical requirement, although it has its own history.

In addition to a general trend toward a secularization of knowledge, three historical developments qua preconditions have helped establish the prerequisites for the unfolding of the human sciences. The first is the Age of Discovery, with its encounter with the "Other," raising in new form the question of human nature and what came to be known as culture. The second is the seventeenth-century scientific revolution and its model of scientific thinking, which, in the so-called Age of Reason, was extended to the human sciences. The third is Darwinian theory, with its insights into humanity's evolutionary nature. I shall be touching only briefly on these episodes, but enough, I trust, to show what is involved.

The Age of Discovery and the "Other"

An informed, complete, and imaginative history of the "Other" would be useful. In lieu of such a work — for it doesn't exist — let us focus on a few highlights. From the beginning, members of the human species have needed others to define themselves. As Christopher Herbert puts it, "Human beings urgently need a sense of boundedness, of definite limits, and . . . they create it collectively by classifying other groups as alien." [30] Almost all nonhuman species have what is other and threatening to them imprinted genetically. For example, one species of ant can recognize an enemy species by smell. In contrast, humans learn by culture what is to be treated as other; and since culture can change, even rapidly, so can the definition of the particular other.

For our purposes, at least as important as the need for others are the significant changes in the very understanding of what is meant by "other." For most of human history and in most societies, "other" has referred to aliens, generally regarded as barbarians and often as nonhuman.[31] Only with the growth of travel literature and a sense of history do hints of a new attitude toward the other surface.

The first major shift in consciousness in this regard appears to have occurred in fifth-century Athens with the rise of philosophy and incipient science. With Herodotus's *Histories,* as Stephen Greenblatt remarks, we are in the presence of the "first great Western representation of otherness."[32] The representation is more dramatic and philosophical than scientific, for we are not yet presented with the concept of culture. That concept, and the rise of what will eventually be the discipline of anthropology, appears to originate in the New World discoveries. Here, Europeans discovered peoples, forms of societies, and ways of living never known to the ancients. In Carl Degler's words, the explorations from 1492 on represented a "large laboratory of human experience."[33] Human nature became problematic in a modern sense, and the nature-nurture controversy took on new specificity.

Not that such topics were entirely new. Timothy Hampton, in an intriguing article on Rabelais's and Erasmus's attitude toward the Turks, shows how, before 1492, Europe was forced to contemplate the barbarians at its gates (literally, for the Turks had besieged Constantinople in 1453) and to understand them in a Christian humanist context. The means for such understanding was to be found in the spirit of charity. Yet charity could go only so far in a time of ideological and martial struggle with an enemy who was perceived as "at once totally alien and strangely familiar."[34] In his closely argued text, Hampton focuses on the expression *Turkish dogs* to illustrate the ambivalent response of Christendom to these others on its borders.

The New World posed a much greater challenge. Here were new creatures, or so it seemed. Were the natives to be treated as humans, with souls potentially to be saved, or as cannibalistic inhumans, potentially to be enslaved? Theologians, especially Spanish ones, tortuously argued the matter. Out of such philosophical musings, and the recorded experiences and observations of conquistadors, missionaries, and travelers, emerged, gropingly, a new sense of what has come to be called culture. This entailed an awareness that humans might be alike as a biological species and yet different because of the cultures in which they lived. Native Americans might be as moral as Europeans and yet their morality take a vastly different shape from that of their European discoverers.

Indeed, some Europeans idealized the natives and saw them as more moral than themselves. The noble savage was born, not on the plains and in the jungles of the New World, but in the libraries of the Old. This newfound awareness and acceptance of the Other both contributed to and benefited from the slow growth of tolerance in a Europe still struggling through its religious wars. A fascinating figure in the emergence of this new attitude to the Other is Michel de Montaigne.

We are still a long way from science, and the discipline of anthropology. As Tzvetan Todorov reminds us, Columbus advanced upon the New World with a more or less closed mind. He knew what he would find: ancient India and a confirmation of the truths of the Bible. In Todorov's words: "Concrete experience is there to illustrate a truth already possessed, not to be interrogated according to preestablished rules in order to seek the truth." All observations were made from a position of superiority and domination, not equality. The Other, for most Europeans, was still inferior. (Conceiving the native as an equal, as Todorov adds in a brilliant insight, was possibly worse because equal status meant the identification of "the other purely and simply with one's own 'ego ideal' [or with oneself])."[35]

The basis for the development of modern anthropology had been laid.[36] The aim of scientifically understanding different cultures gained prominence, even though it was only sporadically achieved. As Christopher Herbert, George Stocking Jr., and other scholars have shown, however, the problems of preconceived opinions and superiority and domination, along with issues of subjectivity, verification, and so forth, persisted for a long time. The West still took the Other as its object while declining to put itself in the reverse position.

From their inception, the human sciences have been overwhelmingly Western. Until very recently, what can anthropology have meant to a Malaysian or a Nigerian other than a Western form of domination? Although new anthropological literature suggests that the supposedly inferior others in fact scrutinized their conquerors as themselves others, it was clear where the power and the "scientific" knowledge lay. If we go beyond anthropology to the human sciences in general, we must ask, What did positivism mean to most non-Westerners or, in any case, non-Europeans?

Our questions are rhetorical. One of the challenges was, and is, to incorporate non-Western experiences and data in a trans-Western science. Recent efforts to revise sociological concepts in the light of new scholarship in the fields of Ottoman Turkish, Mughal Indian, and Safavid Iranian studies show the direction that must be taken. I will not pursue the details of this effort here, but merely rest content with citing it as an example of what must be done in addressing the broad problem.

Ironically, in taking the Other as an object of scientific scrutiny, anthropology tends benevolently to wish to preserve its object's existing culture. In practice, this means denying to natives the increased self and social consciousness that anthropologists seek to derive from their scientific study of "them." The native, ideally, is to remain undisturbed by the Westerner. What is more, the holism and wholeness of native society is often set before the West as an ideal, in opposition to the fragmentation and changeability of modern (Western) society. Again, as a matter of science, there is an irony in measuring modernity in terms of a premodern society whose members can hardly conceive of the type of consciousness required to originate the scientific study of the Other.

Such antagonism to the modern by its own offspring, the science of anthropology, is partly understandable in the light of the history of the discipline. Until very recently, the Other was to be exhibited, like other fauna. From the Crystal Palace Exhibition of 1851 onward, halls were devoted as much to living "primitives" as to the more or less primitive machinery heralding the industrial revolution. At the Louisiana Purchase Exposition of 1904, "the native people of Alaska, the Eskimo; inhabitants of the recently conquered Philippines, some 1,200 in all; the indigenous people of Japan, the Ainu; natives of South America and representatives of fifty tribes of North America; Zulus and Baluba from Africa; and pygmies from the Congo" were all exhibited in their native dwellings — huts, tepees, treehouses, and igloos.[37]

Ethnological museums afforded permanence to these exhibitions, later enhanced by films taken in the field. Statistical measurements were also taken of the natives — of their vision, hearing, and other sensory responses, for example — generally with the pleasing outcome of showing their inferiority to, or at least difference from, Europeans. The "science" of eugenics arose to look at the Other (as well as inferior members of one's own race, the internal Other) in the way a breeder might regard livestock.

There is a further irony: in a number of cases, the classifying and inquiring mentality of Western scientists had the effect of converting the Other into human beings closer in identity to Westerners. In a very important article, "Science 'Gone Native' in Colonial India," Gyan Prakash investigates how this happened for the elite of the subcontinent. As he informs us, as early as the 1840s, in preparation for the Crystal Palace Exhibition, artifacts were collected for local exhibitions in India "as instruments for promoting commerce and advancing a scientific knowledge of economic resources."[38] In 1865 the Nagpur Exhibition attracted thirty thousand visitors in the course of eight days; and visitors poured into numerous succeeding exhibitions; a million went to the Calcutta International Exhibition of 1883.

As part of this process of acculturation, museums sprang up to teach classification to the natives. In a fictional account, we are told how a Bengali schoolteacher learns that knowing about tigers and ostriches as fierce or large animals is not enough. "This is learning things without method. . . . I am so glad that Mr. W. has hit upon this plan of teaching the teachers to value system. In fact he has given us a *second sight*. When I first entered this great hall, I was perfectly bewildered at the vastness of this collection, and had not the least idea in what order and plan they were arranged. I have got at least some notion now of their arrangement."

Ironically, the European's desire to teach science to the Other meant having to recognize the subjectivity of the new student. Only as subjects, not objects, could Indians learn. As Prakash puts it, "If Indians, who were objects of knowledge, were recognized as knowing subjects, the very strategy of hierarchizing and displaying them as objects was invalidated." Once Indians were acknowledged as knowing subjects — an enormous advance in anthropological science — they were also recognized as having intentions and motives of their own. How to read these intentions is a central feature of modern anthropology, as of the human sciences in general. Prakash concludes that we must undertake a "rethinking of our customary notion that the colonial discourse of modernity and science produced nothing other than domination."

Rethinking also forces us to regard the concept of the Other as fundamental to the emergence of the human sciences. "O wad some Power the giftie gie us / To see ourselfs as ithers see us!" Robert Burns intoned. In recognizing the Other, we ultimately come to see ourselves as Other as well. We become an object to ourselves. Yet we know that object to be a subject, endowed with intentions, motives, values, and rules, inherited partly through nature but mainly through nurture, or culture.

The practice of anatomy (a topic I shall discuss further in Chapter 5) made the human species, physically, an object of science. The concept of the Other prepared the way for a "spiritual" anatomy. Although, as has often been suggested, we should study our self and our other selves, that is, other humans, in a value-free manner, in practice, value-free study is entangled with the fact that the object studied is a subject with values.

The New World discoveries not only fostered this inner awareness, but placed the age-old question of human nature in a laboratory of new experiences. Out of this novel mental perception and this body of experimental data gradually came a new human science, anthropology. Born with the caul of Western imperialism about it, the preliminary task of the fledgling science was to break free of its natal condition. Breaking free would mean the freeing of the master as well as the slave, to use the Hegelian metaphor. Only by coming to terms with the Other in our selves, as well as outside, could the human sciences experience further growth.

Seventeenth-Century Science and the Enlightenment

It is hardly necessary to rehearse the achievements of the so-called scientific revolution of the seventeenth century or the Enlightenment of the next century. As it happens, most of what I have to say about the subject will be found in the next chapter, on positivism. Let me highlight here, then, the way these achievements function as a historical precondition for the emergence of the human sciences.

It was the seventeenth-century scientific revolution that effectively taught Westerners, and then humankind as a whole, how to think scientifically. We are all familiar with the litany of shoulder standing—Copernicus, Brahe, Kepler, Newton, with lesser giants attending—that makes for the revolution in knowledge that we call science. It needs to be reemphasized that this science was at first entangled with magic, mysticism, and religious impulsions. What needs further discernment is that these latter forms of knowledge offered competing explanations—understanding—of phenomena over which humans wished to achieve control or at least a sense of order. What Newton and company did was to offer alternative explanations. Even when cloaked in number magic or religious inspiration, their explanations differed fundamentally in being scientific—they appealed to scientific method. Although the heliocentric theory and the gravitational theory were great achievements, what was even more persistently earthshaking was the very conception of a form of inquiry whose boundaries knew no limits other than obedience to the dictates of experience, experiment, verification, and theoretical imagination, among other things.

Whatever the historical origins of modern science, in the end it shook itself free of religion. It had to do so, for religion disdained free inquiry and imposed judgments based on certain and unquestionable (in principle) authority. The special achievement of the seventeenth-century scientific revolution, aside from specific theories, was to free astronomy and physics from the constrictions of religious dictate and to institute scientific method in place of religion in these fields. Geology and biology would remain bound until the nineteenth century.

So successful and overpowering were the achievements of the scientific revolution in regard to natural phenomena that the question immediately arose, Why not extend the scientific method to human phenomena? It is for this reason that the period is called the Enlightenment or the Age of Reason. The results were all too frequently short-sighted, overly ideological, and unsuccessful; still, the touted extension of scientific method to the human sciences must be given full due as a critical precondition for their possible development.

Again, although most of what I have to say on the Enlightenment will be found in the next chapter, a few points should be highlighted here. It must be noted that the initial models for how science should be extended to human phenomena were either Cartesian or Baconian. Hence, for example, when Hobbes wished to establish a political *science,* he took as his model Descartes's geometric method (mediated by Euclid).

In the course of the eighteenth century, however, it became clear to some that human phenomena might be more complicated than the subject matter of the natural sciences. Vico's *New Science* (1725) offered an alternative approach (though largely neglected in the eighteenth century) based on hermeneutics. Human nature was somehow different from the mindless accumulations of atoms in external nature. Humans were moral creatures; and such inquirers into human nature as David Hume and Adam Smith saw themselves as working in moral philosophy or the moral sciences, though still under the inspiration of the scientific revolution, especially Newton.

In a parallel direction, there was the effort, pre-Darwinian in its conception, to establish the grand lines of human development. This was the philosophy of history, to be given much extension in the nineteenth century. In the eighteenth century we find its manifestation in the sketches of the progress of the human mind made by Anne-Robert-Jacques Turgot and Condorcet. Man has become scientific man, and his perfectibility is now conceived in ostensibly secular rather than ostensibly religious terms. A similar but less grandiose version of the effort to give lawlike, that is, scientific, shape to the human career over time is to be found in the work of Scottish sociologists like Adam Ferguson, with their depiction of hunting, nomadic, agricultural, and commercial stages. Such hints as I have given about the way the scientific inspiration of the seventeenth century was extended to human phenomena in the eighteenth are to indicate the general point: that the impulse to the *human* sciences came originally from the triumphs of the so-called natural ones.[39]

In 1789 a great shock was given to those aspiring to apply reason to human affairs. Science as a source of authority was discredited in the eyes of many observers by the excesses of the French Revolution. A romantic reaction against reason and science, whose tremors were already evident earlier, now became stronger. In part, romanticism restored the subject to the human object on whose behalf the science of human beings claimed to be legislating.[40] The romantic movement, fascinated as it was with the exotic, added immeasurably to the openness to the Other. It prized introspection, which differed from the earlier religious version by being largely secular; and it emphasized life, process, and intuition as against the mechanical, the mathematical, and the purely rational. Above all, it entered into dialogue with the dominant scientific thinking of the time and introduced a fresh empiricism — life on the pulse — missing from the arid formalism of some of the previous attempts at human science.

A true history of the human sciences would lay out in great detail, with the necessary additions and qualifications, all of these developments and others. I will end here with the coda coming after the French Revolution. I refer to the work of Auguste Comte and to his effort to save the scientific approach to human phenomena by postulating a new science, sociology, and a new version of scientific method, positivism. Suffice it to say that in the subsequent treatment of positivism I will be fleshing out the postulate that the scientific revolution of the seventeenth century and the consequent Enlightenment were essential historical preconditions for the emergence of the human sciences.

The Darwinian Revolution

If Enlightenment thinkers made it clear that humanity was a fit subject-object for scientific scrutiny, their conception of Man was still rudimentary and restricted. It was Darwin who revolutionized our view of the human species and of nature at large. The theory of evolution by natural selection ended once and for all Man's status as a special creation placed in an unchanging Great Chain of Being. Instead, the human species had now to be viewed as an animal more or less like all others, surviving in a chance universe.

The insights of romanticism about life, process, and intuition were taken up anew but were now given scientific form. The word *biology* itself was coined around 1801 to express a new conception of life. The fact that Darwin took with him Wordsworth's book of poetry the *Excursion* on the voyage of the *Beagle* reflects an affinity with romanticism, not happenstance. The ideas of the ceaseless process and constant strife of the elements were given scientific as well as poetic expression.

The emancipation from religion that had started in the scientific revolution in astronomy and physics was extended to geology and biology. It was geological science that challenged the biblical account and made the habitat of humankind — the rocks, mountains, and seashores — a matter of sustained inquiry, not revelation. It was Charles Lyell who saw the earth as a place of law-like phenomena that were subject to uniform transformations, thus preparing the way for Darwin. Indeed, Darwin began his work as a geologist, studying the processes of the earth's formation and only then becoming concerned about the processes of human transformation.

With painstaking care and a stress on scientific observation and theoretical inference, Darwin traced how, in the natural environment, the human species came into existence naturally. We need hardly recapitulate the details of Darwin's work here; what stands out is how he pursued the injunction to treat Man as an object of scientific inquiry like any other and how, as a result of his inquiries, the object of any human science must now be conceived in evolutionary terms. His work is a fundamental precondition for our subject.

As a result of Darwin's revolution, Man is now seen as an animal more or less like any other animal. The "more or less" is very important. When Darwin himself came to treat of the human animal in *The Descent of Man*, he recognized explicitly that what differentiated Man from the other animals was the possession of conscious morality and religion. Somewhat more implicitly he remarked on Man's nature as a symbolic animal, possessed of a special kind of language. It is this insight that is still being explored by the human sciences.

Darwin's work also played an enormously formative role in the development of anthropology. He voyaged in the Pacific and read widely in travel literature. Evolutionary theory made it possible to classify genealogically the plants and animals, the landscapes and the peoples who inhabited them, and to fit them into an overall context. Although the evolutionary perspective as applied to humans became warped into Eurocentrism, it supplied an ordering principle for the early stages of the new science of ethnography.[41]

The Darwinian revolution also spawned one other important scientific offspring: psychoanalysis. There is little question but that Sigmund Freud's brainchild had Darwin as its godfather.[42] What is especially interesting is that, starting from biology and histological studies in the laboratory, Freud moved, slowly and grudgingly, to inquiring into his subject — humans — by their use of words, with his famous technique of free association.

Freud's attempt at objectivity, at distancing himself from his subject of inquiry, was largely by means of self-analysis. He had a predecessor in Denis Diderot, the encyclopedist and author of *Rameau's Nephew*, who led the efforts both toward increased rationality and toward increased attention to the passions. It was Diderot who wrote in the *Lettre sur les sourds et les muets* in 1755 "that one must be at once inside and outside oneself. One must perform the roles simultaneously of the observer and the machine that is being observed."[43] One need not have assumed the human to be an eighteenth-century machine (noting that the term *machine* had debatable and variable meanings) to accept the call to be both inside and outside oneself; and Freud, placing his work in an evolutionary frame, went beyond Diderot in offering a systematic analysis from both inside and outside the Darwinian human animal.

The Darwinian revolution, following upon the Age of Discovery and the seventeenth-century scientific revolution, was an essential historical and theoretical precondition for the development of the human sciences. The Other has become another self, to be examined scientifically, in an evolutionary perspective. What *scientific* means becomes a key issue. As one of the outrunners of the Darwinian revolution, Freud merits special attention because he illustrates the interplay of positivistic aspirations and hermeneutic outcomes in his effort to establish psychoanalysis as a science, though of a most peculiar kind. Darwin, we can say, establishes the vital context for all work in the human sciences, whereas Freud (whatever judgment we place on his work) is on the cusp of the general methodological problems being addressed. We will be dealing with the intentions of these two men, along with the intentions of many others, as we seek to understand more deeply the nature of the human sciences, initially in terms of positivism and hermeneutics.

2 / Positivism

*P*ositivism often is, or at least has been, viewed as a royal road to certainty in both the natural and the human sciences. Philosophically, it is seen as an epistemology that grounds all knowledge in experience and declares some objects accessible to observation and others inaccessible; the former are objects about which knowledge can be obtained, and the latter are not. In the human sciences, positivism takes the form of a sociology that claims that scientific knowledge about society is possible.

In our search for an understanding of the human sciences, we shall not be interested in an abstract, philosophical definition of positivism. The shapes of what can be called positivism are many; and they are often quite far removed from the stereotypes of the subject. Nor are we concerned with writing another history of positivism. Our intent is, instead, to treat the subject historically, though again with no attempt at full coverage. We shall choose our historical examples with an eye to what can be extracted from them in order to enhance our understanding of the possibility of any human science and to give us an idea of the shape it might or should take.

Francis Bacon

In the beginning is Francis Bacon, whose contributions to our subject deserve an extensive presentation. I shall treat this Elizabethan thinker and statesman, author of *The Advancement of Learning* (1604) and *Novum Organum* (1620), as well as of many other works (though not Shakespeare's plays), as the progenitor, protean in his manifestations, of what we have come to call positivism. As Paolo Rossi writes, Bacon is "responsible for a new intellectual attitude to science . . . which the Enlightenment and Kant — and later the positivists — maintained."[1]

In one view of the history of science, Bacon is taken to represent the "empirical" and René Descartes the "rationalist" basis of scientific thought. In a crude way, this may be so. (In fact, empiricism and rationalism — and both are slippery terms — will later be joined in the unified methodology of positivism.) My comparison here will emphasize the similarities and only then the differences in Bacon's and Descartes's approach. After this, we will concentrate on Bacon alone.

Both men took as their task the achieving of certainty, or as much certainty as was possible, in the area of scientific knowledge. Both rejected the authority of the ancients and their written texts. Both called into doubt human custom as a basis for truth — Descartes spoke of belief being different on the two sides of a mountain, and Bacon of the misleading idols of the mind. In short, they were engaged in a common enterprise: freeing the mind from its false constraints so that it might attain a knowledge about the phenomena surrounding it that could be called scientific.

In sharing this aim, both Bacon and Descartes have been seen as "modern."[2] Some scholars chose to emphasize their medieval elements; modernity does not emerge all at once. Others caution that whatever their modernity, it must not be confused with our later modernity: "Full exploration of the evidence indicates that any understanding of the seventeenth-century worldview necessitates attention to the character of religious motivations, of adherence to philosophical tenets which have subsequently been discarded. . . . In other words, while a number of principles were adopted in the seventeenth century which are congruent with modern scientific knowledge, the general conceptual framework in which these principles were maintained was entirely different."[3]

With this caution in mind, we can return to our comparison of Bacon and Descartes. As is well known, Descartes sought to achieve a solid basis for knowledge by doubting his senses as if he were being deceived by an evil being. Retreating to the rock of the self, which he could not doubt, he constructed, deductively, a world of complete certainty.

Bacon's way was different. His firm base was sense impressions. As he said of the Skeptics, "their chief error lay in accusing the perceptions of the sense, and thus plucked up the sciences by their roots." But Bacon was not naive or simplistic in his empiricism. He was fully aware that the senses could also mislead, or even be insufficient if not aided. He continued, "For though the senses often deceive or fail us, yet, when industriously assisted, they may suffice for the sciences, and this not so much by the help of instruments, which also have their use, as of such experiments, as may furnish more subtile objects than are perceivable by sense."[4]

Instruments like the telescope and microscope can extend our senses. Experiments as envisioned here by Bacon seem less to confirm than to expand the world available to our senses. Both instruments and experiments, in the form of the mechanical arts, are a continually developing rather than a once-and-for-all solution to the problem of acquiring scientific knowledge. In Bacon's words, "The mechanical arts . . . seem full of life, and uninterruptedly thrive and grow, being at first rude, then convenient, lastly polished, and perpetually improved."[5]

Descartes had held forth the vision of a one-man science. Just as civil laws were better coming from the hand of a single legislator — for example, Lycurgus (or God) — than from a motley group of legislators, so, too, were scientific laws. Conceived by a single mind, such laws were both certain and more or less complete.

Bacon saw things otherwise. Grounding science on the senses, aided as they must be by improving instruments and experiments, he anticipated the notion of science as a steadily accumulating body of knowledge. Bacon's science would stand on the shoulders of giants, would be cumulative, but would advance because of humble mechanical advancements.

For Bacon, the superiority of the present rested squarely on its pursuit of knowledge by the correct method, available to all. Descartes, too, claimed that correct method was the key: he laid down four simple rules that could be followed by all who possessed common sense. As we have noted, however, only one person needed to follow out the Cartesian rules in order to erect the edifice of science. With Bacon, the effort had to be a collective one.[6] Institutions, for him, were essential if science was to be a continuing, collective enterprise. A Solomonic cohort, assembled in scientific societies, was necessary for the development of knowledge as a dynamic, cumulative affair. Such societies were to be public, unlike the private, mystical circles of the magicians.

The result of this collective effort would be the famous triumph of the moderns over the ancients. Who, Bacon asked rhetorically, was the older, we or the ancients? The answer depended on perspective, foreshadowed as it was by the earlier developments in painting of linear perspective. If one looked from the perspective of firstborn, then the ancients were older. If, as with the age of the earth, one looked from the other direction, we moderns are older.[7]

With this momentous reordering of time, Bacon had a view of history that was supportive of his notion of cumulative science as the work of many hands, aided by inventions and institutions and proceeding by the correct method of experience and experiment. Here, then, we find some of the fundamental tenets of positivism emerging from his far-ranging and farsighted views on the nature of scientific knowledge and the means of acquiring it.

Was such scientific knowledge, so acquired, certain and final? Bacon was repetitively clear on this matter. All that we could obtain was "degrees of certainty" or "fitting certainty."[8] Perhaps more provocatively, Bacon declared that Man was not the measure of all things. He who pointed science resolutely in the direction of "things," not words, was aware of the distorting effect of what is in the mind on our perceptions.

Here we encounter Bacon's famous four idols, or false notions or dogmas.[9] As he remarks of one of them (and the same is implied for the other three), "The idols of the tribe are inherent in human nature and the very tribe or race of man; for man's sense is falsely asserted to be the standard of things; on the contrary, all the perceptions both of the senses and the mind bear reference to man and not to the universe." He concludes, "The human mind resembles those uneven mirrors which impart their own properties to different objects from which rays are emitted and distort and disfigure them."[10]

The mirror image is central to Bacon's conception of the mind and thus to his epistemology. He seems to believe that there is a reality outside the mind, which the mind must seek to mirror. This is hardly a new idea; it can readily be found in medieval thinkers and is not an idea to which we are very receptive today. It is intimately tied to Bacon's idea of God, as when he announces that "God has framed the mind like a glass, capable of the images of the universe, and desirous to receive it as the eye to receive the light."[11] God sees all things, directly and intuitively, in their true essence. Man, encumbered by a body, distorts them. "For the mind," Bacon intones, "darkened by its covering the body, is far from being a flat, equal, and clear mirror that receives and reflects the rays without mixture, but rather a magical glass, full of superstitions and apparitions."[12]

This is the inspiration behind Bacon's conceit of the four idols. In more secular terms, Bacon is stating that Man can approach reality only through the distorting filters of sense perceptions and especially of language. As he puts it later in the *De argumentis*, the idols deriving from language are "intrinsic to human existence . . . for man cannot live outside society and must therefore of necessity use language."[13]

Thus, given our necessarily imperfect nature, all we can do is to seek knowledge in the best way available, Bacon's recommended method, recognizing the siren calls of the four idols and persevering in our quest for a kind of cumulative knowledge, science, that will always be less than certain.

In setting forth this laborious quest for knowledge by basing inquiry on experience and experiment, Bacon took account of the limitations of the empiricist approach. He remarks about "mere experience, which, when it offers itself, is called chance; when it is sought after, experiment." In language that is all too rare in his works, and too rarely attended to by his critics, Bacon goes on to say, "But this kind of experience is nothing but a loose fagot; and mere groping in the dark. ... the real order of experience begins by setting up a light, and then shows the road by it, commencing with a regulated and digested, not a displaced and vague course of experiment, and thence deducing axioms, and from those axioms new experiments." [14]

Could Isaac Newton have said it better? Is not Bacon's "light" an adumbration of the concept of "theory" or "hypothesis"? With this allowed, Bacon has only himself to blame for the common view that in his inductive method the facts — things — seemed to speak for themselves, bypassing the distortions of theory and of language innate to Man the observer. So obsessed was Bacon with the need to clear away the errors of logic, of words, and of the idols of the mind that blocked whatever certainty was vouchsafed to errant Man, that he tended to pass by the theoretical impulses of the mind as it shaped its mirroring of and experimenting with a reality given by God.

In fact, while rejecting Aristotelian forms, though they lingered in his mind, Bacon retained them as laws: "The form of heat or light, therefore, means no more than the law of heat or the law of light." [15] Whether forms or laws, only God can recognize them instantaneously; Man needs slow and uncertain method, in which experience holds the center of a stage illuminated by light, or theory.

Aristotle had written an "Organum" to show men the way to such knowledge. According to Bacon, that way was faulty and "ancient," and he wrote his *Novum Organum,* literally, "a new machine for the mind," to correct the Greek philosopher and set Man on the true path to science. This new machine seemed to be purely empirical, but, as we have noted, Bacon in his better moments did realize that it had to be guided by an inner light. So guided, Man could, through experience and experiment, acquire new knowledge and do so in a continuous, dynamic fashion.

In Bacon's view, such uncertain, but advanced, science could be achieved with both natural and human phenomena. Unlike Descartes, Bacon did not split knowledge and its accessibility to scientific method. As he wrote at the age of thirty-one to his uncle Lord Burleigh, "I confess that I have as vast intellectual ends as I have moderate civil ones; for I have taken all knowledge to be my province." [16] Attempting to take full possession of his province in the "Novum

Organum," after remarking that he seeks only "fitting certainty," Bacon re-
states his original ambition. "Some may raise this question . . . whether we talk
of perfecting natural philosophy alone according to our method, or the other
sciences also, such as logic, ethics, politics. We certainly intend to comprehend
them all."[17] Paolo Rossi summarizes the matter: "Bacon substitutes his theory
of the congruity of natural and artificial phenomena for Aristotle's theory of
art as imitation of nature."[18]

In short, Bacon vindicates the possibility of a human science pursued by his
new method. For the first time, we see a positivist version of the future human
sciences *avant la lettre*. Although a full development of this claim would have
to await development of social art (to use Condorcet's phrase) or sociology (to
use Comte's) in the eighteenth and especially nineteenth centuries, the pronouncement has been set before us.

Separating the knowledge of Man into Human and Civil philosophy, Bacon
said, "We now come to the knowledge of ourselves . . . since the knowledge of
himself is to man the end and time of the sciences, of which nature only forms
a portion."[19] Self and social knowledge were critical, for, although Bacon rejected Man as the measure of things, he also realized that things were perceived
by Man only through the distorting lens of human nature. One danger was
anthropomorphism, which Bacon explicitly recognized and wanted to avoid.[20]

To avoid it, he advocated the study of human phenomena by the same
method advanced for the study of natural phenomena. The problem is that
when we look at Bacon's actual recommendations, we see how fragile and
superficial they were. The following quotations illustrate the problem. He begins, "We come now to those things which are within our power and work
upon the mind, and affect and govern the will and the appetite; whence they
have great efficacy in altering the manners." So far, no caveat. But Bacon continues, "And here philosophers should diligently inquire into the powers and
energy of custom, exercise, habit, education, example, imitation, emulation,
company, friendship, praise, reproof, exhortation, reputation, laws, books,
studies, etc.; for these are the things which reign in men's morals. By these
agents the mind is formed and subdued."[21]

How disappointing this long, disorganized list is, and in the pages that follow Bacon offers only a series of aphorisms and one long disquisition on "the
Military Statesman." These are in a traditional mold and remind us more of
Lord Chesterfield and Renaissance humanists (Bacon himself being, in part,
one of them) than of modern social scientists. The fact, not to be held against
him, is that Bacon did not even glimpse what came to be modern sociology (a
fact that unintentionally allows us to measure the progress in this area); he saw

only a recommended way, a method, by which to reach it. Perhaps enough for one man in the seventeenth century?

One clue as to why Bacon could go no further lies in the provinces that he treated as out of bounds to his inquiry. Two were economics and political science, about which he explicitly resolves to "impose silence upon myself." [22] Although he feels superbly qualified to comment on these areas, he declines to do so out of prudence.

Another off-limits province was theology and religion. We are obliged, he tells us, to obey divine law and the word of God even "though our reason be shocked at it." [23] "Belief is more worthy than knowledge," he says humbly, thus removing the subject of belief itself beyond the reach of his scientific method. He holds the same view concerning the moral law, which "is too sublime to be attained by the light of nature." [24] As can readily be seen, much of what will occupy future efforts toward a human science is missing and indeed banished from Bacon's plans for the advancement of knowledge.

One other province of great importance is also absent in Bacon's scheme: mathematics. It plays almost no role in his organum, although we know how largely it figures in the scientific revolution of his time. We must turn to Galileo and Descartes for this part of the advancement of knowledge, for it was Galileo who spoke of the universe as being written in the book of mathematics and Descartes who glimpsed the mathematical nature of the universe in his famous dream of November 10, 1619. There is no such awareness in Bacon. When discussing heat, for example, he sounds more like the Greek philosophers about whom he complained than a modern scientist seeking to measure the phenomenon and emerging with mathematically expressed laws. [25]

Bacon's ignoring of mathematics had two important consequences. First, it left him with a gaping hole in his scientific method. It is for this reason that so often he hardly figures in histories of science, where he is passed over with scant reference to his one-sided empiricism. Moreover, Bacon made few contributions in practice to empirical science. Thus, even in regard to the natural sciences, Bacon's method, though enormously powerful, seemed also to have serious deficiencies. Future positivism would have to wrestle seriously with these problems.

Second, his neglect of mathematics in the natural sciences yielded an unexpected bonus in regard to the human sciences. It was a glimpse of this aspect of Bacon that led Giambattista Vico, in his *New Science,* to number the English thinker among his four major inspirators. Although later positivists would try to apply mathematics not only to natural but also to human phenomena, human phenomena tend to be less receptive to this approach. Most of what is interesting about the human species and its societies and cultures does not lend itself readily to quantification. By declaring, however, that his scientific method, nonmathematical as it was, applied equally to human and natural knowledge, Bacon vindicated *in principle* the effort to achieve, for example, a sociology that was scientific in some sense yet not quantitative.[26] What that sense was remained to be defined by others.

Why have we spent so much time on Bacon? My argument has been that in spite of all his drawbacks and omissions he stands at the beginning of the sciences of Man. He elaborated a method of inquiry that would later be known as positivism (and that subsequently required a suitable place to be found for mathematics). His method did not exclude human phenomena from its scientific reach, even though he failed to apply his method successfully. Nevertheless, he opened the way for others. Above all, Bacon tried to free the human mind from its encumbrances of past authority and present custom and to make it initially concentrate on things rather than mere words. Even allowing for his omissions, he helped immeasurably to make mind conscious of its own workings and its forms of idolization and to suggest a better method by which to proceed to the understanding of reality. He believed firmly that man could come to know reality. Admittedly, the knowledge would be neither total nor completely certain. It would, however, represent an advance. And the advance, laboriously achieved through the methods of empiricism and experiment, would continue — so Bacon believed.

Bacon shared the millennial aspirations of many of his contemporaries. His advocacy of a return to the original — experience of nature, religion, whatever — in a Puritan England carried the tones of reformation. Indeed, he was calling for a scientific reformation as well as for a scientific advance. By this call, and by changing our time perspective concerning what is ancient and modern, Bacon opened the way to continuing progress. That progress, though for him based on uncertain knowledge, was to be achieved by the self-correcting methods of science, as he had come to define them — in short, by experience and instrumental experiment. The basis of progress was cumulative science in both the human and natural provinces.

Boyle, Hobbes, and the Air Pump

We can see how Bacon's ideas were put into effect and with what new consequences by comparing Robert Boyle with Thomas Hobbes in a prototypical seventeenth-century episode. It seems almost a replay of some of the differences between Bacon and Descartes. One outcome of observing this dramatic conflict should be a greater understanding of the experiential roots of positivism and a sense of how and why it has come to be socially and historically constructed.

The problem faced by the disciples of Bacon in the mid-seventeenth-century, as well as by thinkers such as Hobbes, was a breakdown of both society and its intellectual underpinnings. The English Civil War was the most dramatic representation of this state of affairs, but the problem was endemic. The challenge was how to bring a new order, social and intellectual, out of disorder. Ancient authority could no longer serve; new grounds for legitimacy were needed. In short, the same inspiration that drove Francis Bacon was still being pursued.

Answers to the problem were forthcoming, on one side from Robert Boyle and various members of the Royal Society and on another side from Thomas Hobbes and his followers. Both sides agreed that neither private judgments nor authoritarian fiats would suffice; what was needed were "demonstrations," which could be observed and followed by all who "witnessed" them. The intent was to rise above passions, and the aim was to achieve impersonal knowledge.

Here, their agreement ceased. Hobbes, enamored of Euclid's geometry, espoused deductive "demonstration," whose certainty could be proven by anyone willing to follow the necessary steps in reasoning. Nothing more was needed (as Descartes had shown). Boyle's was an experimental "demonstration," requiring innovations, that is, a changing material technology, as well as experiments, public observations, and a scientific community of witnesses. Its aim, too, was certainty, but only the "appropriate" certainty claimed by Bacon.

The two solutions came into actual conflict over pneumatic phenomena, and Steven Shapin and Simon Schaffer have examined that disagreement in a brilliant book, *Leviathan and the Air-Pump.*[27] Let me extract from their account, as well as others, various points of significance for our pursuit of positivism. My emphasis will be on Boyle and his supporters.

The battle raged over the existence of the vacuum, which Hobbes sought to determine by deductive reasoning and Boyle by the construction of an air-pump. The air-pump can symbolize for us the experimental commitment. It is a technical innovation and can be built by others, enabling them to repeat the experiment.

The problem is best illuminated in terms of witnesses. Boyle's experimental science, one foundation of positivism, requires a social contract whose major clause establishes a scientific community where disputes can be worked out peacefully and in piecemeal fashion. (A comparison with the very different social contract established in Hobbes's *Leviathan* leaps to mind.) Only members of such a community, accepting its rules and methods, can serve as proper witnesses, either to Boyle's particular use of the air-pump or to others reiterating his experiments.

The experiments were to be public, open to the inspection of all qualified witnesses. Experimental science, however, was much more than this. It also included reports, written up by the experimenter, so that others, in either practice or imagination, might carry out the experiment again, perhaps vary the conditions, and build on previous attempts.

Shapin and Schaffer speak of "virtual witnessing" and explain that "the technology of virtual witnessing involves the production in a *reader's* mind of such an image of an experimental scene as obviates the necessity for either direct witness or replication."[28] A "novel" form (my words) of reporting, which permits witnessing at second hand—virtual witnessing—thus arises: the scientific report, conveyed in the transactions and other publications put out by the learned societies.

Narrative is generally regarded as foreign to positive science. It is important to note its presence at the start of experimental natural science. As Shapin and Schaffer inform us, "The text of Boyle's *New Experiments* of 1660 consisted of narratives of forty-three trials made with the new pneumatic engine."[29] The narrative is the story of how the experiments were carried out, allowing virtual witnesses to pass judgment on their validity. Failed experiments were to be included in the narratives, too. In principle, these were as critical to the advance of science as the more successful ones. The word *trials* also suggests legal trials, with the amassing and testing of testimony, evaluation, and a passing of judgment.

The prose style of the narratives came in for much attention. If the narratives were to secure assent, their writing was almost as importance as the experiment itself, at least for the scientific community. The issue of rhetoric, the concern with persuasion, haunts the scientific revolution from its beginning.

Thomas Sprat's famous injunction to the members of the Royal Society is worth careful attention. Attacking the "Luxury and Redundance of *Speech*" and recommending that "*Eloquence* ought to be banish'd out of all *civil Societies*," the good bishop exhorted his fellow members to "reject all the Amplifications, Digressions, and Swellings of Style; to return back to the primitive Purity and Shortness, when Men deliver'd so many *Things*, almost in an equal Number of *Words*." Concluding, Sprat urged "bringing all Things as near the mathematical Plainness as they can; and preferring the Language of Artizans, Countrymen, and Merchants, before that of Wits, or Scholars."[30] In Sprat's language, allowing for some differences, we hear the echoes of Bacon. Go to things, not words, we are told, for the received words are those of ancient and now illegitimate authorities. Words must be replaced with experiments, replicable and observable by all "citizens" of the new world of science. Ideally, the plain words used to describe the experiments will carry their own meaning with them, their weight due not to eloquence but to the experience—the narrative—itself. "The 'vacuum' Boyle referred to in *New Experiments*," Shapin and Schaffer declare, "was a new item in the vocabulary of natural philosophy: it was an operationally defined entity, reference to which was dependent upon the working of a new artificial device." [31]

Innovations, experiments, a language in which to report them—all were aimed at witnesses. Witnesses are central to the positivist enterprise. And exactly at this point, a major "test" of a different nature arises. If the witnesses are to evaluate the experiment, who is to evaluate the witnesses?

While Boyle was empaneling his witnesses, some of his fellow members in the Royal Society were adopting the same technique to prove the truth of witchcraft. Joseph Glanvil and the American Cotton Mather prided themselves on their scientific approach to spells and familiars; Mather, for example, wrote of how the judges at the Salem trials, "carefully causing the Repetition of the Experiment, found again the same event of it," and he keeps appealing to other experiments and the evidences of "learned men." [32] Much of this was a tribute to the Baconian experimenters, who were setting the terms of discourse. As Henry More, the Cambridge Platonist, explained to Glanvil, "Such fresh examples of Apparitions and Witchcrafts," when subjected to proper trial, must win over "benummed and lethargick minds." [33] If hypocrisy is the tribute that vice pays to virtue, imitation may be the homage paid by such religious experiments to scientific ones.

The witchcraft trials were also the trials of a whole religious belief system, as well as the belief system of the new Baconian science. Central to both was the institution of witnesses. Indeed, for the religious system, witnesses are there at the beginning, to testify to the truth of early Christian events. Should we believe, for example, the early disciples of Jesus who attest to his rising from the dead? What if thousands swear to the miraculous powers of a king, the example given in the eighteenth century by David Hume? To take an even later and less institutionalized example, what if reputable witnesses swear that they have seen Uri Geller, a Hungarian-Israeli prestidigitator, bend spoons at a distance by mere thought?

In short, the invocation and swearing in of witnesses does not constitute science. Witnesses can be blinded by suggestion or prior belief; they can be fooled by appearances, misled by wrong interpretation. What is different in the claim of positive science is that such idols of the mind can be guarded against rigorously.[34] The procedure for doing so is complicated, although it can be summed up in the concept of correct scientific method: the design, where possible, of special instruments — innovations — used in special experiments, with accompanying checks, reiterations for either verification or falsification, judgment by competent witnesses. The definition of a competent witness is that he or she be a member of the scientific community, that is, one who has accepted the tenets of scientific method.

There is a certain circularity here, which cannot be avoided. Science is a social construct, and its proper witnesses must also be constructed. Why they should be believed more than their adversaries is a question to be addressed more fully later. One way of more or less avoiding the issue, however, is Hobbes's. His right reason needed no special skill; all that was required was the common ability to follow a chain of deductions. The whole problem of witnesses is thrown out of court. Leviathan makes the decision, with neither trial nor experiment to influence its finding.

The recommended Hobbesian solution is final and timeless. Boyle's experimental science is open-ended, looking toward correction and aiming only at an asymptotic certainty. I want to conclude this treatment of Boyle by stressing two final points. First, his eulogy to the use of instruments brings in its train the extension of the human senses and the enhancement of human power. The telescope and the microscope permit us to see farther and finer; the air-pump allows us to control pressure and to evacuate space. In using tools, the human being becomes a different person, in a different world of perceptions.[35]

Second, the scientific community called into being by this continuing development is necessarily, in principle, a democratic one. There is no personified authority, only the authority emerging from the series of experiments. All who subscribe to the scientific social contract are of equal worth. (As we shall see, in Chapter 6, practice may differ, historically, from principle.)

The correlation between scientific democracy and political democracy, it has been argued, is high. In the seventeenth century, events in England — the Puritan Revolution, with its introduction of a commonwealth — seemed to give historical veracity, a kind of experimental proof, to a positive assertion of this connection. Would later events allow for this particular correlation to be made into a generalization? We shall see as we follow the story of positivism that the answer is hardly simple.[36]

Varieties and Stages of Positivism

Positivism is not one thing, but many things seeking a unity. It is a form of philosophy, and it is a form of sociology. It is an epistemology, and it is also a historical perspective. In any and all of these shapes it is a multihued affair; but it is often treated as a monolithic shibboleth and as a single figure for whipping.

I shall not be attempting to write a history of positivism—such histories tend to start with Comte, ignoring Bacon and Boyle and all that they symbolize—but to hold up for examination a few of the more significant phases in the development of positivism. I will be emphasizing the varieties and stages of positivism, the fact that it manifests itself differently at different times and places.

I will treat of positivism in regard to the Enlightenment, to Comte, to the post-Comtean challenge, to logical positivism, and to twentieth-century sociology and behavioral psychology. I will be picking and choosing eccentrically, like a magpie building a nest, partly in order to construct a kind of ideal type for our efforts to understand the nature and meaning of the human sciences and partly in order to emphasize that positivism is a changing historical accomplishment. As such, it lends itself to a renewed effort to see what it has of worth to carry forward into the human sciences.

The Enlightenment

Although positivism was foreshadowed in Bacon, it did not emerge as a more or less fully developed concept until the Enlightenment, consequent to the scientific revolution. During the Enlightenment, the notion that the power of scientific thinking, so successful in regard to natural phenomena, could be extended to the phenomena of humans, the social world, became pervasive. We are at the birth of social science.

Historians are fond of repeating that nothing is new under the sun. Anticipations, prophets, predecessors, can always be found. Thus, Niccolò Machiavelli can be numbered among those who sought to carve out "a new route" in the affairs of men by paying attention to what they actually did, not what they piously said they should do. Such realism, in principle, is a step toward social science. Yet Machiavelli's new route was limited to the political in its objectives; and even there, his science proceeded in a very halting manner.[37] Bacon has a much securer claim on our attention.

Ideas move by hops and skips. Descartes, in the middle of the seventeenth century, taking a position opposite to that of Machiavelli and Bacon, denied the possibility of applying scientific method to human affairs, to problems of morality, or to mores (although, in his *Passions of the Soul,* these do come into play).[38] His contemporary, Baruch Spinoza, however, sought to co-opt the great Frenchman. In his *Ethics,* he declared his intention "to treat of human vice and folly geometrically." "There should be one and the same method of understanding the nature of all things whatsoever, namely, through nature's universal laws and rules. Thus the passions of hatred, anger, envy, and so on . . . follow from this same necessity and efficacy of nature. . . . I shall consider human actions and desires in exactly the same manner, as though I were concerned with lines, planes, and solids."[39] We have already encountered this tone in Hobbes. Though representing a very narrow and restricted version of positive scientific method, it catches one part of the positivist quest.

In any event, by the time of the eighteenth century, both the tone and the effort had expanded greatly. Throwing off the restraints that had encumbered even natural scientists, the philosophes refused to leave the supposedly sacred affairs of men outside the bounds of science. Their bible was the well-known *Encyclopedia.* Many of the articles on the arts and crafts were Baconian in spirit. Mixed in with these articles, and their wonderful illustrations, were others on the subjects being treated in the emergent social sciences. The spirit animating both concerns was the same. As the philosopher Stuart Hampshire writes, the "*Encyclopedia* was a statement about the expanding domain of human knowledge, its present frontiers and future possibilities. The great empire of the new sciences and applied arts called for a charter and a written constitution, and this was the time, the 1740s, for a many-volumed manifesto summarizing its numerous conquests and their mutual dependencies."[40]

What representative figures shall I instance? Diderot, as the editor (with D'Alembert) of the *Encyclopedia,* is surely one. Again to quote Hampshire, "Diderot took his place in the line that passes from Democritus through Lucretius to Spinoza. . . . Philosophy was for all of them demystification; they were the metaphysicians of fearlessness, who refused to be haunted by anything that could not be felt or seen."[41] We have seen how apropos this remark was also as applied to Bacon. The long revolt against traditional authority and the substitution of science in its place was constantly under way.

Like Bacon, Diderot thought that the way to go forward, in his case, in the human sciences, was by other means than mathematics. In "Of the Interpretation of Nature," Diderot declared that "we are just coming to a great revolution in the sciences. In the inclination that intellects have toward ethics, literature, natural history, and experimental physics, I would be bold enough to guarantee that in less than a century one will not count three great geometers in

Europe." In a letter of 1758 to Voltaire, he made clear what he had in mind. "The reign of mathematics is no more. Taste has changed. It is now the hour of natural history and *belles lettres*." In support of his position, Diderot could cite the Comte de Buffon, whose *Histoire naturelle* (1766) quickly became a landmark.[42]

Diderot was wrong, or at least one-sided. A few decades later Condorcet was propounding his social art based on the mathematics of probability. At the same time, Bentham was advancing his felicific calculus (though without much actual mathematics to back it up) as the scientific solution to the problems of morality and legislation. In fact, many roads led both to and from the positive spirit. While Condorcet and Bentham pursued the inspiration of mathematics in regard to the human sciences, others followed the muses of a more sociological, political, psychological, and biological nature: thus, Montesquieu could speak of "the Spirit of Laws," and Hume could write that "the science of Man is the only solid foundation for the other sciences. Human Nature is the only science of man."[43] Neither had in mind a mathematical approach.

Only a full history of Enlightenment thought could do justice to this topic.[44] But let me take one final example: Turgot and his *Tableau philosophique sur des progrès successifs de l'esprit humain*. Here, positivism takes on the shape of a philosophy of history. The progressive accumulation by scientific inquiry is placed in a larger context of historical development, as part of the expansion of civilization. The *Tableau* offers an abstract form of historical sociology. Indeed, in his unpublished manuscripts, Turgot advanced the law of the three stages whereby humankind moved from the theological to the metaphysical to the scientific. Progress and positivism are thus tied ineluctably together.

Condorcet would carry forward Turgot's ideas on both fronts. It was, however, Auguste Comte who made them famous and gave them the weight and expansion attached to the terms *sociology* and *positivism*. Behind him, he had the accumulated thought and work of many of the scientists and philosophes of the seventeenth and eighteenth centuries.

Auguste Comte

Comte wrote of positive philosophy and positive polity; although he did not use the word *positivism*, it is indissolubly attached to his name, and his name to positivism. In his work, Comte always employed a historical perspective; we must follow his example and remind ourselves yet again that positivism, or variations thereon, existed before Comte and changed its coloring after him. Still, he is central to our general notions of the subject.

Comte acknowledged his many predecessors, ranging from Descartes through Montesquieu to Condorcet. In a full treatment, innumerable others — ideologues like Dr. Cabanis and Destutt de Tracy, astronomers and physicists like Joseph Louis Lagrange and Laplace, biologists like Xavier Bichat, and conservative thinkers like Viscount de Bonald and Joseph de Maistre — would also have to figure. Of these, I wish to single out Condorcet, whom Comte called "my immediate predecessor." It was Condorcet's *Eloges des Académiciens* (whose five volumes Comte read), cast in the form of funeral orations, that serves as perhaps the earliest modern attempt at a history of science, and it was his *Esquisse d'un tableau historique des progrès de l'esprit humain* that is an early philosophy of history. What Comte may be said to have done in his *Cours de philosophie positif* was to make the history of science into a philosophy of history.

Comte was uniquely suited to his task. His schooling was in the newly established Ecole Polytechnique, the premier scientific educational center, set up in 1795, at the time of the French Revolution. Comte was a gifted mathematician, and when not lecturing on positivism, he tutored in mathematics. Thus trained and equipped, when expelled from the école and required to earn a living, which he did as a secretary to the Comte de Saint-Simon, Auguste Comte was well able to give detail to Saint-Simon's law of the three stages.

Saint-Simon was preceded by Turgot, but because Turgot had not published his ideas, Saint-Simon gets credit for enunciating the idea that knowledge in any field goes through three stages, the theological, the metaphysical, and the scientific. The exposition and working out of this idea in the historical minutiae of the various sciences was beyond the self-educated nobleman. It was Comte who could give body to the assertion; the quarrel over primacy between the two men should not obscure the fact that fully developed positivism emerged from Comte's extensive exemplification of what may not have been, technically, his original idea.

Ranging over the fields of astronomy, physics, chemistry, and biology, Comte amassed data illustrating the way each field passed through the three stages of its development, each becoming in the end a positive science, where *positive* denotes "establishing lawlike regularities by scientific means." In contrast to his fellow mathematician, Descartes, who saw only one right method of conducting reason — the geometrical method — Comte believed that each science develops by a logic proper to itself, a logic that is revealed only by the historical study of that science. The logic of the mind cannot be explained in a priori fashion, but only in terms of what it has done in the past.

Now a new science was about to emerge: sociology. It was on the point of moving from the shadow of theology and metaphysics as a result of man's, that is, Comte's, awareness that social phenomena, like natural ones, obey laws. In the forty-seventh lesson of his *Cours,* Comte proposed the word *sociologie* for what the Belgian statistician Quetelet had named *physique sociale.* Comte did so because of his conviction that each emerging science had to develop its own proper method and that mathematics was not suitable for either biology or the new field of sociology.

Saint-Simon had sought to reduce the social sciences to mathematics, of which he knew little, believing that from the law of gravitation all knowledge could be deduced. Comte, believing that each science must develop its own method—and that the more complex the phenomenon, the more repugnant it would be to mathematical treatment—argued that sociology, like biology, could not be a deduction from some universal concept but could only be a development in the historical progress of the human mind. The preferred method in biology and sociology was the comparative. For the former this could take the form of comparative anatomy; for the latter, the form of comparative synchronicity and diachronicity, or, put concretely, of social statics and dynamics. Although Comte's detailed treatment of social statics and dynamics, as well as his theory about them, left much to be desired (his political views, for example, influenced his "facts" unduly), his overall notion of sociology as mainly historical sociology emerges clearly.

Comte assumed that a science of sociology was possible. An investigator could achieve positive knowledge about social phenomena. With such knowledge, society could then be organized on a positive basis, including ethics. Scientists did not debate whether 2 plus 2 equals 4, his mentor Bonald had argued. And Comte felt that moral certainty was equally amenable to scientific knowledge.

The events of 1789 had overturned traditional legitimate society and knowledge. They could not be reconstructed on the old foundations. To save society, Comte insisted that a new basis for legitimacy in both thought and action had to be established. Fortunately, sociology was on the brink of becoming positive and thus of offering positive knowledge with which to renovate social and moral life. Comte tried to show "scientifically" that altruism and self-love could be reconciled and that "what is" and "what should be" could be collapsed into one self-consistent whole.

Even religion could be made positive. Although he had lost his belief in God and the church, Comte recognized the emotional needs behind such beliefs and sought to provide for them in a Religion of Humanity. This part of his work was rarely taken seriously as part of his positive sociology; rather, it formed part of his polity—his effort to establish a stable society. More secular adherents pushed aside the Religion of Humanity as a regression that should not be allowed to obscure the progressive movement toward positivism of all thought and its embodiment in society.

As we can see, Comte's positive philosophy was born out of the post-1789 upheavals. It reflected specific needs and aspirations of mid-nineteenth-century European society. One of those aspirations was for timeless, universal knowledge. We can deny Comte credit for achieving this latter aspiration while still giving him credit for his embryonic history of science, his "sociology," with its attempt to look at society scientifically, and his overall espousal of the belief that positive knowledge is achievable for both natural and human phenomena.

Numerous criticisms can be made of Comte's work. Among them is the complaint that he fosters the notion that what cannot be made subject to the positive spirit is not worthy of attention and certainly cannot be considered a form of knowledge. The whole world of human fancy and fantasy is dismissed. Further, Comte's positivism, itself an unreflective form of metaphysics, rejects any inquiry into metaphysics and substitutes for it a philosophy of history.[45]

On a different level, Comte can be severely criticized in his use of historical data. Much of that use is warped by his polemical needs. Using erroneous data, he makes sweeping general statements. Much of what he has to say can only be described as naive. Having dismissed psychology as a discipline, his own psychological sense is unable to fill the gap adequately. I could go on, but enough has been said to illustrate Comtean positivism's vulnerability to attack.[46]

The criticisms are just. We must nevertheless try to understand the appeal he had for his contemporaries and for many thereafter. Typical was George Henry Lewes, who rhapsodized that Comte's *Cours* would "be the most memorable work of the nineteenth century. He will have founded a science and furnished its fundamental law. He will be at once the Bacon and the Newton of the nineteenth century." Even a more sober judge, John Stuart Mill, could describe the *Cours* as "at once the most profound, the most complete, and the most masterly in its exposition of any work on the subject."[47]

Yet by the second half of the nineteenth century, faith in positivism had been deeply shaken. Its vaunted claim to ground morality and social order in science sounded hollow. A revolt against positivism took place. Part of that revolt went forward under the banners of romantic aestheticism. Another part took place, surprisingly, on ground that positivism had built on: biological and physiological science, where findings seemed to show that consciousness, the seat of reason, is inseparably bound to intuitive and emotional life.[48]

Lewes, for example, turned skeptical of Comte's positivism as his own researches proceeded. Mill, disturbed by Comte's dictatorial implications — science eliminated moral choice — and his lack of any place for psychology as a form of knowledge, came gradually to turn against his French friend. Though not giving up the effort to achieve scientific knowledge of human affairs, Mill sought it in terms of his Ethology, an effort that, itself partly positivistic, got nowhere.

Others turned to what came to be called pragmatism. We glimpse the latter as an alternative to positivism when George Eliot, an early admirer of Comte, has one of the characters in her novel *Daniel Deronda* argue that "even strictly-measuring science could hardly have got on without that forecasting ardour which feels the agitations of discovery beforehand. . . . In relation to human motives and actions, passionate belief has a fuller efficacy. Here enthusiasm may have the validity of proof." [49] One of Eliot's commentators identifies her "sociology," as opposed to Comte's, as follows: "The images of unity constructed by this mental energy are spontaneous, aesthetic 'previsions' of order which may or may not have a correspondent reality. Their 'reality' lies, rather, in their causal effect on individual and social behavior." [50] William James would say it more forcefully and frequently in his attempt to establish a philosophy of pragmatism. Although pragmatism can be viewed in this context as an offspring of positivism, in James's case it was an offspring that turned on its progenitor.

Others, vitalists and intuitionists, ranging from Friedrich Nietzsche and Henri Bergson to aestheticians like Charles Péguy, excoriated the materialism, naturalism, mechanism(to use what were intended as synonymous terms), that they felt comprised the reduction of life to science.[51] In the end, then, positivism, certainly in the Comtean version, was vulnerable to the charge of not offering a scientific basis sufficient for the moral life, as well as being a restricted and ultimately untenable view of the nature of science.

The Vienna Circle

It was in the early twentieth century, and in Vienna rather than France or England, that a renewed effort at positivism arose. If there had been a revolt against positivism slightly earlier, perhaps we could now speak of a return to positivism. (In fact, of course, as in all such happenings, actions and reactions are constantly occurring at the same time and battling with one another.) Before sketching this "return," let us go back to the late seventeenth century and to the thinker who takes the first approximation of the epistemological position that came to be called logical positivism: John Locke. Locke is the patron saint of empirical philosophy.

We see him stating in his *Essay Concerning Human Understanding* (1690) that no ideas are innate and that knowledge can come only from our sense perceptions. Ideas are simply their reflection in the mind, where they are suitably associated and where simple ideas are made into complex ones. With this as given, Locke goes on to state that knowledge is "nothing but the perception of the connection and agreement, or disagreement and repugnancy, of any of our ideas"—basically, that is, of our sense impressions.[52] Like Bacon and Boyle before him, Locke embraces the "historical, plain method." Like the philosophes who come after him, he speaks constantly of letting light in upon the mind. Like the scientists of his time, he envisions knowledge of God, not as innate but as a reasoned deduction from sense perceptions.

Locke believes he can demonstrate the existence of God with reasoning "equal to mathematical certainty." With more mundane matters, Locke argues that we must "not peremptorily or intemperately require demonstration, and demand certainty, where probability only is to be had, and which is sufficient to govern all our concernments. If we will disbelieve everything, because we cannot certainly know all things, we shall do much-what as wisely as he who would not use his legs, but sit still and perish, because he had no wings to fly."

Variations of some of these Lockean ideas (although not his proof of God) persist through much of what follows under the heading of positivism. The two constants appear to be the emphasis on sense impressions as the origin of all knowledge, certain or uncertain, and the disdain for metaphysics (leaving aside God perhaps).

Lockean ideas persist as we jump to the twentieth century and the Vienna Circle. One of the key figures in this development was Ernst Mach. A member of the Circle, Moritz Schlick, describes his revered colleague's position as follows: "Since all our testimony concerning the so-called external world relies only on sensations, Mach held that we can and must take these sensations and complexes of sensations to be the sole content of those testimonies, and therefore that there is no need to assume in addition an unknown reality hidden behind the sensations."[53]

Immanuel Kant, with his categories of the mind and his defense of metaphysics, which claimed to encompass sensationalist knowledge as well, was the philosopher who had sought to go beyond Locke's empiricism. Mach was at first enamored of Kantianism. Then, as Leszek Kolakowski describes it, he came to believe that "there is no foundation for believing in any a priori conditions of experience whatever. . . . According to [Kant], the history of science shows incontrovertibly that there is no clear-cut boundary between everyday experiences set down in ordinary language, and the theoretical constructions of modern science. Science is a continuation of the same shorthand, symbolic

systematizing of experience that people have pursued spontaneously throughout history." [54]

Positivism was on its way to becoming logical positivism (*logical empiricism* is the other term that could be used here), with an emphasis on language.
Words and things were again to be put into direct correspondence, just as they
had been for the Royal Society, but in a very much more sophisticated manner. What could not be experienced via sensations, experimented with, and
put through a process of scientific verification had no claim to be considered as
knowledge.

The knowledge that could be achieved by the scientific method encompassed all fields, including the social sciences. As Schlick put it, after "the
anarchy of philosophical opinions," order could be imposed on the chaos of
disputing schools by establishing one mode of philosophizing in all domains. [55]
Such a unified science meant, as another member of the Circle, Otto Neurath, argued, that there was no "special realm of empathy (*Einfühlung*) or
understanding (*Verstehen*) and therefore no justification for separate *Geisteswissenschaften* [moral or spiritual sciences]; on the contrary, human activities,
whether individual or group, were deemed behaviours which are fully compatible with physicalist premises." [56]

As early as 1911–12, an *Aufruf*, or Appeal, was issued by an emerging Gesellschaft für positivistische Philosophie, which declared its intention "to bring
forth a comprehensive Weltanschauung, based on the factual material that has
been accumulated by the separate sciences." This Society for Positivistic Philosophy was, as the Appeal noted, to "press forward toward a contradiction-
free unitary conception." [57] Among the Appeal's signatories were not only
Mach, Albert Einstein, and other distinguished physicists but also the psycho-
analyst Sigmund Freud and the sociologist Ferdinand Tönnies.

In the late 1920s, a group in Vienna formed around Moritz Schlick, num-
bering among its members Neurath, Rudolf Carnap, Philipp Frank, and other
scientists and philosophers. This group came to be called the Vienna Circle.
Carnap explained that its purpose was to "banish metaphysics from philosophy, because its theses cannot be rationally justified." [58] He, like Neurath,
wished to build up a unified system of all concepts that would banish the difference between the natural and the human sciences.

The movement toward this Viennese version of positivism was to a great extent inspired by a desire to secure some sober factual grounding in the face of
the medieval obscurantism of contemporary Austrian intellectual and political life. Nazism, with its mysticism and metaphysics, was emerging around the
Circle. In fact, many members of the Circle were soon forced to emigrate.

In the United States, a *Journal for the Unity of Science* was published (as a continuation of the earlier journal *Erkenntnis*), and a series of books was launched. Of a proposed International Encyclopedia of Unified Science, a unit, entitled *Foundations of the Unity of Science,* with twenty monographs, was published. Among the contributing authors were Niels Bohr, Bertrand Russell, Carnap, Ernest Nagel, Carl Hempel—and Thomas Kuhn. A paradox worth noting is that Kuhn's *Structure of Scientific Revolutions* (1962), which has been interpreted as a manifesto for relativism (of the anthropological, not Einsteinian, mode), was commissioned to show the proposed unification of science. The starting point for Kuhn's monograph was his observation that the social sciences were in a sorry state of disarray and uncertainty "concerning the nature of legitimate scientific problems and methods" compared to the natural sciences. In short, Kuhn would attempt to unify the two types of science.

The way he did it was no doubt surprising to his positivist colleagues. He undermined the certainty and positivism of the natural sciences, rather than raising the social sciences to their level. As he stated, "Somehow, the practice of astronomy, physics, chemistry, or biology normally fails to evoke the controversies over fundamentals that today seem endemic among, say, psychologists or sociologists. Attempting to discover the source of that difference led me to recognize the role in scientific research of what I have called 'paradigms.' These I take to be universally recognized scientific achievements that *for a time* [my italics] provide model problems and solutions to a community of practitioners." [59] In one sentence Kuhn seemed to call into question the accumulated positivism of the natural sciences themselves. As one might say, "With such friends, who needs enemies." [60]

It is a short step from Kuhn and paradigms to Foucault and ruptures. Four years after Kuhn's monograph, Foucault's *Order of Things: An Archaeology of the Human Sciences* (*Les Mots et les choses* in the original title) was published in France. Foucault, however, drew his inspiration not from Kuhn but from historians of science like Gaston Bachelard and Georges Canguilhem, and especially the latter, with his emphasis on "ruptures" in the development of scientific knowledge.

Without saying more in detail about either Kuhn or Foucault, it is clear that a new revolt against positivism is to be found in their work and that of other relativists and social constructionists of science. It is not clear, however, I would argue, that they are constructing the issue in the right terms. By focusing on the changing results, the paradigms, they are ignoring the constant feature: the scientific method, which, though it changes and adapts itself in its particulars to the phenomena with which it must deal, remains a consistent method in its essentials.

Can it not be argued, for example, that the break with the view of the geocentric universe and the modes of thinking behind it is precisely what is involved in the scientific revolution? That that revolution, as promulgated by Bacon, Descartes, Boyle, Johannes Kepler, Galileo, and Newton, to name only a few, established a new method for achieving knowledge: the *scientific* method? Although, as I keep stressing, this method has to adjust to new materials, a view espoused especially by Comte, it is a way of seeing the world differently, which has remained constant and has now become a permanent acquirement by humankind. Science may have changed some of its theories (especially about absolutes like space, time, substance), and perhaps these can usefully be described as paradigms, but scientific knowledge is still cumulative and consistent in being founded on a slowly evolving method.

Is that method synonymous with positivism? The question is to be argued elsewhere in this book. Suffice it to say that today, in the fields of philosophy and history of science, positivism has been under severe attack. This is hardly new. Revolts and returns to a positivism of some sort appear to be a constant in the history of modern thought. Either in response to a deep-seated human need for a particular kind of certainty or to a chaotic social situation, or both, positivism, or some version thereof, seems to satisfy some minds while irritating others.

Psychology and Sociology

Comte in his *Cours* refused to recognize psychology as a possible positive subject. After biology there was sociology; psychological phenomena disappeared or were resolved into biology or sociology. Attacked by John Stuart Mill for his omission, Comte suggested that phrenology was the true science — a physiological one — which explained all the phenomena imputed to a so-called psychological domain.

This direction, to phrenology, though very popular in the mid-nineteenth century, was not pursued by later workers in the field of psychology. But one aspect of the positive spirit of Comte's observation was followed. Psychology did tend to become physiology, as in the work of the Russian scientist I. P. Pavlov. He said, "Man . . . is a system (roughly speaking, a machine), and like every other system in nature is governed by the inevitable laws common to all nature."[61] For him, these laws were the laws of the reflex, conditioned and unconditioned.

Even before Pavlov, students of the psychological had turned to the laboratory. Toward the 1870s, moving out of philosophy and away from metaphysics, practitioners insisted that psychology was part of experimental science. Wilhelm Wundt published his *Principles of Physiological Psychology* in 1873 and established the first psychological research institute, at Leipzig, in 1879. His credo was that science could study only external behavior, not imaginary inner states of mind. Introspection, the sickness of romantic minds, was out.

In the early twentieth century, followers of Wundt and especially of Pavlov proudly took the name *behaviorists*. John B. Watson, with his article "Psychology as the Behaviorist Sees It," published in 1913, was the founder of the school. He announced that a new psychology was needed, which made *"behaviour,* not *consciousness,* the objective point of attack." In his textbook, he wrote that "psychology, as the behaviorist views it, is a purely objective, experimental branch of natural science which needs introspection as little as do the sciences of chemistry and physics . . . the behavior of man and the behavior of animals must be considered on the same plane; as being equally essential to a general understanding of behavior." [62] Consequently, the behavior of human beings need not be studied directly; studying rats should tell us most, if not all, of what we need to know about humans.

The connection to positivism was made obvious in the comment of one of Schlick's students, who, back from a trip to the United States in 1930, reported that the behaviorist psychologists were among "the closest allies our movement acquired in the United States." [63] Watson's follower B. F. Skinner carried the positivist approach even further with his theory of operant conditioning (behavior is rewarded, not merely stimulated, as when a rat is given food for running a maze correctly). We ought not to be surprised that Skinner was a self-confessed disciple of Ernst Mach.

The behaviorists had their challengers. Although there were affinities between positivism and pragmatism, William James, for example, launched a savage attack on what he called medical materialism. It "finishes up Saint Paul by calling his vision on the road to Damascus a discharging lesion of the occipital cortex, he being an epileptic. It snuffs out Saint Teresa as an hysteric, Saint Francis of Assisi as an hereditary degenerate. . . . All such mental overtensions, it says, are, when you come to the bottom of the matter, mere affairs of diathesis (auto-intoxications most probably), due to the perverted action of various glands which physiology will yet discover." [64]

James's *Varieties of Religious Experience* (1902) took a different tack. Introspection and self-awareness mattered, he insisted, and could be "observed." Although James had strong reservations about the work of Freud, in this regard the two were allies. Freud, however, sought to deepen introspection, extending it to the unconscious, and to establish a science that took a positivist stance toward the aspects of mind that behavioral psychologists had renounced as a fit subject for "objective" study.

I have merely touched here on a few pieces of the positivist aspect of psychology. Neither behaviorists nor Jamesian pragmatists nor psychoanalysts have had it all their own way. From the nineteenth century on, the field has teetered between the natural and human sciences. The issue is still very much alive today.

What about sociology, which for Comte was the science under which psychology was to be subsumed (or at least those parts of it not fitting under biology)? In its original Comtean form, sociology was given a twin birth with positivism. Its subsequent development, however, was fiercely contested, with some such as Emile Durkheim, even if ambivalently, reaffirming its positivist nature and others such as Georg Simmel taking it in a more interpretive direction. Max Weber appeared to stand somewhere in the middle.

Alongside the great classical sociologists, however, a more naked positivist sociology seemed to flourish. In part, this took a quantitative form, in spite of Comte's cautionary statements. In the United States the sociologist Franklin Giddings, at Columbia University, pioneered the use of statistics in his field, announcing in 1909 that the need was for "exact studies, such as we get in the psychological laboratories, not to speak of the biological and physical laboratories. Sociology can be an exact, quantitative science." [65] Even before Paul Lazarsfeld emigrated from Europe and set up shop at Columbia in the 1930s, the call for an empirical, statistics-based sociology was sounded loud and clear. It was, therefore, only fitting that Lazarsfeld, who had no formal training in sociology whatsoever, chose as his title Quetelet Professor of Social Sciences.

Adherents of statistical sociology were not the only ones to sound the positivist note. In the classic work *The Polish Peasant in Europe and America* (1918–20), William J. Thomas and Florian Znaniecki announced confidently that "the marvelous results attained by a rational technique in the sphere of material reality invite us to apply some analogous procedure to social reality. Our success in controlling nature gives us confidence that we shall eventually be able to control the social world in the same measure. Our actual inefficiency in this line is due, not to any fundamental limitation of our reason, but simply to the historical fact that the objective attitude toward social reality is a recent acquisition." [66] Nothing in their book measures up to this claim, although much of the book has value.

We get an idea of how widespread that claim of sociology to positivist status is in another quotation, this from W. F. Ogburn in 1930: "An idea of value to science must be formulated in some sort of form capable of demonstration or proof. Verification in this future state of scientific sociology will amount almost to a fetish." [67] *Fetish* is a dangerous word. It suggests a faith in the positivist nature of sociology that has not been realized in the actual work of its practitioners. Even as a pious hope, however, it has not gone unchallenged, as a famous *Positivismusstreit*, originating at a conference of the German Sociological Society in Tübingen in 1961, illustrates. Here, the nineteenth-century struggle between classical and historical economists was resumed, and the names of Weber and Wilhelm Dilthey reinvoked. The fight this time was between adherents of the Frankfurt School and some supposed defenders of positivism. In particular, the main disputants were Theodor W. Adorno and Karl Popper (a rather ambivalent positivist).

We catch the flavor of some of the debate in two opposing comments. Adorno declared that "positivism, following Schlick's maxim, will only allow appearance to be valid, whilst dialectics will not allow itself to be robbed of the distinction between essence and appearance." Popper took the position that "although we cannot justify our theories rationally and cannot even prove that they are probable, we can criticize them rationally." [68] In short, Popper's unity of science was concerned mainly with a unity of method, and the objections of the Frankfurt School were primarily to scientism, the view that what cannot be known scientifically is not knowledge. As David Frisby remarked, Adorno "holds up for criticism a naive positivism which is hardly at issue amongst any of the disputants even though it may remain in operation in much social scientific practice." [69]

It is not our task at this point to settle the issue, although I must echo Frisby's comment that in practice, and still often in theory, much of contemporary sociology parades itself in pure positivistic guise. The fetish still holds sway, as it has in earlier times and in different phases of the sciences, natural and human. And opposition to the positivist pretensions still vigorously calls attention to positivism's deficiencies and limitations, both as theory and as a basis for moral and social life. Revolt and return continue their merry, hopefully dialectic, dance.

What Is Positivism?

In the excellent book *Positivism in Social Theory and Research*, Christopher Bryant entitles his introduction "Positivism or Positivisms?" The answer for our purposes is, both. In my short, highly selective account of the varieties and stages of positivist thought — where Bryant starts with Comte, I have gone back to Bacon and the scientific revolution—I have taken account of the various positivisms. Now I wish to take an overview of the subject and extract, or construct, a sort of ideal type: positivism.

In almost all cases, the motive for positivism is a reaction to what is seen as the stultifying and oppressive hand of traditional authority and knowledge. We see this at the beginning, with Bacon and the scientists of the seventeenth century. The opponent in their case is the ancients. With the philosophes, the Christian church, as an ally of the ancients, also becomes a target. Together, antiquity—originally in the Renaissance a liberating force but now become regressive—and religion are seen as enforcing obsolete and erroneous knowledge.

This is a persistent conflict in positivism, one that continues into the late nineteenth century, where we encounter it, for example, in the form of a struggle over the teaching in the universities of the classics, ancient knowledge, versus science and technology. In 1886, Ernst Mach gave a lecture, "On Instruction in the Classics and the Sciences," where he still had to argue "in favor of an education more suited to the needs of the time." Rhetorically he asked, "Are our highest models always to be the Greeks, with their narrow provinciality of mind . . . with their superstitions . . . Aristotle with his incapacity to learn facts, with his word-science?" Instead, Mach suggested, "Is not nature herself our first school-mistress?"[70]

Although rear-guard actions have been fought by the classicists, and a Two Cultures version of the struggle lingers on, the call by natural science for equality or more has triumphed in the schools. Here positivism's insistence that we must go to nature, to direct experience rather than to pronouncements from nonscientific or outdated authorities, has been generally accepted. In this positivism, the emphasis on experience, on empiricism, does not preclude the role of theories. Even in its earliest manifestations, with Bacon, positivism allowed for the "light" of the mind to illuminate the facts. It is only a highly reductive view of positivism that removes this element from its definition of scientific method.

To the emphasis on direct experience, coupled with theory, positivism joins an attention to words, to the language in which the experience is to be described. Hence the positivistic longing for the simplest rendition, in which "the same number of words convey the same number of things." This naive view held by the Royal Society has not lasted, but a concern with linguistics, with logical empiricism, has succeeded it.

Rejecting mere textual authority, of the Greeks or Scriptures, appealing to direct observation of nature and experience, concerned to make language accord as closely as possible to things, positivism hoped to rise above partisan discord and mere assertion of opinion. It sought to offer true knowledge, as close to reality as possible, with as much certainty as was appropriate to the subject at hand.

Overall, in its naive form, positivism has failed. Yet some of its aspirations seem to point in a direction worth taking. As we have seen, demonstration of experiential results in public is a central component of positivism. To overcome subjectivism, experiments, when possible, are to supplement experience, often with the aid of new instruments extending human sense perceptions, and these experiments are to be verified by witnesses, present and future (written reports of the experiment are to ensure the latter possibility). In this way, positivism could attempt to rise above controversy and offer positive knowledge.

A "public" is an essential element of positivism. In practice, this means a community of scientists—that is, of those who accept the principles of an agreed-upon scientific method. The method, in turn, must be strongly disposed to a *form* of positivism. Ideally, such a community is itself part of a larger, democratic community, a polity. It does appear, however, that natural science can suffice with a scientific community, as in the former Soviet Union, separate from a democratic polity. As I shall argue later, the human sciences, in contrast, require ideally a larger, more public community, one accepting of the scientific method, in order to flourish.

Another fundamental postulate of a proper positivism is that the scientific method has to emerge in accord with the nature of the materials under consideration and from the empirical treatment of the phenomenon itself. Comte derived this insight from his study of the history of science; emergent phenomena require emergent methods. These should not come from some abstract idea embodied in a philosophy of science imposed from the outside, and generally derived from one of the natural sciences, such as physics. That would be to revive, in a new manner, the stifling hand of ancient authority in a modern, scientistic mode.

What I have just stated might seem to stand in opposition to another postulate of positivism, that of the unity of science. As Bryant reminds us, "From Saint-Simon's obsession with the law of gravity to Neurath's and Carnap's physicalism, there have been those who base the unity of the sciences upon some aspect of being that connects all their objects."[71] In the nineteenth century, for example, this took the specific form of making knowledge whole, with one formula, the principle of conservation of energy, which was applicable, or so it was argued, to all natural phenomena.

The unity of science does exist, I believe, just as positivism asserts. Yet it is a unity of scientific method (suitably adjusted to particular phenomena) that is at issue, not some single, overarching law. The scientific method can be defined, at a minimum, as being based on the willingness of witnesses to accept public forms of verified experience, acceptable means of logical thinking, and a code by which theory and data can be related and played back against one another. To make this work for particular sciences is a subtle and exacting task, one that still needs to be carried out. Our interest, later, will be in attempting exactly that for the human sciences (see especially Chapters 5 and 6).

One last claim of positivism should be noted here: that positive knowledge is cumulative, that it becomes an increasing heritage of all human beings. Positivism has a definite affinity for the idea of progress and, further, an affiliation with the modern. As we have seen, there is even a strong tendency in some varieties of positivism to assume that positive moral knowledge is identical with natural scientific findings. Without fighting the battle of progress here, it must simply be noted that at least progress in knowledge of certain sorts is a logical consequence of positivism's postulates.[72] Knowledge that is "free" — that is, based on expanding experience — is by its very nature increasing knowledge.

As critics and even some friends point out, positivism is very uneasy with the realm of values and normative statements. Kolakowski voices the complaint that positivism "*denies cognitive value to value judgments and normative statements.* Experience, positivism argues, contains no such qualities of men, events, or things as 'noble,' 'ignoble,' 'good,' 'evil' . . . etc. Nor can any experience oblige us, through any logical operations whatever, to accept statements containing commandments or prohibitions, telling us to do something or not to do it."[73] Positivism too often does fall into the error of denying importance to things — such as values — that cannot readily submit to crude positivist methods. Even if not actively denied, such things tend to become invisible to positivist science. Unquestionably, positivism has had a tendency to reduce all phenomena to physical matter, or, when it does deal with the subject, to base morality on the physical.

In a sense, everything can be said to have a material base, even spiritual thoughts. The issue, however, is that whatever the physical base, part of observed human phenomena shows that people do have religious beliefs, hold superstitions, have neurotic and psychotic impulses, love and hate at the same time, feel a gamut of emotions, and so forth. Any human science that ignores these data cannot pretend to be a positive science.[74]

Humans have fantasies as well as "rational" thoughts. Science, the epitome of rational thought, is a symbolic construction of reality and is thus rooted in human nature. The task of the human sciences is to retain the virtues of positivism, and yet to go beyond positivism by studying the experience that all humans have of what it is to be human. A positivism that ignores the reality of what it is to be human is hollow at its core.

3 / The Human Species as an Object of Study

*A*ny effort at devising and using a suitable form of scientific method must be clear about the subject of study. What are the phenomena into which we inquire? What are their boundaries? Once the field is established, we can ask what techniques (instruments, ways of collecting data, institutional supports, modes of employing evidence, modes of inference) may be suitable to its materials. Involved in this question is the issue of classification, which appears to be a necessary feature of most attempts at science: How can the phenomena under investigation be meaningfully ordered?

The human sciences have these same requirements. Each effort at a human science attempts to define its field; for example, economics concentrates on market activities and institutions, sociology on social interactions and forms. The lines between disciplines are artificial, for economic activity is social, and sometimes vice versa, and so forth.

In the natural sciences the subject matter seems straightforward: the phenomena of nature, such as stars, molecules, glaciers, flora and fauna. Once a subject is identified, specific natural sciences, such as astronomy, physics, geology, and biology arise as specialties, each field being classified and ordered in a distinctive fashion. Crossing of boundaries can later occur, as in biophysics and biochemistry.

What is the subject of the human sciences? The simple answer is, the human species, in its manifold activities. The study of the activities is the basis for specific disciplines, like economics and sociology. As for the overall subject, the human species, its study is shared with the natural sciences, but it presents features, such as consciousness and culture, that are not captured in the nets of natural science.

A Human Species?

The starting point for the human sciences must be *Homo sapiens* as an animal. But what sort of animal? Again, we cannot accept the answer as a given but must confront it as a problem: What is the human species? If the question is simple, the answer is not. Although the species may have remained largely unchanged physically over a relatively long period of time, it may be more useful, at least as a thought experiment, to think of *Homo sapiens* as a changing species (leaving this term ill defined for the moment).

Let us start with the physical aspects. If humans had different senses—if they were all as blind as some moles are—their perception of the world would be different.[1] Many actions and observations would be unlike anything now known to humanity. Thus, the starting point for the human sciences must be in the natural sciences that investigate the physical basis of this particular species.

We need not resume this task here; we must simply be aware of the work of biologists, chemists, physiological psychologists, and the rest (echoes of Comte, without accepting literally his schema, are to be found in this suggestion, although he had scant sympathy for psychologists). What must be stressed for our purposes is that the senses can be said to have changed over time. If we cannot see in the dark "naturally," we are able to do so "artificially," by means of infrared radiation. If we cannot see very far, we extend our telescopes, and if we cannot see very small, we adjust our microscopes.

The extent of humanity's extracorporeal extensions need not be rehearsed here. The important point is that although the biological nature of our species is our starting point, that nature itself evolves in terms of culture—hardly a novel idea. But we can draw a less obvious inference: that we are in a sense studying different species over time and space, although they are grouped under the single heading "humans."

Let me clarify what this involves for the human sciences. These sciences and their subject matter can be said to coevolve. Not until the 1850s, in the West, was the span of humankind's existence extended from the mere six thousand years or so laid down in the Bible to untold millennia. As a result of the work of paleontologists, archaeologists, anatomists, and others, the long evolutionary past came into sight. We now know that we are an old, not a recently created, species. This old species, newly observing itself, has become, so to speak, a new species.

What we observe from this vantage point is a species that has existed in a changing form—*Australopithecus afarensis, Homo habilis, Homo erectus,* and *Homo sapiens.* The last of these manifests itself as Neanderthal and Cro-Magnon, the latter only in the last thirty-five thousand years or so. The early humans were hunter-gatherers who lived in small bands of fifty to five hundred persons and who, until recently, numbered no more than ten million individuals. (One author sees early humans as more scavengers than hunters; as hunters, he also considers humans "butcher[s].")[2]

As hunter-gatherers, early humans were not homogenous, a single type. "For instance, Australian hunter-gatherers with their elaborate kinship and mythical systems are very different from the cognatic and mythless African bushmen; or forest-dwellers in the Amazon are enormously different from the Inuit of Canada."[3] Such variety should hardly be surprising; it can be found in later agricultural and industrial societies as well. Nor does it undercut the fact that for an estimated 99 percent of human existence, hunting-gathering has been the condition of the species.

It is not our purpose to go into the details or the arguments over the details. Rather, I want to suggest some consequences for the effort at human sciences, and to do so in the form of rhetorical questions. Does it matter that we are talking of a species that numbers ten million or five to six billion, on its way to doubling? Do the larger numbers entail conditions of living—overcrowding, pressure on resources—that make for a different life, so that a hunter, an agriculturist, and a city dweller ought best to be thought of as different life forms? Do the numbers mean that the pace of cultural evolution—the number of inventions, the refinements of thought, the sheer brain power available—quickens to such a degree that we should think of separate, if connected, species (or, less threateningly, varieties)?

Should we be studying something called the human species as a whole or segments of it? In the eighteenth century, it was fashionable in some circles to conceive of the human species as a single individual, a single mind, that could be abstracted from actual societies. Thus, Fontenelle, Turgot, and Cordorcet wrote as if humanity were one person that could be studied as a single abstraction. As Fontenelle put it, "A good cultivated mind contains, so to speak, all minds of preceding centuries; it is but a single identical mind which has been developing and improving itself all the time."[4] That such conflation is now recognized as a gross simplification is a measure, I believe, of the distance that we have come in the human sciences. If we do take the human species as our subject matter, we must recognize its evolutionary variations and complexities. Darwin is an essential correction to the narrow linear perspective of the Enlightenment thinkers.

Again, are we studying a species or segments of a species? Is our focus to be on human universals or on describing Other's lives? Or both? This is an important issue.

Provisionally, let us recognize that analyzing, for example, the Nuer of the Sudan, may entail analyzing a different species from modern humans. To say this is to enter a mine field. Am I saying that the Nuer are not fully human or are less evolved than the individuals of a "developed" society? The Nuer are as human as we moderns are, and as individuals, are as evolved, *in the sense that if reared in our society, they would operate on the same level as our members now do* (and vice versa). The Nuer as a group are not as evolved, however, in terms of their society, if we are emphasizing evolution in a cultural sense.[5]

Cultural evolution, however antithetical to current egalitarian ideals, may be an essential postulate for the human sciences. Whether the species being studied consists not of millions or billions but of groups of fifty to five hundred, whose organizing principle, as numerous anthropologists have informed us, is in terms of kinship, must make a difference in scientific approach. Though kinship persists in modern societies, it can hardly be regarded as the most crucial element of social organization.

Even if we regard kinship as an issue of perennial importance, its function may vary significantly within a fixed society. As Eric Wolf observes about certain chiefdoms, "The function of kinship changes from that of ordering similarly organized groups . . . to that of drawing a major distinction between one station and another."[6] Such a change in behavior on the part of ants might entice us into saying that we are studying different species.

Similarly, it is no accident that the concept of exogamy, the requirement by custom to marry outside the immediate kinship or clan group, emerges out of the early ethnological studies devoted to so-called primitive societies (the term *exogamy* was invented by John McLennan in the 1860s). In a modern industrialized society, the term is almost useless. Indeed, as Marvin Harris has suggested, exogamy is not one of the "essentials" in developed societies, although vestiges of it may persist.

In societies that are close to the hunter-gatherer pattern, as Eric Wolf informs us, members do not transform nature but rather "gather up and concentrate for human use resources naturally available in the environment."[7] Although recent studies in environmental history would qualify this statement—primitives do transform nature to some degree—the basic difference is evident. Again, does such transformation mean that we should think in terms of a different species?

And what of the role of women? In hunter-gatherer societies, the type prevailing for most of human existence, men have done almost all the hunting and women the gathering. There are good reasons for this state of affairs. "Men are much better than women to track and kill big animals [men are also stronger on average], for the obvious reason that men don't have to carry infants around to nurse them."[8] Providing meat is therefore a marked function of human males, a trait to be found in only a few other species of mammals.

The roles of males and females in any species are distinctive, as much as markings, size, and other physical attributes. If groups of lions were found in which the male and female roles were changed, we would be tempted to speak of a different species, or at least of a major variety. Changes in gender roles in the human species appear to be of this nature, with truly revolutionary shifts presently impending.

The overall point of the examples is that the subject matter of the human sciences — the human species — is not a given but a reality that has changed. We are led to this conclusion by taking seriously the view that physical evolution for humans has largely been superseded by cultural evolution. We readily accept that physical evolution can result in new species; why not expect the same, mutatis mutandis, from cultural evolution? I have dramatized this point by making the apparently outrageous claim that the human species must be regarded as both unitary (there are human universals) and multiple (we are studying different species) — an assertion that can be understood only in an evolutionary context.

Symbolization

The ability to use symbols developed in the context of the human animal's evolution. Symbol-using ability then opened the way to the evolution of culture; and cultural evolution displaced physical evolution in importance. Symbol use, it must be emphasized, involved both nature outside and nature inside; that is, it served survival purposes in regard to both the physical environment and the social environment.

On the basis of some such analysis of the human species as a symbol-making animal much of the argument concerning the human sciences rests. A starting point is a physical development. The larynx makes human speech possible. (In fact, physiology and behavior tend to evolve in multicausal fashion.) Language, which is first spoken, then written, becomes the carrier of information, taking its place alongside genetic transmission. The story of human evolution, and thus of the species, is the story of the shift in the balance between un-learned, genetically determined forms of conduct and learned forms.

Human language consists of symbols, which serve as representations. As Norbert Elias says, "Symbols are not pictures or mirrors of the world; they are neither windows nor curtains. They have not an imitative, pictorial, but a representational function. They represent objects of communication within a language community."[9] Humans use symbols to represent the world, physical and social, because symbols are a successful means of adapting to the environment in which they must survive. Almost all other species are dependent on innate, genetic instructions on how to cope.

When a human learns to speak, that is, learns a language, such learning is a form of acquiring knowledge. It is social knowledge, because a language is mutually understandable only when it embodies the meanings agreed upon by all who speak it.

Knowledge can advance or regress, expand or disappear. Nevertheless, that we can acquire enough knowledge of the world to survive is a fact of human evolutionary nature. Survival in the struggle for existence casts doubt, in turn, on philosophical doubt as to the possibility of reality-congruent knowledge. If we were truly out of sync with our environment, we would not survive to know this "fact."

The point is worth expounding. If human orientations to reality were fun-damentally flawed and their communications full of misunderstanding, the species would not have lasted. As Elias recognizes, however, "The language [congruent as it is, to some extent, with reality] can also contain fallacies. Thus language can serve as a highly accurate representation of reality and as mis-representations." He goes on: "Human beings alone, as far as is known, have the gift of regulating behaviour in accordance with fantasy knowledge."[10] It is this unique gift, which will lead us later into an analysis of fantasy, myth, and religious belief, that requires the human sciences to be investigated as much in terms of hermeneutics as in terms of positivism. The human species is truly fantastic in a way that goes far beyond the fantastic creatures, such as Mino-taurs and centaurs, that are conjured up in our imaginations.

The human sciences need to inquire into the quagmire of normality: What is normal in the species? Is it made up to a large extent of madmen and mad-women? The evidence suggests strongly that reason is matched by unreason, or irrationality. In the words of Rousseau's tutor to Emile, which might as well be about the human scientist: "He will live not with wise men but with madmen. Therefore, he must know their madnesses since they wish to be led by them."[11]

The fact that an estimated 1 percent of the population is hopelessly schizophrenic (irrespective of culture, though embedded in it) must give us pause. So, too, the numbers of paranoid and psychopathic psychotics, not to mention neurotic obsessives, hysterics, and the rest, must be taken into account in any effort at understanding the species. According to some estimates, at any one time in developed societies (where such figures are compiled), 10 percent of the population is in mental asylums, and at least one out of four members of the society will experience a nervous breakdown requiring treatment during the course of his or her life.

The subject matter of the human sciences is thus a symbolic animal with deep flaws in its view of reality or in the appropriateness of its symbols. A breeder of race horses, finding equally serious flaws, might well give up on the stock. Humans do not have this luxury. Our very capacity for knowledge brings with it the knowledge of our own symbolic nature, with all its flaws, with which we must live. It also brings us, as we evolve culturally, the wish for greater understanding, or scientific knowledge, not only of the world's but of our own condition, even if that condition is frequently one of madness. Such knowledge, for us, will be mediated primarily through symbols rather than obtained by the more common genetic means of transmission found in the rest of nature.

Emergent Phenomena, Culture, and Consciousness

It is not just the human species that is evolutionary in nature but its society and culture as well. The consequences of this fact are enormously important for the human sciences, constituting one of its major problems. We can get at this topic by taking up anew the emergent nature of our phenomena. The concept arose most prominently in biology, in regard to organic systems. In Ernst Mayr's words: "Systems almost always have the peculiarity that the characteristics of the whole cannot (not even in theory) be deduced from the most complete knowledge of the components, taken separately or in other partial combinations. This appearance of new characteristics in wholes has been designated as *emergence*."[12]

The *Webster's New World Dictionary* definition of *emergent*, in its philosophical and biological usage, is "appearing as something novel or unpredictable in the course of an evolution." In the Darwinian universe, variations are random and basically unpredictable. All that can be predicted is that some random change will take place according to the dictates of natural selection and whatever other influences obtain, such as sexual selection or mutation. That animals will emerge and that, for example, one, the *Pikaia,* will have a backbone, eventually making humans possible, cannot be predicted.[13]

Emergence can also characterize inorganic systems, and Mayr notes that T. H. Huxley had the concept in mind when he asserted that water's "aquosity" could not be deduced from our understanding of hydrogen and oxygen. Here, we shall be most concerned with extending the concept to social systems, such as the agricultural and the industrial — an extension involving certain special difficulties along with ambiguities. For example, the ordinary usage of the term *emerged* (as in "X emerged as a leader") complicates our usage of the term *emergence.* To avoid such difficulties, we must remember that the word *wholes* is central to the concept of emergence.[14]

In one sense, *all* evolutionary products must be construed as emergent: the ant and its nest have not always existed. Nevertheless, although minor changes may occur, the basic nature of the nest and the behavior of ants, once emerged, have altered relatively little in the past forty million years.[15] Matters are different with the human species. Although certain features remain constant — all along humans have had to secure nourishment and shelter — the conditions of their life have been drastically transformed. The cave has given way to the city, and food today is cooked by means of fire and, more recently, by processes, such as microwaving, undreamed of a couple of millennia ago. In short, the human species is emergent, and the emergence has been accelerating recently. The acceleration has taken the form of culture. The emergence of human culture, in turn, has been based on the development of certain language forms, symbolic manipulations, technological developments, and social interactions, to name a few of the constituent elements.

How do we know all this? The reason is that emergent knowledge accompanies the emergent cultural phenomena. The conditions facing humankind, requiring adaptation to survive, call forth renewed cultural changes, which carry with them a kind of self-knowledge about the new knowledge involved in the material and social adjustments. The discovery of fire brings with it not only technological and social changes — eventually encouraging the forging of weapons and utensils, along with new ways of enlarging the food supply (smoking meat, for example), thus permitting the survival of larger groups of people — but also new mentations: myths about Promethean figures, philosophical musings about fire as one of the four elements of the universe, a science of thermodynamics.[16]

All of this may still seem distant from the formation of the human sciences. It is not. In culture, which is the accumulation of symbolic activity, we witness the emergence of something new, unprogrammed, and basically whole.

Emergent social structures, correlative with changing human nature, pose new problems of understanding and new possibilities for achieving such understanding. It is essential to realize that the emergence of the new phenomena and the possibility of understanding them must proceed in multicausal fashion, with changes in human nature (as we have defined it, emphasizing its evolutionary character) being part of that multicausality.

The idea can be made clearer and starker if we take up again the point that until the emergence of industrial society, preceded as it was by the emergence of philosophy and the natural sciences, scientific knowledge about such a society was unthinkable. The change in the life situation of large numbers of people, starting in the late eighteenth century (and comparable in its transformative power to the agricultural revolution of twelve thousand years ago), must be seen as creating phenomena that then required new understanding—in other words, the human sciences.

There can be no economic science until a market economy has emerged. There can be no sociology until the concept of society emerges out of the changing forms of human cohabitation, as in the shift from the feudal to the industrial. Hunter-gatherers, no matter how bright, could have no ideas about industrial society before the fact.

Modern humans are different from ancient humans. As Francis Bacon and others recognized at the beginning of the modern period, the use of ancient science in a world that the ancients never experienced was of limited value. Bacon's contemporaries, he believed, were different kinds of human beings from their predecessors (though sharing certain universal attributes of being human); they responded to a different environment, physical as well as social. A new "species" may be said to have emerged, both individually and in its group existence. Modernism, then, is in large part an interpretation of the experience brought about by the scientific and industrial revolutions.

Let me try to make this notion of the emergence of human phenomena even more concrete by taking up imperialism again. Darwin's theory of evolutionary biology would not have been possible before the emergence of modern imperialism—or its equivalent—brought the entire globe into the purview of the Western world (let me quickly add that the scientific result need not be seen as a justification for the atrocities committed in the colonial process). Without the examination of varying ecologies, the widespread collection of species, the observations of an enormous range of flora and fauna, and all the other results of European expansionism, the materials for conceptualizing evolution would not have been available (any more than the world of microscopic phenomena would have been available before the microscope). Darwin's evolutionary theory required, for example, the collections and classifications made by Linnaeus's "missionaries," along with those of a host of other imperialist researchers.[17]

Another illuminating example links modern imperialism with the science of anthropology. Other imperialisms did exist before the European, and undisciplined attempts were made at what came to be called anthropology. The difference is in degree. The European imperialism of the late nineteenth century finally filled in the empty spaces on the map of the world. The confrontation of the fully modern with the "traditional," "primitive" small societies that were on the verge of either perishing or being absorbed in emerging nations gave rise to the new discipline of physical and cultural anthropology.

Two observations can be made. First, at least until after World War II, most modern anthropology, issuing out of unprecedented change in the human condition, largely ignored the problem of change and the meeting of cultures. It is as if the early fieldworkers intuited that their subject was, indeed, cut off from modern people in some fundamental way; in some cases, they nostalgically hankered after the "simpler," premodern life. Second, much of modern anthropology has tended to ignore the emergent nature of humans and to hold to the conception of anthropology as uniquely positioned to find out the essentials of social life "otherwise disguised in complex, modern societies."[18]

There is a large grain of truth in the second observation: human universals do exist, and comparative anthropology is one way of identifying them.[19] This view, however, if held one-sidedly, ignores the fundamental fact of emergent human phenomena. Neither primitives nor the ancients have experienced the environment created by modern humans and requiring the understanding — the knowledge — that can be derived only from that experience. The task of the human sciences requires that increasing weight be given to the emergent new.

What shape can such understanding take? One consequence of our observations on emergent cultural phenomena is to recognize that the burdens imposed on the human sciences are much heavier, and different in form, than those faced by the natural sciences. Just when we think we may be able to secure positive knowledge of some aspect of human behavior, the behavior changes in ways that cannot have been foreseen.

In this light, we can entertain the notion that what changes most is consciousness. I will only hint at what is involved, reserving an extended treatment for the last chapters. The fact is that during 99 percent of human existence, the hunting-gathering phase, the species had no history, that is, consciousness of a past in "scientific" terms. Oral accounts, mostly of a mythical nature, existed; not until the fifth century B.C. and the emergence of written texts can we speak of history proper.

What this means is that primitive societies, including those existing today, can be said to have a magico-religious way of thought that bears some similarity to science in the effort to understand and control phenomena. What they have not had is a developed consciousness, which would allow them to conceive of alternatives to their established beliefs. Developed consciousness (which is not the same as logical ability) is an emergent phenomenon, a product of evolution. One other facet of what is here being called developed consciousness is that it has to be a social construction. It requires a community or social structure that both embodies and supports such a development. Human sciences, like natural sciences, cannot develop in a vacuum.

The Evolutionary Perspective

To repeat a point made earlier, without tumbling into sociobiology, the biological is the essential starting point for the human sciences.[20] Comte's intuition was correct: biological knowledge, the last of the natural sciences to emerge, was the basis for a human science that he labeled "sociology." We need not accept his positivism, or his rigid schematization of the natural sciences, or his actual sociology, to recognize the power of his insight.

Unfortunately, in Comte's scheme evolutionary theory is absent. And we have established that it is a key to the human sciences: on the portals of our house of human science should be emblazoned the slogan "Humanity, an evolutionary species with emergent features." Humans are creatures whose ancestors, millions of years ago, swung down from the trees and assumed an upright posture. That posture had innumerable consequences—among others, bipedalism, the freeing of what now became hands to use tools, and frontal sexual engagement with attendant psychological implications.

More to our immediate purposes, it allowed humans to raise their eyes to the heavens and to contemplate them steadily. As Hans Blumenberg puts it, what bipedalism made possible was "lifting one's gaze out of the sphere of biological signals and drawing something inaccessible into one's range of attention." He perceptively adds: "To see stars epitomizes the surplus that man was able, as a by-product of his upright carriage, to superadd to his pressing, everyday concerns."[21] In short, a biological change, not known to us until the nineteenth century—the emergence of a new species a few million years ago—opened the way a few thousand years ago to what was to become the first natural science: astronomy.

For the earth to be the setting for life required a delicate balance in which the cloud covering allowed just enough sunlight through to foster growth but not enough to burn everything up. Similarly, the medium in which humans lived could not be, as Blumenberg notes, "so opaque as to absorb entirely the light of the stars and block any view of the universe," as is the case with Venus and Jupiter. These conditions were requisite if any species were to emerge, and especially so if a human species, gifted with eyes of a special sort, able to see three-dimensionally, and to lift its gaze to an observable heaven, was to evolve.

To say that biology is the foundation on which any knowledge, natural or human, must be erected, is not to say that the elaboration of astronomy or physics required biological knowledge; quite the contrary, as the historical record shows.[22] It is to say, rather, that without the evolutionary biological developments, the animal that thinks scientifically, the human species, would not be in any position to do such thinking.

My purpose until now has been to establish the general evolutionary perspective with a few chosen examples. Let us continue to explore the subject of human evolution by taking up again what was touched on in Chapters 1 and 2: the fact that 99 percent of the human species' existence has been in the form of hunting-gathering groups.

Somewhere around twelve thousand years ago, the shift to agricultural cultivation began to take place, and a consequent increase in population sizes and social groups resulted. As Herbert Spencer (so wrong in so many other ways) pointed out correctly, one discernible direction in the evolutionary record was toward increased differentiation along with increased integration.[23] On this basis, the first civilizations began to arise, with specialization of functions and the emergence of different social groupings within the overall group — a change in habitat and behavioral patterns that we might well consider to have created a new human species (as I have argued and will argue again). The process is nicely summed up by Johan Goudsblom: "As a direct result of agriculture there was a trend towards higher production of food in increasingly more concentrated areas ... leading to an *increase in numbers* of the human population and to an increasing *concentration* of people in ever more densely populated areas. Both within and among these areas of dense settlement there were processes of *specialisation* as to social functions and of *organisation* in increasingly large units such as states, markets, and religious cults. As specialisation and organisation proceeded, they gave rise to increasing differences in power, property and prestige, in other words to a process of social *stratification*." [24]

Humankind as a Whole

If the object of the human sciences, whether conceived positivistically or otherwise, is all too obviously the human species, the obviousness of the statement may prevent our looking closely at the phenomenon itself, or even ignoring and certainly underestimating the subject. It must be said again: what we study in the human sciences is not, in the first instance, wars, or trade relations, or cultural exchanges, but the human species as a whole, an evolutionary animal whose characteristics are such and such and which must be studied in particular disciplined ways.

What I am suggesting here must be distinguished from what earlier I described as the philosophes' efforts to study humankind as a whole. Basically, they had in mind a new kind of cultural history, an incipient history of ideas or history of science seen as the history of the human species.[25] Whatever specificity such studies had and have, they do not have the kind of specific, detailed research that must underlie the inquiry into the human species. Such research has to be predicated on the theory of evolution by natural selection. Only with evolution in mind can it build resolutely on the commitment to scientific method in the human sciences that emerged in the eighteenth century.

What of scientific instruments in the study of the human sciences? There appears to be an obvious difference between the natural and the human sciences. The former were to a large extent made possible by the development of such instruments as the telescope, the microscope, the thermometer, and the barometer. These tools enormously expanded the scope of the human examination of nature, creating the observations and concepts to be understood scientifically by astronomers, physicists, biologists, and the rest.

In the human sciences, however, the fossils have been dug up by pickax, the cadavers secured from the cemeteries, all in old-fashioned ways. The modern instruments subsequently brought into play, such as carbon dating, are applied to the phenomena, not used primarily to create them. In cultural anthropology and comparative history, so-called scientific instruments are even less in evidence (the use of computers, for example, is to massage the data, rather than, as with the telescope and the microscope, to bring forth worlds previously unknown). The tools of the human sciences, for example, free association, seem largely to be of a different nature from those of the natural sciences.[26]

The human sciences appear mainly to resemble evolutionary science. As with Darwin's early evolutionary biology, few instruments play a decisive role. Other means, other shapes, of the scientific method seem to be called for. Such are the consequences of taking seriously the fact that our subject of study is fundamentally and initially the human species as a whole, increasingly viewed in terms of cultural as well as physical evolution.

A Species?

At this point, are we not in danger of reducing human science to a form of natural history or sociobiology? Not at all, because humans are a species, *Homo sapiens,* and we are therefore obliged to understand this biological assertion. Untidy and ill-bounded as it is, the notion of biological species is essential as a classifying concept, allowing us to order usefully an enormous phenomenological diversity.

Here is Wilson's definition: "*A species is a population whose members are able to interbreed freely under natural conditions.*"[27] *Homo sapiens,* then, is a unitary species whose members can interbreed freely. The various so-called races (in fact, many scholars see only a continuum, with racial categories a mere invention)—Caucasian, Negroid, Mongoloid—are unified by the potential for successful breeding. In contrast, humans cannot interbreed successfully with, for example, apes and monkeys (myths and legends to the contrary).

Wilson advances in this regard the interesting notion of "natural conditions." The example he offers is of lions and tigers. In captivity, they can be crossed, creating tiglons (where the father is a tiger) and ligers (where the father is a lion). In natural conditions, however, they do not appear to hybridize. Wilson's explanation is suggestive.

> First, they liked different habitats. Lions stayed mostly in open savannas and grassland and tigers in forests. . . . Second, their behavior was and is radically different in ways that count for the choice of mates. Lions are the only social cats. They live in prides, whose enduring centers are closely bonded females and their young. . . . The adult males and females hunt together. . . . Tigers, like all other cat species except lions, are solitary. The males produce a different urinary scent from that of lions to mark their territories and approach one another and the females only briefly during the breeding season. In short, there appears to have been little opportunity for adults of the two species to meet and bond long enough to produce offspring.

What implications, if any, does this have for the human species? What are the natural conditions and reduced opportunities that may have limited potential hybridization? The historical record seems to show relatively widespread human "hybridization," a number of cross-"racial" fertilizations, yet without overwhelming effect on the major "racial" categories. The object of evolutionary biology in this context, it must be emphasized, is populations, not individuals. Thus, potential changes in the gene pool from individual matings become reabsorbed in the existing categories.

Yet at some time in the more distant past, the major categories emerged, probably as a result of natural conditions. And before that, separation into different species was involved in the long evolution from *Australopithecus afarensis* through *Homo habilis* and *Homo erectus* to our present *Homo sapiens*. What boundaries to interbreeding were necessary, and how and when did they arise, so that the new racial varieties could evolve? In a more limited version of this question, what caused the replacement of Neanderthals by their cousins on the branch of the human tree, the Cro-Magnons, with their somewhat different anatomy and enlarged brain size? Might the answer lie in natural conditions, keeping the gene pools separate by artificial, that is, culturally created, means?[28]

To sum up: the definitional question concerning the human species forces us to focus on the natural conditions, now to be thought of primarily in cultural terms, which surround human breeding, that is, the behavioral patterns that make up the societies in which this species exists and may or may not interbreed. Courtship, territoriality, stratification, with all their attendant signaling systems, and other traits for ethnological study now enter claims upon our attention, but as cultural, not physical, attributes. In taking this stance, we both define our object of study, this particular species, and the means by which we observe and seek to understand it—the cultural behavioral patterns that match up with our human science disciplines.

Population and Stratification

Three points require further reflection. The first is that the number of humans counts. Population—the study of which first took on a conceptual and semi-scientific form in the seventeenth century—matters. The amount and kind of differentiation possible in a population of fifty to five hundred is so different from one of a hundred thousand or a billion, to instance China, as to arouse interest in whether a qualitative and not just a quantitative change in the subject of our study has taken place. Again, are we studying the same human species? When does a variation become a new species?

The second point concerns classification. Natural sciences classify the flora and fauna and, within each, different families, genera, species, and varieties. Human beings also consciously classify themselves, by status positions, kinship ties, ethnic affiliations, and so forth. Must the human sciences use this self-classification as the basis for a scientific classification?

The third point concerns the nongenetic feature of social stratification: humans, unlike ants or bees, are not genetically destined for their positions in the emergent social structures that they construct. This nongenetic feature entails, among other things, the recognition of heterogeneity as an essential characteristic of this particular animal, the human, with all the consequences for the human sciences that follow from this fact.

An illuminating connection can be made among these three points by returning to cultural evolution. For most of human existence, the species mostly resembled, say, wolves in its organization. In this hunting-gathering phase, the human "pack" that numbered around fifty was only about five times the size of a wolf pack. Its social structure also seemed similar. Then, as humankind entered into agricultural enterprise, its numbers increased. More complicated social groupings, in cities, emerged, and these produced stratifications and divisions of function more akin to that of the hill of the ant and the hive of the bee. Human ranks and stratifications, however, differed from those in anthills and hives by being socially conferred and constructed—in other words, nongenetic. The essential point is that the human species can be conceptualized as having changed in fundamental ways because of changes in population size. A different classification of the species came into play, itself evoked by the changed stratification system brought into existence by human cultural evolution. This is why the notion of social stratification lies at the core of human science: it leads us to what basically constitutes the human species and human society.

Biology, evolution, emergent phenomena, population, classification, nongenetic stratification, heterogeneity—these are some of the key words underlying the effort to construct the human sciences. It is to be hoped that we have now built up a broader context for further explorations into the meanings and possibilities of the human sciences. Let us proceed by turning to the presumed alternative to positivism, the hermeneutic method.

4 / Hermeneutics

*C*ulture is the key concept in much (though not all) of human science. Yet it is more useful to think of humans first as symbolic animals, rather than cultural animals, for symbolic abilities underlie the existence of both cultures and societies. The species, at the same time that its brain physically evolved, adding a cerebral cortex to its limbic core, increasingly replaced immediate, instinctual responses with delayed, thoughtful actions, mediated through symbols. We need not examine in detail the familiar findings of recent research that tell us of the development of symbolic language, with denotative and connotative characteristics, especially as evinced in spoken form (which presupposes a correlative development of the larynx and vocal cords); the ability, through symbols, to think in abstract and time-related terms; the capacity to pass on such thought-through knowledge by nongenetic means, that is, by learning; and ultimately, and only comparatively recently, the recording and retaining of such knowledge in written form.

As a symbolic animal, the human being adapts to conditions by inquiring after knowledge. Only in the past few hundred years or so, through the acceptance of a particular kind of method, the scientific, by a small number of investigators who make up a scientific community, has such knowledge been characterized as positive, or certain, meaning that it is beyond the challenge of nonscientists. Thus, scientists tell us that the earth revolves around the sun, and not vice versa, and we have come to believe them. Enticed by the giddy prospect of achieving scientific knowledge, from the seventeenth century on, claims to positivity were extended to human as well as natural phenomena. Social sciences were conceived that promised to be certain, verifiable, and predictive, offering the possibility of control over human behavior equal to that over physical forces.

Although the aspiration to positive knowledge has much to recommend it, its claims were both simplistic and unrealizable, especially in regard to the human sciences. Even before this became apparent, a different approach was being recommended by many thinkers. Human phenomena, it was argued, are separate in kind from natural ones. They are filled with human meaning and have to be understood in essentially those terms. Not positive science, with its claims to objectivity and repeatability, but interpretation, emphasizing subjectivity and uniqueness, was called for. A chasm was declared to exist between the natural and the human sciences.

Two names illustrate the nature of the quarrel. Bacon argued that the natural and the human were parts of the same nature and intertwined with one another. A century or more later, Vico argued that humans (along with God, of course, who also made man) can really understand only what they themselves make, that is, culture and society, while God alone can have certain knowledge about nature. Anticipating work in philology, mythology, anthropology, and many other fields, Vico argued for what is now known as hermeneutic rather than positive knowledge.

Both positions, Bacon's and Vico's (with Bacon being acknowledged by Vico as one of his four major progenitors), claim to offer understanding, but in one case it is scientific and in the other presumably something different (although Vico called his book *A New Science*) and more appropriate to the human sciences. Without going into the long history of this division, it behooves us now to give as much attention to the interpretive method, which I shall treat under the heading of hermeneutics, as earlier we gave to positivism.

To anticipate the final argument in this chapter, in my view, interpretation is fundamental to both the natural and the human sciences. Further, their inquiries are pursued with a similar intent—the achievement of understanding—and, basically, by a similar use of reason. Reason itself is rooted in the common human use of symbols. Not a chasm but a misunderstanding is what too frequently separates work in the human and natural sciences.

How are we to approach this putative divide, with its huge and varied set of subtopics? Let us first take the historical route and, as with positivism, examine how the notions about Geisteswissenschaften, or moral sciences, and hermeneutics arose.[1] Then we can review what is involved, for our purposes, in the hermeneutic method itself. At that point, it becomes helpful and perhaps even necessary to say something more about the Other, a subject that we took up originally in Chapter 1, and about madness. We should then be prepared to examine the subject of narrative and its possible claims to being a form of scientific understanding. At the end of such a treatment, we need to step back from the subject and deal with the issue of criteria: What are the criteria by which the validity of any interpretation can be judged?

Geisteswissenschaften

As noted earlier, the name of the human sciences has been a matter of debate and conjecture: *social sciences, moral sciences,* and *Geisteswissenschaften* have been offered as alternatives (although not satisfactory ones). In taking up anew the subject of Geisteswissenschaften, now in conjunction with hermeneutics, we have a chance to review the question of terminology more closely.

The German term apparently arose as a translation of John Stuart Mill's *moral sciences.*[2] In his *System of Logic,* which he started writing around 1830 and published in 1843, Mill made a major effort to define the correct method for both the natural and what he called the moral sciences. If we are to understand the origins and meanings of the term *Geisteswissenschaften,* introduced as the translation for "moral sciences," we would do well to probe Mill's original account.

In fact, a clearer and more succinct version of his ideas on the subject is found in his essay "On the Definition of Political Economy; and on the Method of Investigation Proper to It" (written in 1831 and thus with the *Logic* in mind but not published until 1836). In this essay, Mill declared that "the real distinction between Political Economy [which can be taken as a stand-in for any and all of the moral sciences] and physical science must be sought in something deeper than the nature of the subject-matter; which, indeed, is for the most part common to both."[3] Although what Mill has in mind by "common to both" differs from what I have suggested to be in common — that is, symbolic reasoning by means of an appropriate version of a shared scientific method — he still presumes a kind of unity of science.

In his next sentence that unity is rudely shattered. "If we contemplate the whole field of human knowledge, attained or attainable, we find that it separates itself obviously, and as it were spontaneously, into two divisions, which stand so strikingly in opposition and contradistinction to one another, that in all classifications of our knowledge they have been kept apart. These are *physical* science, and *moral* or psychological science." Driving his message home, Mill continues. "The laws of matter are those properties of the soil and of vegetable life which cause the seed to germinate in the ground, and those properties of the human body which make food necessary to its support. The law of mind is, that man desires to possess subsistence, and consequently wills the necessary means of procuring it. Laws of mind and laws of matter are so dissimilar in their nature, that it would be contrary to all principles of rational arrangement to mix them up as part of the same study. In all scientific methods, therefore, they are placed apart."

What Mill has placed asunder ought no one to restore? Mill's own mentors, his father, James, and Jeremy Bentham, thought otherwise, with Bentham convinced that a felicific calculus equal in power to Newton's differential could be applied to the subject of legislature and morals. Many of the philosophes, as we have already seen, also emphasized the common method and the material-based nature shared by the natural and moral sciences. We have already evoked the names of Turgot and Condorcet in the chapter on positivism, and their views can stand for many others. Condorcet, honoring his mentor in his *Vie de M. Turgot,* rhetorically asked his readers, "Why shouldn't politics, grounded like all the other sciences on observation and reasoning, be perfected accordingly, as more subtlety and exactitude are brought to its observations, more precision, profundity and accuracy to its reasoning." Addressing the French Academy in 1782, he drove home his point. "In meditating on the nature of the moral sciences, one cannot indeed help seeing that, based like the physical sciences on the observation of facts, they must follow the same method, acquire an equally exact and precise language, attain the same degree of certainty."[4]

I am using Condorcet as a stick figure — he was vastly more sophisticated and complicated in his thought — in order to set up the stereotype perceived by many of his contemporaries and the generation that followed. Subsequently, in the face of what was seen as their hubristic, arrogant assertion of the dominance of scientific ways of thinking in both the natural and moral worlds, a reaction set in, which we think of as "romantic." Against a presumed reductionist reality, reached by the withering tendencies of abstract reason and calculation, opponents of such a way of thinking proclaimed another reality and a different method of reaching it.

That reality was spiritual, and the method was hermeneutics. To understand exactly what is involved, we need to continue to examine the German word for "spirit," *Geist* (which can also be translated as "soul" or "mind"). It is most famously used by Hegel, whose Spirit makes its way, as a manifestation of Reason, through time and history. This is the Hegelian usage best known to most of his readers.

Hegel, however, also spoke repeatedly not only of Spirit but also of the "Spirit of the Times." By this he meant "*one* specific being, character, that permeates all aspects and represents itself in political affairs and in all else as in various elements; it is *one* condition that is coherent in all its parts and whose various aspects, as manifold and random as they may appear, as much as they seem to contradict one another, nevertheless contain nothing heterogeneous to the basis."[5] (That is, there is unity amid the heterogeneity.)

Thus, two meanings inhere in Hegel's use of the term *Spirit*. The first is peculiar to him and his philosophy of history and basically means "mind," or "Reason." In this usage, he is inventing a dialectical version of the Fontenelle-Turgot-Condorcet conceptualization of the single human mind. It is more fertile than theirs because Hegelian Spirit carries with it an implication of developing consciousness lacking in the philosophes' rather mechanical progress of the human mind.

The second meaning of *Spirit* in Hegel, as in his "Spirit of the Times" usage, is time bound, restricted to an existing society. It resembles anthropology's future emphasis on holism, on the notion that a given society, usually primitive, is animated by a single spirit that suffuses all the institutions and mores of the society and culture. In Hegel's own time, however, it echoed, for example, his compatriot Herder's emphasis on *Volksgeist*, or "national genius," laden as is this usage by future racial overtones. As Hegel expressed the idea in his lectures on the philosophy of history, "The common stamp of its [society's] religion, its political constitution, its ethics, its legal system, its morals, even its science, art, and technical skill" is imprinted by the national spirit.[6]

In his writing on the Spirit of the Times, Hegel was reflecting additionally his own spirit of the times in the sense that the usage itself was widespread. John Stuart Mill, for one, had written on "Spirit," as in his essay "The Spirit of the Age" (1831). There he declared that "the 'spirit of the age' is in some measure a novel expression. I do not believe that it is to be met with in any work exceeding fifty years in antiquity. The idea of comparing one's own age with former ages, or with our notion of those which are yet to come, had occurred to philosophers; but it never before was itself the dominant idea of any age."[7] What is embodied in Mill's usage is the same notion that lies behind the conceptualization of "society": the awareness that society is a human construction, subject to change, possibly guided by a newly developed consciousness. So viewed, society is an emergent phenomenon; and its spirit is a unique and shifting one.

That the concept of society and the notion of spirit are related and very much in circulation, is made evident by the work of Mill's older friend Thomas Carlyle, who has a large claim to being the first to use the term *society* in its modern meaning, as in his phrase *industrial society*. In his 1829 essay "Signs of the Times," Carlyle writes of "the spirit of these times" and speaks of the "Age of Machinery," just as Mill in his own essay speaks of "the signs of the times."[8] Spirit is the shifting cultural pattern that animates the changing societies constructed by the human species.

What makes the history of the term *Geisteswissenschaften*, whose counterpart is *Naturwissenschaften*, so puzzling is the confusion between an Anglo-French Enlightenment tradition, reflecting the positivist-tinged spirit of the age (to which Mill belonged) and a mainly German tradition, bestowing an idealist cloak on the term *spirit*. Mill is the paradoxical link between the two traditions, for, while opposing the unity of the natural and moral sciences espoused by the positivist-minded philosophes, he embraced their positive view of science. Indeed, when he came to attempt a separate moral science, it was mainly on the Anglo-French model: his "Ethology, or the Science of the Formation of Character," which was part of his *Logic*, ended up as a vague mixture of general laws of mind conjoined with particular circumstances. Ultimately, Mill had to admit failure; and he had no immediate inheritors of his enterprise, which has largely vanished from the history of the moral sciences.[9]

Not so his idea of the moral sciences as separate from the physical ones. This was a more enduring bequest, becoming a major inspiration for the development of the Geisteswissenschaften. My argument is that in translating Mill's *moral sciences* into German by using the term *Geisteswissenschaften* a spiritual ambiance was imposed on Mill's idea — an "interpretation," so to speak — that is foreign to his own usage of the term *spirit*. Between the German and English words stand two different understandings, or traditions, which separate them.

In the particular German tradition of which I am speaking, reality is primarily spiritual, not material, and the method necessary to reach it, hermeneutical.[10] By *spiritual*, moreover, the German hermeneutic thinkers generally meant the domain of religion and myth, rather than the positivist-tinged spirit of the age proclaimed by French and English thinkers of the Enlightenment. By *hermeneutics* the Germans meant understanding rather than explanation, and interpretation rather than analysis. In short, their hermeneutics can be seen as an alternative to positivism.

The Origins of Hermeneutics

As with positivism, we must inquire into how the hermeneutic method came into being — how the lens was ground in the mills of history by which we then look at that history itself.

Hermeneutics has a long history, although it is only with romanticism that it becomes a general method rather than a technique practiced in particular disciplines. The name itself is associated with the Greek god Hermes, who served as herald for the other gods. He is a highly ambiguous character. He seems first to have been a god of fertility, represented with phallic images. In the *Odyssey,* however, he appears mainly as the messenger of the gods—hence the name *hermeneutics.* He also conducts the dead to Hades and is seen on occasion as a dream-god, or "conductor of dreams." Elsewhere, his trickery and cunning are emphasized, and he is accorded numerous other attributes; he is a many-sided character. (See the entry on Hermes in the 1911 *Encyclopaedia Britannica.*) When I add that as Hermes Trismegistus, a later Egyptian version, he is seen as author of the sacred books, known as hermetic, and thus of hermetic, or sealed and magical, knowledge, we can readily see how well chosen he is for the amorphous approach known as hermeneutics.

Two fundamental connotations, therefore, are attached to the word *hermeneutics:* a mythical-religious origin and the notion of a translator or carrier of messages. The word itself comes from the Greek for "interpretation." *Interpretation,* in turn, comes from the Latin for "agent, negotiator, or the act of giving meaning." The word *meaning,* to take the last step, derives from Old English and signifies the thing that one intends to convey, especially by language; it carries with it also the sense of something intervening or intermediate. These are the key terms, with *hermeneutics* and *interpretation* sometimes being used as synonyms for a method whereby meaning results. The notion of "exchange," of a negotiated act, is central. The end result is a form of understanding that is distinct from and independent of ordinary scientific explanation.

Although the notion of hermeneutics was known to and used by the Greeks, it was not for them a significant method by which to achieve understanding. Rational argument was preferred. Only with biblical exegesis did hermeneutics come into its own. The movement had a number of outriders: the development of classical philology in relation to humanistic as well as religious texts, and the application of hermeneutical techniques to matters of jurisprudence. But by the seventeenth century the central and enduring aspect of hermeneutics was its attachment to religion and the interpretation of the sacred texts.

Hermeneutics in its preliminary modern form emerged at the same time as epistemology became the major focus of Western philosophy, yet hermeneutics seems detached from that philosophical concern. It resumed the God connection found in its Greek origins, with Hermes, but now solely in regard to the Christian religion, specifically Protestantism. One result was that unlike its emerging counterpoint, the Western scientific method, it did not grow in importance, at least initially, or spread in use in non-Western cultures. Its claim to universality was embedded in its religious doctrines and not its method. (One further implication is that the elimination of a belief in God also would abolish the possibility of messages being carried from him by the hermeneutic method.)

In the seventeenth century, then, hermeneutics was tied to specific disciplines — theology, jurisprudence, and philology — and was essentially Christian. It was the accomplishment of Friedrich Schleiermacher, while retaining the Christian element, to proclaim hermeneutics a general method, applicable to the mental world of all humanity. As he asserted in 1819, "Hermeneutics as the art of understanding does not exist as a general field, only as a plurality of specialized hermeneutics."[11] Inspired by romanticism, he implicitly extended hermeneutics to include all of humankind, even exotic peoples and texts outside the realm of Christianity.[12]

Although implicitly accepting Descartes's division of mind and body, Schleiermacher was hostile to the reductionist, analytic science that emerged from the Frenchman's work on the world of body — in other words, the world of the natural sciences. Descartes had disparaged the possibility of certain or reasoned knowledge about customs and morals — the realm of the moral sciences. Vico had tried to redeem the world of myth and moral from Cartesian skepticism, claiming even greater certainty for the world that humans had created. Nevertheless, Vico's world remained that of the classical-Christian, and his hermeneutical method was overshadowed by his grand theory of cycles.

Schleiermacher, unintentionally, inspired by romanticism, went beyond Vico's evident parochialism. As Hans-Georg Gadamer tells it, "Schleiermacher's idea of a universal hermeneutics starts from this: That the experience of the alien and the possibility of misunderstanding is universal."[13] We are, I would suggest, on our way to the interpretation of the modern Other (to use my terminology).

The full extent of Schleiermacher's achievement is laid before us as Gadamer continues. "Through critique, he rejected everything that, under the rubric of 'rational ideas,' the Enlightenment regarded as the common nature of humanity, and this rejection necessitated completely redefining our relation to tradition. The art of understanding came under fundamental theoretical examination and universal cultivation because neither scripturally nor rationally founded agreement could any longer constitute the dogmatic guideline of textual understanding. Thus it was necessary for Schleiermacher to provide a fundamental motivation for hermeneutical reflection and so place the problem of hermeneutics within a hitherto unknown horizon."

The task was to understand the Other, whether in one's own tradition or in another. This meant entering the Other's world, mainly via texts, and then returning to one's own. The experience has to be transformative, one in which the interpreter and the text interact. The interpreter is not a Baconian mechanical observer but a particular human with a personal horizon. As such, the interpreter must combine that horizon with the text's.

Schleiermacher referred to the means by which this is accomplished as "the hermeneutical circle." One must understand the whole—holism is a vital part of hermeneutical thinking—in order to understand the parts; but one can understand the whole only by understanding the parts and building up a composite picture, which is itself requisite for understanding any of the parts. In short, hermeneutical understanding requires a continuous shuffling back and forth between text and context (or between fact and theory, to put it in a way less congenial to Schleiermacher but suggestive).[14] As Schleiermacher himself remarked, "Complete knowledge always involves an apparent circle that each part can be understood only out of the whole to which it belongs, and vice versa. All knowledge that is scientific must be constructed in this way."[15] Effectively, Schleiermacher had established the foundations of modern hermeneutical thought.

If I were writing a history of hermeneutics, I would need to pay sustained attention to the contributions of Kant, Hegel, and many other figures. Kant, for example, with his division of reality into noumena and phenomena seems to prefigure the separation of the Geisteswissenschaften, based on hermeneutics, from the Naturwissenschaften, based on causal science (science that positivistically seeks causal explanation). And his idea of the active mind, participating in the knowing process (though with fixed, universal categories), is an essential forerunner of the hermeneutical circle. Hegel, in turn, with his own philosophical interpretation of history as the movement of mind—in the form of Reason, or Spirit—toward increased self-consciousness and awareness, provided a vital piece of the continuing construction of hermeneutics.

Instead of entering into these details, let us skip ahead to Wilhelm Dilthey, who effectively transformed Schleiermacher's generalized hermeneutics into a method of understanding history. Where Kant had offered a critique of Pure Reason, Dilthey saw himself as supplying a critique of Historical Reason, the only kind that could truly underlie the moral sciences.

In undertaking his task, Dilthey was attempting to go beyond the religious origins of hermeneutics. Schleiermacher was the son of a Moravian chaplain and himself became a professor of theology at the University of Berlin in 1810. Dilthey, who, incidentally, wrote a *Life of Schleiermacher,* was also the son of a Protestant clergyman, and he was at first destined for a theological career.

Passionately interested in history, however, he eventually became professor of philosophy at the University of Basel in 1867.

With Dilthey, Geisteswissenschaften can be said to have taken on modern scientific form, to have accepted the standards of scientific method as found in the natural sciences. Unlike Schleiermacher and other early proponents of hermeneutics, Dilthey was broadly in favor of science and the scientific method. His argument, trailing implied contradictions behind it, was that human, or moral, phenomena required a different kind of science from that directed at the natural world. Its aim was to be understanding, not explanation, and its method was, in fact, hermeneutical.

In terms reminiscent of Vico's, Dilthey declared that the range of the Geisteswissenschaften "is identical with that of understanding and understanding has the objectification of life consistently as its subject matter. Thus the range of the human studies [Geisteswissenschaften] is determined by the objectification of life in the external world. The mind can only understand what it has created. . . . Everything on which man has actively impressed his stamp forms the subject matter of the human studies." [16] Specifically, Dilthey included in human studies such disciplines as history, economics, politics, psychology, and the study of religion, literature, architecture, and philosophical worldviews and systems.

In all of these studies we seek to understand an inner world, constructed by humans. We are to do this not by means of either introspection or pure reason but by the method of interpretation, by hermeneutics. Because all human beings have elements of common humanity, at the same time that that humanity is expressed in particular cultural ways, we can penetrate to the inner life of others. In Dilthey's words: "Interpretation would be impossible if expressions of life were completely strange. It would be unnecessary if nothing strange were in them. It lies, therefore, between these two extremes. It is always required where something strange is to be grasped through the art of understanding." [17] Here we can recall the etymology of *hermeneutics*, starting with Hermes as the intermediary to the gods.

Dilthey's intent, however, is secular. He aimed at objectively valid interpretations of expressions of inner life. We reconstruct the spirit of the age animating a text and thus enter into the meaning of the text by the methods of scientific history. Like the physical sciences, history is a method of inquiry (the Greek word *historia* means "inquiry"), which uses observation, comparison, classification, and testifiable hypotheses. Although it lacks crucial experiments, has little in the way of quantifiable material, and so forth, for Dilthey history is still basically an empirical science.

Yet to overstress this aspect of his thought is to miss the tension between the empirical method and hermeneutics. *Verstehen* (understanding) is the key word in this context. Do we understand the other by entering into the other's thoughts, by reconstructing the other's mental processes?[18] Is the other's *Weltanschauung* (worldview) penetrated by means of our own leaping intuition (*Anschauung*)? In general, Dilthey seems to say that it is the lived experience of history that allows us to come to know ourselves — and thus what I am calling the Other. Throughout his work, Dilthey wrestled with such problems. He never really put his ideas together succinctly in one full book. As his *Einleitung in die Geisteswissenschaften* of 1883 suggests, his work was an introduction (*Einleitung*), which he strove to flesh out in the rest of his untidy writings. Max Weber and his use of Verstehen, R. G. Collingwood and his philosophy of history — these are a few of the attempts following upon Dilthey's to work through the nature of the Geisteswissenschaften. The problems expressed by Dilthey still stand at the heart of efforts to conceptualize the human sciences, and I shall say more about them shortly.

For the moment, let us finish our brief historical treatment of hermeneutics. Rejecting Dilthey's attempt to subject hermeneutics to the dictates of scientific reasoning, a number of theologians and philosophers sought to establish its independent validity, subject only to its own imperatives. I shall touch on two of them as examples.

The first is the theologian Rudolf Bultmann, working in the 1920s. Bultmann squarely faced the problem of understanding the Bible as a text completely outside scientific judgment. He waived aside challenges to the validity of biblical cosmology by asserting that the cosmology was "only the context of a message about personal obedience and transformation into a 'new man.'"[19] Real understanding, therefore, comes from the correct interpretation of the Bible's message as an inspiring one, making for self-transformation. Only a true believer, one already transformed by the reading, it is implied, can understand and be a judge of the Bible's meaning. That meaning transcends both ordinary and scientific inquiry.

If Bultmann tended to remain within the walls of theology, Hans-Georg Gadamer in the 1960s sought to elaborate a philosophical anthropology that claims all knowledge for its province. He forms part of a school, which includes such other philosophers as Helmut Plessner and Arnold Gehlen and is connected on one side to Edmund Husserl and on another to Martin Heidegger. Gadamer's own history of hermeneutics in his book *Truth and Method* is encyclopedic.[20]

Gadamer's text is infused with meditations on thinkers like Kant and Hegel and on subjects like *Bildung,* as well as on hermeneutics. His starting point is that "the problem of hermeneutics goes beyond the limits of the concept of method as set by modern science. . . . It is not concerned primarily with amassing verifiable knowledge, such as would satisfy the methodological ideal of science—yet it too is concerned with knowledge and with truth." But, he asks, "what kind of knowledge and what kind of truth?" Lest we misunderstand him, he continues, "Hence the human sciences are connected to modes of experience that lie outside science: with the experiences of philosophy, of art, and of history itself. These are all modes of experience in which a truth is communicated that cannot be verified by the methodological means proper to science."[21]

It is not science that stands as Gadamer's ideal of knowledge; rather, it is art or nonscientific history. In fact, he seems to disdain scientific knowledge as such,making almost no reference to it or any of its achievements. Where, for example, I have postulated that the perspective of Darwinian evolutionary theory is essential for efforts in the human sciences, Gadamer offers one trivial mention of Darwin and ignores evolutionary theory completely.

Embracing the idiographic view of history, Gadamer declares that "its ideal is rather to understand the phenomenon itself in its unique and historical concreteness. However much experiential universals are involved, the aim is not to confirm and extend these universalized experiences, in order to attain knowledge of a law—e.g., how men, peoples, and states evolve—but to understand how this man, this people, or this state is what it has become."[22] The question of why not aim at both sorts of knowledge does not really interest Gadamer. Not evolution—evolving—but "Bildung," or "forming," is what Gadamer has in mind.

Art is his ideal, and here "the frontiers of reality are transcended." Gadamer is insistent that art is timeless, though always to be understood hermeneutically in the context of a unique time. It is as if with Gadamer we were once again fighting the battle of the ancients and the moderns, in the guise of classic art versus new science, with the victory this time being withheld from the sciences.

History is also an art. As Gadamer puts it, "This is shown by the fact that the great achievements in the human sciences almost never become outdated. A modern reader can easily make allowances for the fact that, a hundred years ago, less knowledge was available to a historian, and he therefore made judgments that were incorrect in some details. On the whole, he would still rather read Droysen or Mommsen than the latest account of the subject from the pen of a historian living today."[23]

If prejudice was abhorred by Enlightenment thinkers, it is defended in Gadamer-like hermeneutics. Gadamer indicates that the question, which he calls a fundamental epistemological question, should be, What is the ground of the legitimacy of prejudices, or, to put it another way, what distinguishes legitimate prejudices from those that are presumably illegitimate? One result of asking the question is that "the possibility of supernatural truth can remain entirely open."[24] As we have seen with Bultmann, such a view implicitly assumes that hermeneutic inquiry can start from the presumption that God speaks legitimately through the text.

What of the secular author of a text? Does the author speak authentically through it? The romantics, as Gadamer interprets them, conceived of understanding as the reproduction of an original production. Thus, interpretation meant re-creating the thought of an author, living through him, so to speak. Yet Schleiermacher, inspired by romanticism, went beyond this formulation and insisted that we become conscious of many things of which the author was unconscious. As Gadamer sums up the position, "Schleiermacher asserts that the aim is *to understand a writer better than he understands himself,* a formula that has been repeated ever since; and in its changing interpretation the whole history of modern hermeneutics can be read. Indeed, this statement contains the whole problem of hermeneutics."[25]

These are strong words. They fit correctly with the construction of the hermeneutic circle, in which the horizon of the interpreter is merged with that of the author, thus going beyond the latter's. A latent danger is that such an approach can also go to the extreme of ignoring the author's intentions entirely.[26] Gadamer's hermeneutics, however, does not, and could not, envision such a move.

His is a serious effort to inquire into the Hermes-like function of the interpreter of texts. Texts are the linguistic expression of human mentality; they are to be interpreted and understood. They are not products of nature, merely to be explained. It follows, then, that for Gadamer, hermeneutics is *the* method of the Geisteswissenschaften, standing in opposition to the scientific method of the Naturwissenschaften.

How valid is this division, where Geisteswissenschaften and hermeneutics are tied together by thinkers like Dilthey and Gadamer? How useful and reliable is the hermeneutic method? These are the questions to which we must now turn.

Interpretation

Many hermeneutical thinkers insist on a sharp division between the natural and the moral sciences and set off interpretation from causal analysis as the unique method characterizing the moral sciences. How valid is this position? Is interpretation — to take up anew a question asked earlier — in fact restricted to the human sciences?

The evidence seems overwhelming that interpretation is a necessary part of the natural sciences, just as it is for the human ones. Common sense would appear to impose this conclusion upon us. That it does not always do so is probably owing to the extreme positivists' claim to complete objectivity. According to this claim, it is possible (and desirable) to eliminate the human element and so to see phenomena as they really are. The unrealism of this claim has given rise to hermeneutic mimicry, though in an opposite direction, which denies interpretation to the natural sciences.

In aid of common sense, let us consider a few supporting quotations. Speaking of physics, Erwin Schrödinger remarks of photometric recordings that "they must be read! The observer's senses have to step in eventually." [27] If we move toward what is often thought of as a slightly softer science, but still a natural one, we find that the Geological Society of London, founded in 1807, sought desperately to impose a positivistic (and Baconian) creed upon its members by writing into its procedures the eschewal of interpretation and restricting discussions to facts alone. The reality was different. As Stephen Jay Gould tells us, "The striking vectors of geological history read literally must be interpreted, by Lyell's method of probing behind appearance." [28] Last, here is a comment on a human science, anthropology. Where Malinowski wrestled with the possibility of eliminating interpretation from his observations, Raymond Firth admits flatly, "Even the simplest record of what purports to be the 'facts' of a native culture has involved a considerable amount of interpretation, and every generalization about what the people do has meant a selection from the immeasurably wide field of their activity." [29]

As these quotations suggest, interpretation is as present in efforts at understanding in the natural sciences as it is in the texts of the human sciences. I would go further and argue that there can be no science without interpretation; what is at issue is how interpretation functions in a particular inquiry and in what form. To say this is not to eliminate the hermeneutic differences between the natural and human sciences but to look toward establishing what they are. Naive empiricism in physics or geology is the equivalent of scriptural literalness in religious studies: both are equally deficient in not recognizing the necessary role of interpretation. In the end we must accept that interpretation underlies all our attempts at comprehension, in everyday life as well as in science. [30]

The central role of interpretation arises from symbol-making activity.

Humans are constantly interpreting each other's actions and the meaning of those actions, as well as interpreting their experiences vis-à-vis nonsubjective nature. Alfred Schutz has underlined this phenomenological aspect of interpretation. In a useful paraphrase, Richard J. Bernstein states Schutz's position as one that sees us as "continuously ordering, classifying, and interpreting our ongoing experiences according to various interpretive schemes. But in our everyday life these interpretative schemes are themselves essentially social and intersubjective. Intersubjectivity lies at the very heart of human subjectivity."[31]

Interpretation occurs in a social matrix. It does not arise as an innate, individual, completely detached activity. It occurs in a cultural context and has meaning only in such a context. Clifford Geertz expresses the idea this way: "The concept of culture . . . is essentially a semiotic one. Believing, with Max Weber, that man is an animal suspended in webs of significance he himself has spun, I take culture to be those webs, and the analysis of it to be therefore not an experimental science in search of law but an interpretive one in search of meaning."[32] Geertz's interpretation recapitulates the sharp divide between Naturwissenschaften and Geisteswissenschaften. I have already suggested leaving open the question of the extent to which the human sciences can also search for law and for causal explanations. With that caveat, we can return to Geertz's emphasis on interpretation.[33]

There is little question as to the connection between symbol-making activity and culture. One makes the other possible and underlies the shift from physical evolution to cultural evolution. Symbol making is also at the bottom of nongenetic social stratifications. It is as much part of *Homo sapiens* as the prehensile thumb and upright posture. It pervades everyday life, permitting interpretation of both the social and the natural worlds initially at the level of ordinary experience. It emerges from culture and constructs that culture at the same time. In this sense, the hermeneutic circle is in constant evidence.

Humans have intentions and motives, and these must constantly be interpreted in terms of the relevant culture if we are to make sense of social relations — and to survive. It seems a natural step to extend this notion to "nature," which is thus anthropomorphized and made a part of the culture (in addition to animism, we need only think of Aristotelian thought and its "nature wishes . . ."). Much of the modern history of humanity, however, has circled around its attempt to free itself from such anthropomorphizing and to look at natural phenomena objectively, that is, nonteleologically. This effort is directly tied to the attempt not to be deceived by appearances but to go beyond them to a deeper reality.

Here yawns the divide between Naturwissenschaften and Geisteswissenschaften. The attempt to look at nature objectively may be an impossible ideal, because all such attempts must necessarily be tied to interpretation and thus subjectivity. Yet in seeking the ideal, natural scientists have been able to construct a scientific methodology—characterized by an abundant use of verification procedures that involve evidence and inference, extension of sensory inputs, use of experiments, and so forth—that permits the community of believers in that method to go beyond appearances to an abstract reality that allows for a large amount of prediction and control. We get a glimpse of what is involved in F. A. Hayek's observation that "this process of re-classifying 'objects' which our senses have already classified in one way, of substituting for the 'secondary' qualities in which our senses arrange external stimuli a new classification based on consciously established relations between classes of events is, perhaps, the most characteristic aspect of the procedure of the natural sciences." The most appropriate language in which to express such relations is mathematics. As a language, mathematics is itself a means of interpretation. It performs its own translations. As Hayek further remarks, "In other words, although the theories of physical science at the stage which has now been reached can no longer be stated in terms of sense qualities, their significance is due to the fact that we possess rules, a 'key,' which enables us to translate them into statements about perceptible phenomena." [34]

The ability to abstract from phenomena, especially by the use of mathematics, and thus to objectify motives as much as possible, has not been feasible to any great extent in the human sciences. Human motives are intrinsic to and inseparable from human actions. They must be understood from the inside in a way that differs from the procedures brought to bear on natural phenomena—this is the message of hermeneutics. In W. H. Auden's wonderful lines:

> Between us and the Insects,
> namely nine-tenths of the living, there grins
> a prohibitive fracture
> empathy cannot transgress.   ("The Aliens")

We may seek to understand the ants and the bees, but we cannot share their feelings or know what form they might take or even whether they have any. Anthropomorphism, it follows, is a false form of fellow-feeling. In contrast, true empathy is at the heart of much of the human sciences. [35] Using it as an investigative tool, human scientists, too, like the natural scientists, have sought, especially in modern times, to go below the surface of appearances, to a deeper and thus truer reality.

Possibly differing on the role of motives but agreeing on the need to go beyond appearances, scientists on the sides of the divide stand eyeing each other. Both use interpretation; but the problem of the human sciences, focusing on motives and intentions, is how to recognize the "correct" interpretation. There is neither a sufficiently agreed-upon scientific method nor a scientific community committed to its clear procedures. The effort to go deeper may eventuate as readily in mysticism; one may well drown in subterranean waters instead of descending to a more fundamental scientific reality. How can we distinguish between the two scuba-diving expeditions after knowledge? Why might one be preferable to the other? And how would we recognize the preferable one?

The Hermeneutics of Modern Science

Let us resume our historical approach and look now at the emergence of modern science in the light of our hermeneutic concerns. Before we did so with an eye on positivism. This time it will be useful to focus on Galileo.

It is Galileo, following upon Copernicus and Kepler, who "discovers" modern astronomy along with Jupiter and its moons. The difference between astronomy and astrology is an important one. As Hans Blumenberg remarks in his brilliant work *The Genesis of the Copernican World*, "Grasping at the phenomena hastily, astrology relates the universe to man, makes it the sum of signs for him, and thus makes him the reference point for all physical processes."[36] In astrology, the heavens are anthropomorphized. In astronomy, on the contrary, Galileo sought to remove human beings as the measure of the planets and to study the heavenly bodies "objectively." This effort at self-removal was the precondition for a scientific astronomy. Although I would wish eventually to qualify his approach—for humans and their symbol-making, interpretive nature must serve as the reference points for any knowledge they may obtain— Galileo's way did lead beyond appearances and false interpretation to a deeper knowledge.

First, the appearances had to be extended and refined. This required a new instrument, the telescope. The instrument also required a new way of looking, or, to put it more correctly, a new way of thinking.[37] Galileo claimed to be amazed at the refusal of some of his learned contemporaries to look through the telescope or, if they did, to acknowledge what it showed. As he wrote to Kepler in 1610, "What do you think of the chief philosophers of our gymnasium, who, with the stubbornness of a viper, did not want to see the planets . . . ? In truth, just as he [Odysseus] closed his ears, so they closed their eyes against the light of the truth. That is monstrous, but it does not astonish me. For men of this kind think that philosophy is a book, like the *Aeneid* or the *Odyssey,* and that truth is to be sought not in the world and in nature, but in the comparison of texts (as they call it)." [38]

This theme, a version of which we encountered in Bacon, resounds throughout the seventeenth and eighteenth centuries. There is a kind of battle of the books — for Galileo also produced books — in which one side compared texts against texts and the other side, in this case, Galileo's, claimed to compare them with "reality." [39] The first uses hermeneutics, or interpretation, but restricts it to what has already been written, which, itself claims to be a correct interpretation of reality. The second seeks to look anew at that reality in a more scientific manner.

In Galileo's science, we need to read the pages of the heavens with fresh eyes, freed from earlier interpretations. We need to penetrate beyond appearances, which is all that the vulgar see. We must operate as virtuosi: "what the mere sense of vision presents to us is like nothing in comparison to the marvels . . . that the ingenuity of those with understanding discovers in the heavens." [40] Galileo's was an aristocratic conception. As he wrote to Grand Duchess Christina, one need not give heed to popular common sense, which paid attention only to appearances, which, moreover, were false and constricted; rather, one should give preference to ingenious understanding.

Galileo's distrust of what is merely written in books extended to the Bible, although he could only express himself on the subject in circumlocutions. The holy text resolved itself into a question of interpretation. Surely, he declaimed, God is no less "excellently revealed in Nature's actions than in the sacred statements of the Bible." Moreover, the Bible could be misread, and figurative statements could be taken literally. In doing the latter, Galileo's opponents, "contrary to the sense of the Bible and the intention of the holy Fathers, if I am not mistaken . . . would extend such authorities until even in purely physical matters — where faith is not involved — they would have us altogether abandon reason and the evidence of our senses in favor of some biblical passage, though under the surface meaning of its words this passage may contain a different sense." In words that have become famous, Galileo quoted an eminent ecclesiastic to the effect that "the intention of the Holy Ghost is to teach us how one goes to heaven, not how heaven goes." [41]

Galileo's opponents were right to distrust him, for though outwardly accepting faith in the Bible he advocated going beyond the knowledge contained in it. "Who indeed," he asked pridefully, "will set bounds to human ingenuity? Who will assert that everything in the universe capable of being perceived is already discovered and known?" Equipped with such a conviction, science as a progressive, accumulative acquisition of knowledge lies before us. In this realm, Galileo argues, one seeks to explore "the difference that exists between doctrines subject to proof and those subject to opinion." Explicitly, Galileo made protestations of submission to papal authority; yet it is apparent that his comments implicitly open the way to positive science. (And we must remember that the preservation of Galileo's life was predicated on the remarks he made and their reception by the church.) Whether science is restricted to the natural or can be extended to the human became, as we know, a central concern of eighteenth-century philosophes–cum–social scientists, as well as subsequent inquirers.

The way is also open to a more critical hermeneutics. Looked at closely, Galileo's expostulation suggests that a different reading of the Bible from that of the traditionalists is possible. From what he says in his letter to the Grand Duchess Christina to the *Historical and Critical Dictionary* of Pierre Bayle the step is not too long. Also, the step had been prepared for by the Erasmian humanists and the Lutheran reformers, who were at pains to establish the original texts and then to offer a more philological and less authoritarian reading. Bayle continued on this path, but with a secular intent. By the nineteenth century, we are on the highway to the Higher Criticism, which originally served religion but which deviated with the Young Hegelians into more secular and revolutionary tracks.

As we have noted repeatedly, hermeneutics has been, and is, frequently connected with religion. It need not be. Although Schleiermacher's attempt to establish hermeneutics as a general science was religiously inspired, his creation could and did turn into an opposition, seemingly from the Devil, undermining by its interpretations both established religion and the state that so often supported it.

Hermeneutics, like positive scientific method, presupposes that something is hidden, that behind appearances, there is a truer reality. At first, dragging the word's etymological origins behind it, many hermeneutical thinkers believed that only God could know truer reality. What hermeneutics can do, they claimed, is to allow us to peep behind appearances to a transcendent reality — the true one — known fully only to God. But in the physical sciences, Galileo and many others came to believe that God revealed himself in nature as well as in the Bible and, with the scientific method, could be read correctly in his physical manifestations. In the moral sciences, the heretical notion arose subsequently that the notion of God had to be examined and that closer interpretation indicated that he was a reflection of humanity's own desires and intentions. As Feuerbach and the Young Hegelians, for example, pointed out in the early nineteenth century, man was not made in God's image, but vice versa.

Increasingly, the conviction has grown that humans construct their own culture and society on the basis of their symbol-making nature, that, indeed, everyday existence is spun out of symbolic construction and interpretation of what I am calling the Other, as when projections onto mythical gods or "foreign" peoples take shape as culture. By either eliminating God, as Nietzsche did, or relegating him to the far reaches of the heavens, radical thinkers have placed responsibility for human actions squarely on humans. In their view, part of that responsibility means that humans must fashion reality. In doing so, their motives and intentions enter into the reality that they seek both to construct and to understand.

Earlier, I suggested that the two terms *hermeneutics* and *interpretation* are synonyms; and so, in part, they are. But if the physical sciences also use interpretation, their version of it has become scientific method. The term *hermeneutics* is not used to describe that method. *Hermeneutics,* consequently, is reserved for the human sciences, where motives and intentions rule and empathy is requisite for much of our understanding.

In assigning humanity responsibility for its own creation (while bearing in mind the role of evolution in that creation), we also place human beings in a most precarious position — one that gives rise to a malaise, expressed by such terms as *alienation, anomie,* and the Weberian *disenchantment of the world.* The position is marked by uncertainty about the very foundations of humankind's and society's existence, especially about the origin and soundness of values. It is why, for some, spurning regression to earlier beliefs, there is a desperate need for human sciences that can offer correct and certain guidance, that can hold out the hope of prediction and control. Alternatively, many seek these goals in a return to religion. Where positive science falters, however, and is found unreliable, can some form of hermeneutics take its place?

The Other

Three more subjects need to be considered before we confront the question just asked frontally: the Other, madness, and narrative. In Chapter 1, I discussed the Other in terms especially of the New World discoveries and the developments from those events that underlay attempts in the human sciences. Now, we need to relate the concept of the Other to hermeneutics. To do this, we must look further at perspective and the ways views concerning it changed as a result of the Copernican revolution. Only by studying the rise, historically, of the new attitudes can we gain insight into the way the idea of the Other figures in our interpretations.

Copernicus not only displaced the earth as the center of the planetary system; he also called into question traditional ways man viewed the universe. The central issue was one of perspective, which in turn invoked the problem of appearances and reality. As Blumenberg puts it, Copernicus "reflects the cosmological differentiation between the parochial perspective of his terrestrial 'corner' and the central point of construction from which the universe cannot, indeed, be viewed, but can be thought." [42]

How can we think beyond our visual "corner"? Copernicus was a loyal Catholic priest, yet his thought leads us to merge with God, not in a mystical embrace but with rational insight. Copernicus imagined himself in God's place, seeing the whole of the universe in a glance. From that perspective, the earth can be viewed by the mind's eye as rotating around the sun, rather than vice versa; one is no longer misled by false sense appearances.

Copernicus' astronomical "instrument" was mathematics, the exercise of reason. A further proof of his theory came at the hands of Galileo, using a telescope. With its use, we see things unavailable to the naked eye. New appearances are substituted for existing ones; we see more deeply, confirming our new, Copernican perspective.

At the other end of the scale—for issues of scale and proportion now become critical—we have the microscope. Galileo's contemporary Christian Huygens compared its use to a new voyage, declaring excitedly that he and other microscopists were "discerning everything with our eyes as if we were touching it with our hands; we wander through a world of tiny creatures till now unknown, as if it were a newly discovered continent of our globe." The times were heady with new vision. Huygens, who was given to writing poems as well as to performing scientific experiments, expressed the viewpoint neatly in "On the Telescope." Ecstatic over humanity's novel ability to see "everything from the highest point of the heavens to the tiniest creatures on earth" (Sylvana Alpers's words), Huygens wrote:

> At last mortals may, so to speak, be like gods,
> If they can see far and near, here and everywhere. [43]

Whether by mind or matter (i.e., instruments), by the seventeenth century humans had achieved a new vantage point from which to view the universe—and, it followed, humankind itself. The New World discoveries had already given new shape to the debate about the Other, to the human who appeared radically not like oneself. From what perspective was one to view this difference? The revolution in science of the sixteenth and seventeenth centuries gave intellectual form to the problem: it was all a question of perspective, and no one vantage point was privileged.[44]

This radical conclusion did not emerge all at once, nor was it followed widely. It required further discoveries, further explorations. For one thing, even the new perspective was a Western one: The explorations were European voyages, made for European purposes, and the science was also mainly European, though claiming universal validity. Emergent European capitalism and imperialism supplied the incentives for the voyages. To take a trivial illustration, the charting of the coasts of the seas was undertaken by Westerners rather than, say, Chinese because their boats, unlike the larger junks and praus of the East, drew as much as twenty-two or twenty-three feet and thus needed deep waters under their keels.

It was on such a charting voyage that Charles Darwin sailed in the *Beagle*. In pursuit of imperialist trade, the British needed to know where their ships could go safely. Science followed the flag, starting with Cook's voyages to the Pacific, accompanied by naturalists, and continuing with the *Beagle* voyage under Captain FitzRoy, and then many others. In Darwin's case, the voyage not only made him a "new man"; it created a new perspective on all men. *Homo sapiens* had become an evolutionary creature around the globe.

Yet the European point of view still initially prevailed. In regard to the Fuegians, for example, Darwin himself was revolted by their savage appearance and behavior. *This* Other he saw as monstrous. In fact, evolutionary theory was used, or misused, to establish a scale of evolution, thereby supporting an existing view of Western superiority.

But the same voyages bred the new human science of anthropology. Again, the initial positions were Eurocentric. The Other was devaluated, dehumanized, made childlike and primitive, or else material to be missionized and modernized.[45] It is true that, for a few, the primitive was superior to the civilized, retaining ancient values in the face of Western degeneracy and the fall into apathy and alienation.[46] For most, however, the primitive was seen as an object for objective study. Such humans were no different in this sense, it appeared, from insects.

We have already noticed in Malinowski the difficulties of such a stance in regard to humans. With the hermeneutical attitude, humans, it was now understood, unlike insects, required empathy and Verstehen. The romantics had earlier glorified the exotic and exalted its unknowability. Now the task was to understand the exotic Other. To do so meant to interpret others in their own terms as well as our own. It meant adopting a new or at least a different perspective.

Anthropologists have continued to struggle with the defining aims of their putative science. Is its intent to establish universals characterizing the human species based on comparative studies? Or is it to study particular tribes, delineating their unique qualities, with anthropologists content to offer us the details of other lives? Or is it some combination thereof? Even positivist-minded British administrators in India needed to understand the intentions of the Others whom they ruled; and anthropologists were forced to pay attention to strange perspectives, utilizing techniques like interviews to discover how the Other saw the world.

Whatever the intent, interpretation stood at the center of the anthropological effort. And interpretation, as Copernicus, Galileo, and those unlikely allies the romantics had shown, was always from a perspective, a point of view. The temptation for many anthropologists was to accept the point of view of "their" natives, to identify completely with the Other. Seeking holistic understanding of the culture being studied, many anthropologists thought of their object of study as a closed circle, shut off from history and other points of view.

The alternative option was to swing toward a total cultural relativism. Any point of view was as good as any other; an observer could not stand outside the phenomena observed, as God-like and objective, and pass judgment. This was true whether in regard to the native's interpretations — they were as valid as Western ones — or the anthropologist's interpretations, for these, too, were subjective and not superior because supposedly scientific.

By the twentieth century, any presumed center appeared to have disintegrated. Ours is a time when we are heir to all perspectives, whether of time or of space. Centuries of imagery, drawn from particular perspectives concerning, say, the figure of Christ, are in our possession. We have roamed the world and taken possession of every manifestation of humans and culture, ranging from the !Kung San to the barrio dwellers. These perspectives are ours simultaneously. No wonder a certain optical dizziness, a blurring, has seized us.[47]

What I am calling the Other is no longer outside us; the Other is now consciously perceived as an integral part of our own being. This development has been fostered by the globalism sweeping our planet. The situation is partly characterized by the term *multiculturalism*. As a result of, among other things, satellite communications, we are with the Other in real time — the Other is no longer in an isolated culture at the other end of the globe. To do business with the Other, or to make war, we need to grasp the Other's perspective (and the

Other ours). In more and more parts of the world the center of our interpretive efforts is no longer the Bible or the Quran but the media. The media is what connects us and makes the Other our immediate neighbor. The book of the world now appears on a screen or monitor.

A major intellectual response to these developments is postmodernism. The deconstructionist critique of the notion of the center has been scathing. Foundational thinking has become anathema. It is no accident that hermeneutics has come center stage, *the* method by which all meanings are made a matter of perspective. Foucault, for example, in the preface to his book *The Order of Things: An Archaeology of the Human Sciences,* declares that "this book first arose out of a passage in Borges . . . [which continues] to disturb and threaten with collapse our age-old distinction between the Same and the Other." Then comes the famous passage quoting from a "certain Chinese encyclopedia," which offers a classification of animals totally at odds with anything resembling the Linnaean order. The implied conclusion is simple: ours is an idiosyncratic, not truly scientific classification, and the other's, that of the Chinese, is equally worthy and representative of reality.[48]

For most postmodernists, everything is a text: books, incarceration practices, scientific findings. Postmodernism appears to restrict itself, in fact, to a discourse about discourses (although these, admittedly, are about practices). Interpretation reigns supreme, and none can derive from a central, ordering perspective. Everything becomes a matter of extreme hermeneutics.[49]

The great *Encyclopaedia Britannica* of 1911 — the classic edition — has a three-line entry under "Hermeneutics," including a definition: "the science or art of interpretation or explanation, especially of the Holy Scriptures." Limited and partial as this entry is — after all, Dilthey's work had been published at the time — it at least notices the subject. In contrast, by the 1967 edition of the *Encyclopaedia,* all mention of hermeneutics had vanished.

Postmodernism has redressed the balance. The postmodern thinkers, whatever their weaknesses in terms of our overall concerns, have unquestionably and rightly restored hermeneutics to our awareness. In the process, they have gone far beyond Schleiermacher and Dilthey, breaking the hermeneutic circle into 360 points of view. The result verges on interpretive nihilism.[50] The Other has been fused with us, but at a price: both of us have lost a sense of identity or intention (texts for postmodernists, too, must be read as if without an author). One psychiatrist likens the situation to schizophrenia.[51]

Our challenge is to recognize the strengths that are in hermeneutics and to evaluate how the method may serve efforts to construct (as well as deconstruct) human sciences. If Copernicus and the New World discoveries "decentered" our existence, making life and culture a matter of interpretation from newly created perspectives, the concept of the Other gave specific forms to this development. Carried to an extreme, however, the Other could call into question our own identity and leave us unbalanced. To understand what is involved in hermeneutics, we need to look further at exactly how balanced the human being has ever been.

Madness

Perhaps the term *madness* is too strong. *Insanity* or *irrationality* might be better terms by which to characterize much of what is to be found among the human species. Madness, however, dramatizes the issue and reflects the way ordinary speech often describes even an innocuous action or thought as "mad," or "hare-brained," or "loopy." As the saying goes, "All the world is mad, but thee and me / And sometimes I am not sure of thee." At the heart of any usage of these terms is the notion of being seriously out of touch with reality.

The paradox is that madness arises from the symbol-making aspect of human nature, which is employed to gain a better, more farsighted relation to reality than is acquired by reflex action or instinct. Madness is more or less a unique quality of the human animal, the price paid for its evolutionary acquisition of symbol-making abilities. We may speak of rogue males among elephants, or foxes turning rabid; these are comparable phenomena but not on the same level as human madness.

Madness results when the complicated human brain misfires, so to speak, and hits the wrong target. Misfirings may be genetically caused, chemically induced, or psychologically produced, or triggered by some combination thereof. Misfirings can be held back or fostered by social conditions—in this sense, R. D. Laing, Foucault, and others grasp a piece of the truth in ascribing madness to the reigning culture (although they note only one part of the etiology) —and its particular manifestation may be culturally determined.[52]

Another, less dramatic, way of speaking about madness is to use the terms *normal* and *abnormal*. By definition, the normal is what most people do; the abnormal deviates from the norm. Different societies will have different norms and different conceptions as to what behavior is conforming and what is deviant. To accept this mode of conceptualizing the problem, however, is to accept that in a madhouse the normal person, as perceived from outside by the majority, is mad.

We need to take a different position to move forward with the human sciences. For heuristic purposes, we must step outside particular majorities and view the human species as a whole (while recognizing cultural differences). Like Copernicus, our perspective will be from on high.

What do we see? Most people, we observe, even without institutionalization, cross the borders between normality and abnormality, to use those restricted terms for the moment, at frequent occasions in their lives—in abusing children, beating their spouses, indulging in paranoid fears of the Other. If the touchstone is reality, innumerable humans have skirted its edges in religious hallucinations and related phenomena, not to mention beliefs.

Schizophrenia, the extreme case, seems to occur in all societies. A genetic base or at least predisposition is predicated. E. O. Wilson refers to it as "the commonest form of mental illness" and describes it as consisting of "various combinations of hallucinations, delusions, inappropriate emotional responses, compulsively repeated movements of no particular significance, and even the death-like immobilization of the catatonic trances." He concludes, "The borderline between normal and schizophrenic people is broad and nearly imperceptible. Mild schizophrenics function undetected among us in large numbers." [53]

The *concept* of schizophrenia, the lens that we use to classify the behavior at which we look, is itself recent. Emil Kraepelin introduced the diagnostic category in 1896, and Eugen Bleuler coined the term *schizophrenia* in 1908 to describe a "specific type of alteration of thinking, feeling, and relation to the external world which appears nowhere else in this particular fashion." Louis Sass, a present-day clinical psychologist, adds that "to judge from what schizophrenics report, they themselves are not spared this unnerving sense of aberrancy and alienation. They may experience the most profound alterations in the very structures of human consciousness, in the forms of time, space, causality, and human identity that normally provide a kind of bedrock foundation for a stable human existence. . . . The objective world may loom forth as a solid but strangely alien presence, or else may fade into unreality, or even seem to collapse or disappear." [54]

Without ignoring Wilson's genetic underpinnings, Sass highlights the schizophrenic's alienation from reality. I would also point out the connection to the Other: there is none. The schizophrenic in effect inhabits a world void of others and their reality. It is not just a matter of a different perspective but of no perspective. It is a kind of total nihilism. The only reality is the person's inner life, shut off from external reality.[55]

The schizophrenic is not neurotic but psychotic, that is, so significantly out of touch with the consensual world as to have lost a true grasp of reality. Again we are faced with a diagnostic classification involving loss of ego boundaries, poor reality testing, and delusions or hallucinations about world destruction. The *Diagnostic and Statistical Manual of Mental Disorders* (DSM III-R) defines the psychotic as an individual who "makes incorrect inferences about external reality, even in the face of contrary evidence." Such individuals "believe in the literal truth of idiosyncratic ideas that, in actuality, are 'false' and grossly 'incorrect.' " [56]

Needless to say, the definitions can be challenged. They may be thought to misconstrue the reality of the mental conditions being described. But common sense, not to mention reason itself, would insist that they point to recognizable phenomena, even if those phenomena are differently evaluated.

Schizophrenia and psychosis under which it falls mark the extreme conditions of what I am referring to as human madness. They may indicate the outside limit of what, on the other side, is the basis for creative approaches to the external world. Fantasies, even hallucinations, may lead as well to scientific and artistic creation as to pathological detachment from reality. Madness is the correlative of human symbol-making activity, which, though evolved to bring the species better in touch with its surrounding reality, may take on a life of its own — a psychotic or a neurotic life.

As we now know, there is meaning and even method in madness. Most prominent in investigating the meaning has been Sigmund Freud. It is fashionable nowadays to denigrate Freud, but this is to ignore his real achievements.[57] He has truly brought about an introspective revolution. Admitting all his weaknesses — a tendency to overspeculation, a rigid assumption of a certain type of rationality, an overemphasis on the sexual, a mistaken psychology of women — he has taken the hermeneutical approach to its furthest point of connection with the scientific method. He has deepened our knowledge of human madness by creating a new science, psychoanalysis, that stands uneasily on the edge between the so-called two cultures of C. P. Snow.

This is not a book on Freud.[58] I will, however, attempt to disentangle from his work a number of themes pertinent to the effort at human science. One of the most important is his notion of a spectrum from normality to abnormality, on which every individual has a place. Thus, in his studies on hysteria, Freud went to great pains to exonerate his patients, in this case women, from the charge of degeneracy. As he insisted explicitly, they exhibit symptoms and causes that are deep within all of us. Indeed, we are all neurotic. We exhibit our neurotic character not only in symptoms but in everyday parapraxes and in our dreams. For Freud, moreover, symptoms, parapraxes, and dreams all have meaning; they are not just somatic events, reflex actions of a sort. Their meaning requires interpretation to unravel them. This makes psychoanalysis the most advanced form of hermeneutics that we have encountered. It is no accident that Freud's first major book is entitled *The Interpretation of Dreams.*

How can we know what is going on in the human mind? What are our intentions and meanings? We cannot find out with an X-ray or CAT scan, which shows just the structure of the brain. We have no telescope or microscope that can mechanically extend our senses as we peer into the inner recesses of mind and strive to detect and interpret meaning. In this perspective, Freud's work can be seen as a pioneering effort to find new instruments with which to examine mind and its meaning.

Hypnosis (which proved the existence of unconscious mental processes) and then free association were Freud's preferred probes. With them, he sought entrance to the mind's depths. His evidence came from his clinical practice, as well as from his own self-analysis. On this basis, he erected theories and models, some good and some bad. He frequently revised them, and his successors have carried on in the same spirit. To take the claims of psychoanalysis seriously, as I believe we should, it is an empirical science that, building on Darwinian biology, eventuates as a hermeneutical enterprise.

In this most human of sciences we are forced to accept the unconscious nature of most of humanity's mental processes. The matter is given theoretical form and analysis. Freud, defining the mind as a dynamic mechanism, tried to give specific form to the "defense mechanisms" deployed, such as projection, repression, displacement, and transference. We cannot help but recognize ourselves in these terms. Even anti-Freudians serve as Freudian exhibits in their daily lives.

The extent of our madness may differ, but in one degree or another we are all mad, as seen from this viewpoint. Who can deny the craziness that exists in the world? A whole chapter could be devoted to the miracles that defy reality and are nevertheless accepted by individuals claiming to have witnessed them; to the Jim Jones Temple worshipers and other cultists; to the conspiracy theories that flourish through the ages, ranging from outbreaks of witch hunting to the Nazi ideology; to the body snatchers and invading aliens perceived by so-called normal people, and on and on.

If the touchstone is reality, then such people deviate from it significantly and can be correctly described, in this aspect of their lives, as deranged. Such people, however, turn out to be alter egos of ourselves. In *Dr. Jekyll and Mr. Hyde*, Robert Louis Stevenson gave classic form to the perception of the "double man," the split personality that lurks in all of us. The abnormal and unfathomable Other is oneself in other shape. Thus, a literary writer intuited what Freud sought to understand by more scientific means (though he appealed to literature for both evidence and confirmation); both were sounding the same caverns of the human mind. As a result, hermeneutics was irrevocably changed and needs now to deal with unconscious as well as conscious meaning.

Narrative

Freud admitted ruefully that his scientific reports on his patients read like "Just So" stories. But whereas Freud's stories are the result of interpretations, most stories are themselves creations demanding interpretation. The first are attempts to take the fictions of the analysands and make science out of them; the second are presentations of the fictions as themselves a direct form of a special kind of truth.

Humans, it appears, have a propensity to tell stories.[59] Stories are a form of explanation. They are generally structured to have a beginning, a middle, and an end in order to impose meaning on otherwise shapeless events. They can take many forms, ranging from mythological to literary to scientific.

Science, may be a surprising place to find narrative. On reflection it should not be. I have already alluded to Robert Boyle in this vein. Boyle, in the prefactory letter to his *New Experiments Physico-Mechanical Touching the Spring of the Air* called his description of the experiments "narratives." [60] He was telling the story of how he went about his experiments, with an eye to helping readers repeat the steps and reach the same scientific conclusion. His narrative allows for the virtual witnessing spoken of in the chapter on positivism. We, too, can repeat his story. The narrative is both a demonstration and a proof.

Such a scientific narrative is quite different from a historical narrative. The former stands (or claims to stand) outside time (and, indeed, place). It can be performed in the next century as well as in Boyle's, in Timbuctu as well as in England, and be unchanged. I cannot put the difference better than Jürgen Habermas does, so I quote him fully. Remarking on historical narrative, he states: "The predicates with which an event is represented in narrative form require the appearance of later events in the light of which the event becomes a historical event. Consequently, with the passage of time historical description of events becomes richer than empirical observation at the time of occurrence permits it to be." In the physical sciences, in contrast, "within the frame of reference of empirical-scientific theories, events are described only in terms of categories that can also be used in making a protocol of the observation of the event. An event that is scientifically predicted can be designated only in an observational language that is neutral with respect to the time of occurrence."[61]

Boyle's use of the term *narrative* is not generally echoed in the reports of physical scientists after him, and narrative as a form of explanation plays little overt part in their accounts of their work. When we approach the geological and biological sciences in the nineteenth centuries, however, we see narrative playing its role in a more pronounced fashion. Lyell, for example, compared early geology to history, suggesting that the explanatory patterns in both fields were similar. Of course, Lyell sought to derive generalizations, uniformities, and laws from his historical accounts of the rocks; but then many general historians of his time sought to do the same (Henry Thomas Buckle is a shining example).

Lyell influenced Darwin. The young man's first glimpses of his theory of evolution came during his five years on the *Beagle*, when he was reading Lyell's *Principles*. He published his findings as a *Journal*, a narrative account of his voyage. The narrative hints at a transition to scientific theory, though remaining a travel story. We see the tension in a typical lyrical description of the flora and fauna of Tierra del Fuego: "I can only compare these great aquatic forests of the southern hemisphere with the terrestrial ones in the inter-tropical regions. Yet if in any country a forest was destroyed, I do not believe nearly so many species of animals would perish as would here, from the destruction of the kelp. Amidst the leaves of this plant numerous species of fish live, which nowhere else could find food or shelter; with their destruction the many cormorants and other fishing birds, the otters, seals, and porpoises, would soon perish also; and lastly, the Fuegian savage, the miserable lord of this miserable land, would redouble his cannibal feast, decrease in numbers, and perhaps cease to exist."[62]

In this passage, Darwin already has in mind the theory of ecology, or what he called the "economy of nature," where each part is connected to every other. His emphasis here is on destruction or extermination, the obverse of origin. In his *Journal*, however, natural selection, a theory, based on variations, is absent, at least explicitly. Instead, we have a narrative of imagined events as they would unroll if the kelp were destroyed. By later introducing his theories, Darwin will go beyond simple narrative to biological science. Even that science, however, will be marked by the narrative mode, for evolutionary theory is a story of origins and extinctions played out over time. The events are unique; unlike Boyle's experiments, we cannot repeat the evolution, say, of the Galápagos tortoise Nevertheless, though a different sort of narrative from Boyle's, Darwin's shares with its predecessor the aim of serving a scientific purpose, one that leads to generalizations and "laws."

Where Darwin was influenced by Lyell, Freud was influenced by Darwin. Freud's psychological human was a product of Darwinian evolution. Nevertheless, as Freud sought to gain understanding of the human mind—and its madness—he was forced, initially much against his will, away from the positive method of organic-minded psychiatry to the will-o'-the-wisp of hermeneutics. His science, psychoanalysis, increasingly became a matter of interpretation and linguistics (while retaining its biological base). It was for this reason that his case histories were exactly that: histories, or narratives ("Just So" stories) claiming to be science.

All science can be viewed as an effort to establish lawlike connections among phenomena. In Freud's construction of psychoanalysis, the connection explaining a symptom is a narrative one, a story. That story is placed in the framework of psychological formulas (laws or theories of a very special kind). The one thing the story cannot encompass, however, is an explanation of how exactly this particular "experiment" can be repeated, for it is the very nature of the relationship between analyst and analysand, with its powerful underpinnings of suggestion and confidentiality, that cannot be minutely replicated. Instead, we must tell other stories, appeal to similar materials, such as myths and dreams, and by the comparative method construct and test our psychological formulas—let us call them theories—which we then apply to new materials (as well as old).[63]

The narrative itself becomes a form of causal explanation: "X happened because Y did this." Freud claims to give this explanation a truer scientific form by, in turn, giving to the connections cited a theoretical underpinning. We understand Y's doing what he did because it stemmed from his Oedipal feelings. The theory of the Oedipus complex comes from a host of evidence: other clinical encounters, Greek dramas, our dreams. The complex, in turn, is understood in terms of a whole array of other theories, based on other evidences and brought together in systematic form.

The whole of the above enterprise is controversial. Eschewing the arguments for and against, I have been trying to see here how narrative as such can, in principle, be viewed as a form of scientific explanation.[64] It need not be dismissed as a mere story or waved aside as pure fiction; it may, in fact, be a powerful and reliable form of truth telling that explains a unique event in approximately scientific fashion.

How do we know when truth telling is in the ascendancy in the narrative mode? We are thrown back, if we insist on escaping Freud-like pulls, to hermeneutics, or interpretation, without the benefit of overt theories. The fact is that most use of narrative is not on the model of Freud, or Boyle, Lyell, and Darwin. Most narrative is to be found in the form of myth or so-called creative literature. It is to these forms that we must now proceed.

Myth

It is often thought that myth is in complete opposition to science and that with the advance of the latter, myth will disappear. This is generally put in terms of mythos versus logos, with logos becoming triumphant. In a typical treatment of Greek philosophy and history, Henri Frankfurt and associates show how mythical accounts of nature were gradually supplanted by philosophical reasonings, as in Thales, Anaximander, and others, and how Herodotus and Thucydides introduced historical reasoning—a weighing of evidence and a use of strict inference—in place of mythical accounts.[65]

A strong case can be made that history shows just such a development, in which logos, that is, science, does emerge out of mythos. One example is Thales' prediction of a solar eclipse; here theory concerning natural causes is substituted for the workings of the gods. Where I would differ from some accounts (and agree with others, such as Blumenberg's) is in regard to the view that science makes myth extinct. The two forms of thinking coexist, although mythos remains relatively unchanging in essence and science develops cumulatively. Indeed, this conclusion is itself a finding of the human sciences. Even if one wished to do away completely with myth—an arguable matter—the phenomenon appears as a constant in the human species. The task is to seek to understand it as fully as possible.

We must recognize that myth is a fundamental means by which humans try to understand, and survive in, their environment, both natural and social. Myths are narrative accounts, whether about the origins of the universe or the doings of heroes (these are the two major types of myth). The myth-making attribute, it should be noted, takes its rise from, and is unthinkable aside from, the symbol-making nature of humans stressed earlier.

In the view of Hans Blumenberg, myth persists because we are continuously faced with what he calls the "absolutism of reality." By this he means that "man came close to not having control of the conditions of his existence and, what is more important, believed that he simply lacked control of them."[66] Instead, control is in the hands of "superior powers." The root cause of this loss of control resides in the adoption of an upright posture, which forces a confrontation with new horizons that offer both novel possibilities and unsettling challenges. Instinct is replaced by symbolic adaptation to the environment. The result is anxiety, a state of feeling that persists in the human unconscious (which remains the same even as logos comes along to assist mythos in assuaging our fear).

On this account, myths are a means of reducing anxiety or at least transforming it into manageable fears. As such, myths may be said to be in the service of a rational desire for explanation. They explain beginnings and possible endings if they are origin myths (the biblical account of God's creation of the world in seven days is an example). The heroic quest myths are somewhat more complicated; they offer us a model of how to behave "manfully" (although they are often about women as well) in the face of nameless terrors, unpredictable events, and the capricious actions of the gods. In this sense, they do exactly what the modern bourgeois novel attempts: to instruct us how to live in unexpected circumstances by offering vicarious examples.

Myths have other functions as well. They justify power and custom. In the words of Malinowski, "Through the operation of what might be called the elementary law of sociology, myth possesses the normative power of fixing custom, of sanctioning modes of behaviour, of giving dignity and importance to an institution."[67] Their power lies in seeming to come from extrahuman sources and being beyond the reach of normal reason. They are once-and-for-all stories, not amenable to examination (except by later practitioners of the human sciences). To their believers they are not matters for debate, analysis, verification, or possible dismissal. They are stories, accounts that carry their own conviction within them.[68]

In the view of Blumenberg, myths are not changeless. They are subject to a selection process in which modifications take place. He defines myths as "stories that are distinguished by a high degree of constancy in their narrative core and by an equally pronounced capacity for marginal variation." In this they differ from dogmatic religious texts, which *claim* inerrancy. "So myths are not like 'holy texts,' which cannot be altered by one iota."[69]

If we accept this line of reasoning, myths are not intrinsically an antithesis of reason but an alternate, nonscientific form of reasoning. They, too, name the nameless and seek to subject phenomena to human understanding and control. They are not a remnant of the human past but a persisting part of the present. Nor do they represent simply a primitive or non-Western mode of thinking; they are, as Gananath Obeysekere argues, "equally prolific in European thought."[70]

Myths, wherever found, have a lasting power. Prometheus and Faust tell us about the human aspiration to knowledge in a manner that positive science cannot match. In explaining the apotheosis of Captain Cook, Obeysekere resorts to the Prospero myth, which takes on an inimitable explanatory power.

Myth is narrative, subject, like any other narrative, to the strictures of hermeneutic interpretation. In fact, the passage to modern narrative occurs by means of myth. For example, humans seem to have an obsession with attempts to cross the boundaries between life and death and visit the dead. An attempt to do so is frequently embodied in a journey to the underground. Orpheus, by the power of music, was able to redeem his wife from the other world and bring her back to the living. This myth is taken up anew by Vergil and then by Dante in their described trips to Hades. For Dante, *The Divine Comedy* is not only a renewed use of mythical material but a biographical account of the sins and aspirations to which mortals are prey. In the great medieval Italian poet, we see myth being transmuted into modern narrative. The link is immutable.

Literature and Narrative

The modern narrative, in its own way, is as much a substitute for myth as are the natural and human sciences. As Peter Brooks points out, in the last century and in Western culture "the making and the interpretation of narrative plots assumes a centrality and importance in literature, and in life, that they did not have earlier, no doubt because of a large movement of human societies out from under the mantle of sacred myth into the modern world where men and institutions are more and more defined by their shape in time."[71] Literature, trailing its caul from myth, can take many forms: poetry, drama, novels (and some would add history). All can have narrative features, but such features are especially strong in drama and novels (leaving aside history); and it is on novels that we shall concentrate.

What does the novel claim to do? Basically, to tell a story. It shares with all other narratives the need to have a beginning, a middle, and an end. As Tony Tanner puts it, "We can no longer ignore that truth succinctly expressed by George Eliot that a 'beginning' is an indispensable 'make-believe' (we might say fiction) — in science, in poetry, and indeed very notably in the novel."[72] In having a beginning, the novel is no different from myth, folktales, or even science, and this in spite of the timeless claims of myth and science.

The beginning can be vague — "Once upon a time" — or it can be precise: "It was the year 1789 when . . ." But a beginning any novel must have. This necessity forces a selection upon the writer; into the seamless cloth of time and event he or she must make a cut and, by this act, start the story.

The literary work must also have an ending, or what Frank Kermode calls in the title of his book "the sense of an ending." Such a structuring makes sense out of time, which is otherwise featureless. We are, of course, constructing a fiction. Such a construction, as Peter Brooks points out, carries with it "a gain in knowledge, a self-conscious creation of meaning." [73]

A beginning and an ending also mean a chosen perspective from which to view the whole novel. Only when readers have reached the end, which the author has already foreseen (or at least contrived), can we look back over the narrative and say, "Ah, so that's what happened." The end, itself an arbitrary cut in time, gives meaning to all that goes before it. And we can mentally project the story and ourselves into the future: "They lived happily ever after."

Much of modern literature has constituted an attempt to destabilize these fixed posts of narrative. By defying conventional space-time structure, sometimes trailing off incoherently or offering multiple endings, some novels call into question the verities of narrative. The novelists seek to do consciously what schizophrenic discourse, it is said, does unconsciously.[74] Even such works, however, start on a first page and end with a final page. In between is the story, even when it claims to be a nonstory.

What does the novel do besides offer a perspective on events and give a meaningful structure to time? Does it claim to offer truth about whatever it describes and whose story it tells? In the development of the modern narrative, history and the novel have engaged in a quaint minuet. History has claimed to tell truth from the time of Herodotus through that of the seventeenth-century European historians (with the Middle Ages mainly given to chronicle) to the scholars of the twentieth century. Its claim to truth telling is based on the weighing of evidence, the testing of testimony, and so forth.

This claim took on special form in a subgenre of history, the travel literature resulting from the New World discoveries. Why believe accounts about cannibals or mermaids? One answer was in terms of firsthand witnessing. Bernal Diaz in his *Conquest of New Spain* claimed to be producing a "historia verdadera" because he participated in the actions; he was there and saw events with his own eyes and gathered the reports of other witnesses. (In fact, there is much in Diaz that he could not have seen but only received from some who did and reported their experiences to him.) Whatever Diaz's veracity, readers of his account are immediately thrown into an interpretive experience, weighing his statements and those attributed to others, for we were not there and cannot repeat Diaz's experience (except in a rereading) the way we could Boyle's experiment.

The problem of credence haunted the new explorers. One, Jean de Léry, "asks how his French readers can be made to 'believe what can only be seen two thousand leagues from where they live: things never known (much less written about) by the Ancients; things so marvelous that experience itself can scarcely engrave them upon the understanding even of those who have in fact seen them?' " [75] Léry was also aware that preconceptions — what one expected to see — influenced what one saw.

Skepticism was heightened by accounts such as those of Mandeville's travels, which turned out to be false. Yet, in general, most travel accounts and histories carried with them a strong presumption of truth, based on either witnessing or a proper use of other's firsthand accounts. Naturally, the accounts needed to be read with an interpreting and testing eye — but the implication was that in the end there was truth.

The modern novel, in its early years, aspired to the same status. *Don Quixote* is the classic attempt to tackle the problem of appearance and reality. Anthony Cascardi reports that it was "Cervantes' project to legitimize fictions by bringing an end to the 'illicit' fancies of romance, and continuing through the techniques of verisimilar narration." [76] Much of the subsequent history of the modern novel details the same effort. Thus, realism in the eighteenth century was a constant effort to spread the cloak of verisimilitude around what was otherwise fiction. A declaration in the preface of Daniel Defoe's *Robinson Crusoe,* which was modeled on travel literature, reads: "The editor believes the thing to be a just history of fact; neither is there any appearance of fiction in it." Henry Fielding called his books on Joseph Andrews and Tom Jones "Histories" and kept assuring readers that he was presenting them with truth; in the preface to *Joseph Andrews,* Fielding informs readers that "everything is copied from the Book of Nature, and scarce a Character or Action produced which I have not taken from my own Observations and Experience." [77]

Indeed, much of the literary history of the seventeenth and eighteenth centuries in Europe revolved around an inconsistent effort to discriminate between history and literature and between fact and fiction.[78] (Ironically, in the postmodernist currents of the late twentieth century the boundaries are under attack by some historians and anthropologists, who proudly proclaim their work as mere literature, more fiction than fact.) In general, the novel today is left to stake out its own version of truth.

What is that version? The original meaning of the term *fiction* (from *fictio*) is "something made," "something fashioned." It does not mean "something false." If one wants to know about human relationships, about the formation of people's character, the novel may be a better way of arriving at understanding than, say, sociological surveys.[79] In dealing with subjectivity, subjective methods are called for.

Subjectivity does not necessarily mean the lack of any attempt at objectivity. We have already noted Fielding's invocation of "my own Observations"; Brooks, remarking on Balzac's "penetrating powers of observation, his capacity to delve beneath the surfaces of other people's lives, to discover the stuff of their hidden stories," then quotes him on how this power of observation "gave me the faculty to live with the life of the individual on whom it was exercised, allowing me to substitute myself for him." [80]

Observation is a fundamental part of any attempt at scientific knowing, and no less so in narrative. What is observed, however, is the shifting, tortuous reality of the human condition. In the process, observation can become empathy, and vice versa.

Human life is ambivalent, ambiguous, inconsistent, and more or less unfixable. A few quotations will show what is involved. Here is one: "I look upon good novels as a very valuable part of literature, conveying more exact and finely-distinguished knowledge of the human heart and mind than almost any other, with greater breadth and depth and fewer constraints. Had I not read Madame de la Fayette, the Abbé Prévost, and the man who wrote *Clarissa,* that extraordinary feat, I should be very much poorer than I am; and a moment's reflection would add many more." [81]

Another remark goes: " 'Middlemarch' has frequently been described as a web, a hall of mirrors, an echo chamber — all terms that seek to convey its density, its refractive and reflective qualities. No character in 'Middlemarch' can ever be seen alone; all are subject to a constant process of qualification and modification as they interact with and comment upon one another, or as we compare them, or as the narrator leads us toward one set of judgments, and then, with a peroration, or a scenic flick of the wrist, turns those judgments inside out and upside down." [82] Here, I would argue, we have life in its reality, though presented in a fictitious setting.

There are more ponderous ways of stating what the novel shows. Take the following: "The language of rules and models, which seems tolerable when applied to 'alien' practices, ceases to convince as soon as one considers the practical mastery of the symbolism of social interaction — tact, dexterity, or savoir-faire — presupposed by the most everyday games of sociability and accompanied by the application of a spontaneous semiology, i.e., a mass of precepts, formulae, and codified cues. This practical knowledge, based on the continuous decoding of the perceived — but not consciously noted — indices of the welcome given to actions already accomplished, continuously carries out the checks and corrections intended to ensure the adjustment of practices and expressions to the reactions and expectations of the other agents."[83] Is it not questionable whether this codification conveys more knowledge than the lived presentation of the "spontaneous semiology" in *Middlemarch?*

What the novelist is doing is interpreting her characters and the milieu in which they move and then presenting us with a textual representation that we, in turn, must interpret. Characters in literary texts are themselves engaged in interpretive actions. *King Lear,* to take a play for an example, is about a misinterpretation of feelings and motives. King Lear misreads his daughters, taking deception for truth — he is misled by words — and taking his daughter Cordelia's words amiss. The play is about his discovery — a self-discovery and at the same time a discovery about his three daughters — and his coming to know the truth. In the process, the words *love* and *loyalty* take on new meaning for the king — and for us, the readers, as well. We realize that only in his deranged state does Lear work his way through to the true meaning of the events that he set in train by his decision to divide his kingdom while still alive. In hindsight, we see that his true derangement was in that beginning, where he was beguiled by appearances of love and loyalty. All of this is a matter of interpretation, Lear's and ours. The end result is a profound knowledge of human realities, of the truth of human relationships.

In a way, this is scientific knowledge befitting its subject matter. It cannot be conveyed by a model, where the constant, subtle series of adjustments would escape capture; the metaphors of cybernetics and feedback are too gross. Knowledge given by the narrative is not codified and cannot be. The material is too varied and shifting, and only an Eliot-like or Shakespeare-like mirror can catch the shapes and colors. Yet, given the shared nature of human beings — their propensity to rivalries, jealousies, loves, self-deceptions, trusts, misapprehensions, and madnesses — we the audience are in touch with a form of universal knowledge. Particular cultural overlays may obscure the knowledge for some, but the potential is there for all who pierce the local presentation.

Literature takes up a task similar to myth's.[84] It seeks to put humans in relation to a social and natural reality mediated by symbolic means. It attempts to allay our anxieties about an unknown and threatening world, to provide examples of how to navigate successfully through such a world, and to offer ways of comprehending our place in the universe. It is the same task as that undertaken by the natural sciences but now carried out by literary means. There can be paraphrases of what literature does; there can be no substitute for its actual doing.

Criteria

Even this brief survey of myth and literature should show the limitations of the strict positivist approach as applied to the human sciences. In most cases, naive positivistic science is not possible and, in many cases, not even desirable in describing the realm of human interaction. We must seek truthful knowledge by other means.[85] But there is not one correct way of thinking in regard to natural phenomena and another in regard to human phenomena. We must search, instead, for the different ways an appropriate scientific method may be applied in regard to all phenomena. Indeed, earlier, I argued that this is the situation in the natural sciences, where each branch may require a different variant on essentially the same method.

That method, at a minimum, requires an effort to establish a public reality that is constructed with careful attention to observation, verification, open debate, constantly checked theory, careful inference, and so forth. To do this consistently and faithfully, there must be a scientific community, that is, a group of trained specialists who are willing and eager to abide by the requirements of general scientific method, suitably modified in regard to their particular disciplines.

There is no reason why a similar demand should not be placed on hermeneutical interpretations. With this demand accepted, we can agree that the interpretive problems of hermeneutics, though of a different magnitude from those encountered in the natural sciences, can be dealt with by appealing to scientific method.

The crucial problem for any interpretation is criteria: How do we differentiate a true from a false interpretation, a better from an inferior one? To insist that all interpretations are equally valid is to give up hope of knowledge and to fall into total, nihilistic relativism. Such a stance negates the idea that human symbolic activity has evolved to help adapt to the environment, both natural and social. Like much else in evolution, adaptation can go astray and defeat the very aim originally intended. The antlers of elks can become too large; human symbolic activity can get out of hand.[86]

Let us not accept such a despairing conclusion. Humans still seek to acquire both human sciences and natural sciences. In doing so, they must confront in

ordinary life, as well as in more formal texts, the question of criteria in interpretation.

Interpretation is not one single thing, in spite of Schleiermacher's attempt
to establish a general hermeneutics. Moreover, we must recognize that an uncertainty, an ambiguity, inheres in the notion of interpretation, for, alas, we
must interpret whether or not an interpreter explains correctly and conveys
the meaning appropriately.

In this light, let us consider, as an example, jurisprudence, theology (or religion), and, a domain already glanced at, literature. Although all are expressed
in forms of narrative, each appeals to a particular, rather than a general, hermeneutic practice. What links them may be the necessity of persuasion by
rhetorical, not scientific, means.[87]

As I remarked earlier, modern hermeneutics took its rise in part from
seventeenth-century juridical debates. In a legal trial, we are forced to choose
among competing interpretations. "X says . . ." and "Y says the opposite."
Whom do we believe? How do we weigh the testimony grain by grain, reach a
sum total, and come to a final judgment? The hermeneutics of jurisprudence is
unique in having a legal code that constrains interpretations. It approximates
scientific method in establishing criteria for our judgment.[88]

With this said, however, we are forced to recognize how much is still a
matter of unstructured interpretation. Witnesses are notoriously unreliable.
Though sworn to tell the truth, they may lie, they may forget, or they may tell
only part of the truth. A life may hang on the outcome. A jury must go through
a constant series of shifting adjustments throughout the trial, suspending judgment if possible until the final bit of evidence is in. And interpreted.

The trial, in the end, becomes an exercise in persuasion. In this one way, it is
no different from a discourse about a scientific experiment or theory. What is
different is the role of rhetoric in the evaluation of evidence. Lawyers appeal to
our emotions as much as to our minds. They tap into a different level of what
goes on in interpretation, seeking to convince us not by the evidence but by the
irrational components of our judgment, our prejudices. Stanley Rosen, though
speaking of philology, not jurisprudence, but with the same point in mind,
says, "The practical results are produced by the interpretation. If the measure
of validity in interpretation is power to persuade, then philological hermeneutics is a testimonial not to objective methods but to the will to power."[89]

The only perfect "trial" is the detective story, which ideally proceeds by objective methods of interpretation. As Peter Brooks says, the detective story is
a "nineteenth-century invention . . . which claims that all action is motivated,
causally enchained, and eventually comprehensible as such to the perceptive

observer." Sherlock Holmes solves the mystery in a manner that permits no further doubt. He has interrogated the informants, rightly interpreted the clues, read correctly the motives involved, and come to the one right, rational conclusion. Any further juridical procedure is unnecessary (though, as we know from real life, clever lawyers can press for a trial and achieve a false verdict). This happy conclusion, we must remind ourselves, is in a work of fiction. Nevertheless, as Brooks claims, "The clearest and purest example of the hermeneutic would no doubt be the detective story."[90]

What of the situation in theology or religion? Here we encounter the major source of both ancient and modern hermeneutics: the need to interpret the sacred documents. Unlike in jurisprudence, opposing witnesses are not generally allowed. There is no ongoing debate. We are presented with the texts as coming from a power greater than human beings. Yet the hermeneutic problem cannot be avoided.

To rehearse a familiar story, argument can and has taken place over the authenticity of texts. Where more than one version of St. Matthew exist, how do we decide which is the ur-text? If parts of the Bible speak "with the tongue of angels" how do we translate their song into language that we can understand? Indeed, how do we translate from Hebrew or Aramaic—and which came first?—into Greek, then into Latin, and then into the vernacular, and be sure that we are being faithful to the "original"? Even if we think we know what is being said, how do we correctly interpret it? Which Hermes, coming from which God or gods, do we trust to deliver the message accurately?

Theologians have split hairs, and their supporters have sometimes split skulls, arguing over such points. Satirists and skeptics, such as Pierre Bayle, have had great fun mocking the religious texts and pointing out inconsistencies.[91] Yet most of the philological and hermeneutic challenge to the texts has come from within the religious camp itself. The Higher Criticism of the nineteenth century arose, after all, in the German theological schools.

Most of the Higher Criticism remained in the domain of theology, where it started (in debt already to Luther and his fellow reformers). With the Young Hegelians, it moved out to the wider world. Hegel, a theological student before turning into a philosopher, attempted in his philosophy to save religious belief by means of Reason. His effort was carried on by David Strauss, Bruno Bauer (inadvertently), and Ludwig Feuerbach, all Young Hegelians and all students of theology.[92]

With Feuerbach, theology turned against itself. Employing reason rather than Reason (Hegel's synonym for an all-knowing God), Feuerbach plunged religion into the chilling streams of rational thought and declared, as I have noted, that it was not God who had created man in his image but man who had created God or the gods. Theology and religion were now transformed into anthropology.

To finish this brief treatment of the way theology, that is, the Higher Criticism, turned into social theory and even social science, we must turn to Karl Marx, who is, among other things, the great secular hermeneutist of the nineteenth century (followed by Freud in the twentieth, another admirer of Feuerbach).[93] Marx declared that Feuerbach's critique of religion was the basis for all other criticism, and he proceeded to extend it to politics and then economics. (Marx, incidentally, began his university studies in jurisprudence—our previous example of hermeneutics—before shifting to philosophy, under the influence of Hegel, and then to economics.)

Marx is the great interpreter, peering past the appearances of society, especially bourgeois society, and revealing the reality behind or beneath. His prose is filled with images of "tearing veils" and "exposing" the naked truth. Narrative, along with analysis, is his chosen form, and, like Dante entering the inferno, Marx descended into the nether regions of capitalism, taking us with him on his terrible journey.

What is worth noticing further is how Marx introduces the notion of representation, so central to present-day hermeneutics. What the bourgeois claim to be depictions of reality turn out only to be their representations of it (with the implication that these are false). Such representations, Marx argues, are merely "ideological." By "reading" capitalism correctly, penetrating its disguises, Marx claims to bring us in touch with reality. In his eyes, his presentation is no longer merely one interpretation among others—hermeneutics—but positive scientific knowledge.

Earlier, we spoke of perspective as central to the interpretive enterprise. The term *representation*, stemming from both Marxist and non-Marxist traditions, appears today to be the focus of discourse.[94] The argument seems to have shifted to the criteria for judging correct representation, rather than for assessing proper perspective. In whatever terms, the same problem persists, and we must continue to examine it.

The immediate task in most of hermeneutics is to establish correct texts (jurisprudence has a different challenge, although it seeks to establish a written record as well). We cannot interpret until we know what we are interpreting. In earlier times, all texts were transmitted orally. In this sense, a text was not stable, altering, even if minutely, in each telling. When oral texts came to be written down, it became a constantly reproducible product. Repeated interpretation of a single supposedly fixed text was possible.

A related problem involves forgeries, and I could set forth a long and sometimes fascinating history of the subject. Steering clear of the details, let me stress the obvious: interpretation of a text's meaning is affected by whether it is authentic. Even a forgery has a meaning, of course, which requires further interpretation — but the meaning is different from that which would obtain if the document were authentic.

Here we have obtained one criterion for a correct and truthful interpretation: Does the text originate as claimed? The text is a fact, and we must get our facts right. A Jewish conspiracy may exist, but it cannot be argued from the Protocols of Zion, which is a forgery manufactured by those with a case to make. On the other hand, the masses of authentic documents showing the existence of the Holocaust cannot be negated by those who wave them aside and declare the Holocaust to be a fiction.

The question of texts is far more complicated. We are dependent on what survives. The burning of the library at Alexandria, in 391 A.D., eliminated dozens of plays by Aeschylus and Euripides and probably other playwrights, as well as most of the poetry of Pindar and Sappho. What would be our view of Greek literature if they had survived? Among Greek epics, only the *Iliad* and the *Odyssey* remain. We know how overwhelmingly they have colored our perspective on Greek mentality and culture. If we had none of Homer's epics what would be our interpretation of Greek life?

Here is a more modern example. Until about 1932, many of Karl Marx's writings were unpublished and unknown. Any interpretation of his thinking was conditioned on what existed in print. Then came the *Economic and Philosophic Mss., The German Ideology,* and the *Grundrisse,* and suddenly we had a different, much more Hegelian Marx. If yet another manuscript were discovered, Marxist scholars would have to scurry to reinterpret their subject.[95] In short, we are dependent on the selection accidentally handed down to us.

Selection is a key to interpretation. It happens at many levels — not only accidental selection in terms of textual survival but also our own selection among the texts that do exist — do we ignore Marx's "On the Jewish Question"? — along with the emphases that we place on the texts we single out and then on the passages and phrases that we emphasize in our choice of what to interpret.

Then there is context. We must decide in what context to place the chosen text — what version of Homeric Greece, what spirit of the age that gave rise to Marx? Context, in turn, is established by judgments and interpretations concerning other works of the period, in short, by a constant "warping and woofing" of history's fabric. And there is our own context, which we bring to the interpretive encounter. The hermeneutic circle is not easily and securely drawn.

Should we give up in despair? Many a teacher of literature has been shaken by a student's asking, "Why isn't my interpretation right?" and pointing to a few bits and pieces of supporting evidence in the text. The teacher's task often is to further explicate the text, pointing to counterevidence and then offering a different interpretation. Yet, as any good teacher knows, there is no one right interpretation; and occasionally a bright student comes up with an insight not to be found in the existing expert accounts.

So, are some interpreters better, more expert, than others? If knowledge of reality is a matter of perspective, are some perspectives better, more true, than others? Modern thinkers have been preoccupied by these questions. Within psychoanalysis, for example, Freud wrestled with the question of whether the assent of an analysand to an interpretation proved its correctness. He concluded that a simple yes by the patient to the analyst's construction may be valueless "unless it is followed by indirect confirmations, unless the patient, immediately after his 'yes,' produces new memories which complete and extend the construction."[96] The criterion for correct interpretation, in this case, is its correspondence to further evidence in the form of additional symbolic actions. As for the overall Freudian model itself, this, as Peter Brooks suggests, "must essentially find its justification in the illumination it can bring to texts."[97]

The sociology of knowledge arose to deal with the same issues by identifying the positions—social, economic, political, philosophical—from which interpreters viewed the world. Taking its origin in Marx's notion of ideology, sociology of knowledge in the hands, for example, of Karl Mannheim sought to take its stand beyond ideology. By postulating a sort of unattached, free-floating intellectual, Mannheim hoped to find an impartial, semiobjective observer.[98]

Recently, sociology of knowledge has been overtaken by notions about the social construction of knowledge, including natural science, and by the ill-defined field of cultural studies. In these accounts, any claim to a general interpretive science—hermeneutics à la Schleiermacher—appears to have disappeared in an emphasis on local knowledges. Many Others crowd the epistemological stage, with all voices more or less equal.

Conclusions, Tentative and Otherwise

Is there any firm footing in the morass of human intention and behavior? Can we come to any conclusions? At this point in the discussion of hermeneutics, there appear to be only a few clumps of ground upon which we might stand securely. We can insist upon the distinction between public and private knowledge. An individual may claim to have a private experience—say, communication with a god—but it is in its very nature incommunicable. It can not be displayed before us, nor can we repeat that exact experience. Although such experiences are part of reality — they happened (or may have happened) — they cannot become part of public knowledge.

Some form of public verification seems an essential part of any claim to science, natural or human. That verification must be by means of a suitably adjusted scientific method agreed to by a scientific community. The best features of positivism must be preserved even in those sciences that are mainly hermeneutical in nature.

What of the hermeneutical method? We must constantly bear our subject matter in mind. It is composed of a species — the human — that is characterized by a superabundance of madness and irrationality; that is heterogeneous in motive and behavior; that occupies nongenetically assigned positions in society; and that must be understood as a symbol-making creature. The possibility of human sciences exists in the fact that the heterogeneity of such creatures is homogenized in the form of culture. Culture orders and disciplines the symbols that humans use and prevents their use from being completely idiosyncratic.

Yet perspectives do differ, and interpretations may vary widely, especially as societies develop and become more differentiated. Interpretations, in such settings, become less and less a matter of right and wrong; authoritarian traditions break down. To replace or stand beside these traditions, natural sciences have arisen with their attendant scientific claims. In the domain of humanity (still a part of nature), an additional claim arises: a demand for depth of explanation.

A major criterion, then, for hermeneutical knowledge is the accumulation of more and more interpretations. They pile up as we view the texts from more and more different contexts and perspectives, as befits our increasingly complex nature. The interpretations must still pass the test of an agreed-upon method, though many flout such an agreement, saying proudly that Geisteswissenschaften or cultural sciences are exempt from public requirements. But when methodical strictures are insisted upon and are observed, interpretation becomes a matter of depth. This is the form accumulation takes in many, probably most, parts of the human sciences.

Let us glance at one simple example drawn from history. How do we explain the origins of modern Western society? Did it take shape because of certain technological developments, such as the stirrup or the crossbow, giving a military edge? Because of population movements and plagues, making for a shortage of labor? Because of the discovery of the New World and its metallic riches? Because of the Christian religion and especially the rise of a Protestant Ethic? Because of conceptual changes, such as the coming of humanism and the scientific revolution? Surely it is clear that the answer is all of the above.[99] The factors cited, taken together, amount to a causal explanation—all those "becauses"—but hardly a unitary or simple one; and the problems of verification are horrendous. Nevertheless, in principle, such problems, however difficult, can be dealt with.

What is more difficult is the problem of weighing. How much weight should be assigned to each of these causes? How disentangle them one from another? The answer is that we cannot do so scientifically, anymore than we can with a single individual, who has many motives when entering into even the simplest of actions. Shall I get married? Charles Darwin asked himself—hardly a simple action—and then proceeded to list the pros and cons. His Q.E.D. was yes; but it is doubtful that even he knew what weight each pro and con had in the final, gut decision.[100] When we approach something as complicated as the origins of modernity, this issue becomes compounded manyfold.

No less is true in the hermeneutics of jurisprudence, theology, or literature. What goes into a decision? What motives were involved? What intentions did the various actors have? Whether concerning real people or fictitious characters in a novel, the interpretive task must be carried out with increasing depth of understanding until finally, we, in our turn, must make a decision, a "final" interpretation, one that is necessarily partial and unfinished. Such interpretations are part of our continuous creation as self-interpreting and Other-interpreting human beings.

Even if we, as human scientists, have firm grounds for believing that our interpretation is correct, what are we to do with it? Let us assume, for example, that much of American politics revolves around racial anxieties, and let us suppose that we can understand the workings of such anxieties. How can we bring such knowledge to bear on the actors involved? To tell a madman that he is mad does not make him sane.

Putting knowledge—hermeneutics—into action is a problem that I shall not grapple with here, although it is a most important one. It must follow, however, upon the need in the realm of the human sciences to establish the grounds of knowledge by means of interpretation. All I hope to have accomplished in this chapter is to have given the general outline of the hermeneutic method, the areas in which it is most applicable, and the insights that it offers to balance those offered by positivism. In the next chapter I shall outline the implications and consequences of a knowledge of hermeneutics for the uncertain, human sciences.

5 / Some Achievements to Date

*T*he future of the human sciences appears unpromising. On one side, we are faced with a positivism that functions in faint imitation of its relatively successful use in the natural sciences. On the other side, we are confronted by phenomena that call for a hermeneutic method. The status of that method, however, is problematic. In sum, the use of an effective scientific method along the lines of positivism is handicapped, and the employment of the interpretive method highly unreliable.

The human sciences are uncertain, and, at the same time, the natural sciences no longer assume the certainty of past times. The confident Laplacean belief of the eighteenth-nineteenth centuries in determinism, prediction, and certainty has been irremediably shaken by recent work in physics, biology, and the philosophy of science. Now the natural sciences are seen as only relatively certain (though vastly more so than their humanistic counterparts); or, to put it another way, they are also on the spectrum of uncertain sciences.

It is in this context that we must place our evaluation of the past developments and future possibilities of the human sciences. Our first assignment, the topic of this chapter, is to assess what these sciences have achieved to date. We shall do this in two ways: by a general estimate of the situation and by a look at existing disciplines — in particular, five representative social sciences: economics, sociology, anthropology, political science, and history. These I take to be representative of the social sciences, though others — for example, social psychology — could be considered as well. The purpose of the exercise is to evaluate their achievement.

The question immediately rises, What standard should they be measured by? One such standard is positivism — a standard that all the social sciences have at one time or another set for themselves. However, we have already noted its limitations. We need, therefore, a fallback, one that accommodates the hermeneutic needs of social science phenomena while retaining the virtues of positivist aspirations. In other words, we must fit our standards to the historical development of the human sciences. Whereas controlled experiments and predictions may be desirable, they are not the only way scientific knowledge can be acquired and tested. Varied phenomena require various approaches. The validity of these approaches as science must be established sui generis while remaining faithful to the injunctions of scientific method.

Accomplishments

The debate over how scientific the human sciences are may obscure the powerful and solid achievements that have been made in the areas they address. Of enormous importance are two major shifts in perspective, two revolutions in conception. The first involves time and the knowledge of the span of human existence, a topic already touched upon in a different context. In the Christian world (and potentially everywhere else), a revolution occurred in the nineteenth century; only then did the view that humankind has been around for more than the biblical six thousand years or so take hold.[1] By the 1850s, even before Darwin's evolutionary theory, geology had extended the chronology of the human species far back into the past. Now, it is certainly true that mythical accounts spoke of a long-ago, dim past of uncertain date. What was accomplished in the nineteenth century in the West was a specific dating, giving humankind an extended existence subject to the more or less precise measurements of modern science, that is, of geology and paleontology. I have already spoken of the discovery of Neanderthal man in 1856–1857. Precision has increased unceasingly since then, enormously forwarded by carbon dating and other new techniques.

There is no need to rehearse the details of that expansion of knowledge here. Instead, we must focus on the overall extraordinary achievement: the securing of true scientific knowledge of the chronological existence on earth of the human species, its full extent. (It should be added that knowledge of the spatial extent of the human habitat also became scientifically established, starting with the Age of Discovery expanding the contours of the known world and continuing up to the present day and our ability to stand outside our globe, in space, and to measure and perceive it whole.)

We can catch a glimpse of the enormous change involved if we consider that even advanced, far-sighted thinkers of the eighteenth century, such as Kant or

Condorcet, while aware of an extended past, knew it only in a shadowy sense. They had little knowledge of its actual dimensions or its particular denizens. They were, literally, living in a different universe from that of the late nineteenth century. Though they might speculate on a time on earth beyond the biblical limits, they had no or little solid data — no real scientific knowledge, based on continuously revised facts and theories that were cumulative in their weight.

We, today, possess a reasonably precise and certain knowledge, however debatable in the details, of the existence of humans in a geological-biological setting of 2–3 million years, and of the changes in the human condition occurring during that time to the present; and of the coming of civilization, and all that that entails in the way of increased scientific knowledge and consciousness.

One example will suffice. In the second century A.D., the Roman poet Lucian wrote,

> Why as to Nineveh, it is gone friend, long ago,
>   and has left no trace behind it,
> There is no telling whereabouts it may have been.

By the mid-nineteenth century, Austen Henry Layard and other European archaeologists were busily engaged in excavating Nineveh and other cities of the Assyrian empire and shipping the wonderful stone sculpture reliefs back to museums in London, Paris, and Berlin. The unknown had become known, with great though gaping certainty.[2]

Underlying and yet part of this momentous novel knowledge and consciousness is a changed perspective as regards our own nature, our second revolutionary (in its consequences) development. In a shift over the past four or five centuries it is now accepted, certainly among scientists, that mankind is itself an object of scientific inquiry. This has not always been the case; again, we must stress what an enormous accomplishment (though it may also carry additional, unpleasant consequences) is involved in this shift, underlying, as it must, any and all possibility of a human science(s). The critical step in the human sciences is thus, to repeat, the recognition that mankind is itself an object of scientific study.

We have already touched on this change. Now it must be highlighted. Perhaps it best begins with anatomy and the awareness that a human is, in fact, a body (as well as a mind), and can be examined scientifically. In earlier days, the practice of anatomy was considered monstrous. A. J. Ayer captures the matter brilliantly in the opening of his article on "Man as a Subject for Science." He begins by a quotation from the novelist Thomas Love Peacock as follows: " 'Oh, the monster', exclaimed the Reverend Doctor Folliot 'he has made a subject for science of the only friend he had in the world.'" As Ayer continues, "Those who catch the allusion in my title may remember that the science in question was anatomy. The devotees of science to whom Peacock was referring were body-snatchers who provided medical students with corpses for dissection: for in 1831, when *Crochet Castle* was published, it was still thought improper, mainly on religious grounds, to treat even lifeless human beings as subjects for experiment."[3]

Anatomy, of course, merely helped pave the way. Once humans were deemed a fit subject for scientific inquiry, not only the body but the mind of man became a proper subject for such examination. Travel narratives, with their attention to climate and/or custom as shaping human beings, catered greatly to this development. The way was open to the establishment of modern anthropology, a subject that we shall pursue further shortly. Though the "science" of anthropology at first stressed physical form—the result of work in comparative anatomy and in physiology—even from the beginning the cultural loomed large.[4]

Quetelet, the Belgian statistician and one of the founders of one branch of sociology, saw clearly the possible link of physical anatomy and cultural science. Defending the analysis of moral phenomena, which included such disturbing subjects as crimes, he asked, "But is the anatomy of man not a more painful science still?—that science which leads us to dip our hands into the blood of our fellow-beings, to pry with impassible curiosity into parts and organs which once palpitated with life? And yet who dreams at this day of raising his voice against the study? Who does not applaud, on the contrary, the numerous advantages which it has conferred on humanity?" Then he adds, "The time is come for studying the moral anatomy of man . . ."[5]

The new attitude marks an extraordinary jump in consciousness. Mankind has always anthropomorphized external nature. Humans have tended to see all experience in terms of their own desires and fantasies. They have projected their hopes and fears onto nature and nature's gods. A de-anthropomorphization, so to speak, was needed, which made not only nature but man a subject of "inhuman," that is, objectivized, study as a necessary first step (and then, paradoxically, turned that study into anthropology, which in its subsequent and more recent developments has become a re-anthropomorphization of man and his knowledges).

With man as an acceptable object of scientific study, the way was prepared for Darwin's treatment of man's "animal" nature, where he is an animal like others, produced by the forces of evolution. On this physical basis Darwin could then, as he did, speculate on what might be involved in cultural evolution. One must read *The Descent of Man* and *The Expression of the Emotions in Man and Animals* to see how Darwin pursued this effort.

By remarking on anatomy, travel narratives, and Darwin, for example, I have tried to hint at the data supporting the "naturalization" of man as a scientific subject for himself. As we know from what has been said about hermeneutics, the effort runs the perils of reductionism, false analogizing, and inappropriate application of scientific methods better suited to other fields. Such perils, however, should not obscure the overwhelming change in perspective that first envisions the human species as a legitimate object of scientific knowledge, whose methods of inquiry must then be suitably modified.

The disciplines that address such knowledges are themselves very recent. For example, biology is a term first used around 1801, marking a specific area of study. Archaeology appears in the mid-nineteenth century. Sociology emerges around the 1840s in name, and as a discipline around the 1880s. In fact, almost all of social science, in the form of academic departments, professional associations, journals, etc., in other words, their institutionalization, first manifests itself in the period 1870–1920.[6]

Stated in a few paragraphs, the gains in human science may seem small and simple. They are certainly subject to controversy and challenge, especially in their details. In some cases they are vaguer than we would like. If we step back, however, we can see that they represent a change in knowledge and perspective that is truly overwhelming. Indeed, we can hardly overstate the transformation that has occurred. It is breathtaking. Our situation is not that we don't "know," and know with sufficient scientific guarantees, the facts and theories stated above; *it is that we have not readily incorporated such knowledge into our behavior and beliefs.*

Of course, it must be added that such knowledge animates our societies in a subterranean manner. In the chapter on Positivism, we alluded to Bacon's failure to pursue his own admonition that "philosophers should diligently inquire into the powers and energy of custom, exercise, habit . . . for these are the things which reign in men's morals." Since his time, we have come far in these matters. Even unconsciously, knowledge of the sort outlined above is widespread in present culture.

Nevertheless, to repeat, it is little acted upon consciously, in acknowledged fashion, by the great masses of people. In only small part is this because the status of the knowledge described above is insufficiently "scientific." However shaky at the edges, the claims to scientific knowledge cited are backed by sufficient data, and arrived at by acceptable scientific means, to persuade any "reasonable" person.

How can we explain the alleged disparity between the accomplishments of the human sciences and their conscious, active acknowledgement? I shall be arguing—and this is an extremely important point—that what is lacking in the human sciences, above all, is an acceptable "scientific" account that can explain the anomaly just cited. In short, a major deficiency in the human sciences is its inability to explain why many, perhaps most, humans do not accept and act upon consciously the findings hitherto achieved, and how, if at all, one might go beyond this state of affairs.

This is not to say that major problems of scientific method in regard to the human phenomena do not remain; and may always remain, as will be discussed in the final parts of this book. It is simply to call attention to the disparity between the actual accomplishments of the human sciences in the last few hundred years, and the awareness and acknowledgment of their achievements. The fact is that especially in what I have been calling human science, whose object of study is the species as a whole, the advance has been spectacular.

The Social Sciences

I have made a distinction between a human science, devoted mainly to the species as a whole, and the more specialized disciplines, the particular human sciences, especially as they relate to the social sciences. It is now time to take a closer look at the human sciences, plural, for they are among the lenses by which we look at various aspects of human phenomena. To know what we know, we must also examine the ways we have tried to know. We shall be flying high above the actual terrain of the particular social sciences under consideration: economics, sociology, anthropology, political science, and history. I shall be sketching some of their prime contributions to our understanding and evaluating their claims to be scientific. Experts on each of the fields are requested to be patient; they will be more than familiar with the details. The aim, however, is to have all readers share a common understanding of some of the accomplishments of all the social sciences under review.

Economics

Economics, often accused of "physics envy," claims to be the hardest of the social sciences, the most scientific. It speaks of laws and uses mathematical models. It pretends to an accumulation and refinement of its theories denied to the other social sciences. Here, if anywhere, practitioners of a social science believe that what they do approximates the positivist ideal.

Economics as a science arose in the seventeenth and eighteenth centuries in the West. Everything, of course, can be found in Aristotle, and it should be no surprise that he, too, wrote on economics. But his economics referred to the household and its management (the word *economy* comes from the Greek *oikonomia*, meaning "household management"). At best, the term could be stretched to cover the provisioning of the polis, the Greek city-state. In other cultures and civilizations, economics could be found in the concern with management of provisions, trade goods, and government revenues.

What is new in the West is the attempt at a rational science of economics, modeled on the achievements of the scientific revolution of the seventeenth century.[7] That effort was not, however, a pure intellectual aspiration but, to a large extent, a result of the emergence of a commercial-industrial society, with novel features both practically and intellectually. Both the science and the phenomena that economics seeks to understand are emergent: the coming into being of free markets; the exchange of commodities produced for those markets; the use of money; the creation of "free" laborers working for wages; the division of labor, making for increased productivity; the turning of land into a commodity, used for extraction of rent and production for profit; the existence of capital rationally seeking profit; and the division of the product into wages, rent and profit—all these were required for the emergence of a science of economics.

The science sought to establish abstractions drawn from the complex economic transactions of peoples, that is, the facts on the ground, and to set up a model that simulated the process, producing the abstractions. The process itself was a result of complex historical developments, generally treated under the rubric "the rise of capitalism," and is a subject for economic history. The abstractions, theories, and models devised to deal with the results of this process are a subject for the history of economic thought. The science of economics is the product of these histories but claims to lead an autonomous existence.

Underlying economics is an extraordinary human invention: money. Money abstracts from goods and serves as a neutral medium of exchange. It allows for multiple transactions, with the original commodities completely lost sight of in the end. It permits accumulation that is, ideally, not subject to wear and tear and that can be invested to bring profit or interest. Its potential to bring good and evil to the human species is enormous.[8]

Not until modern times did money become the foundation of capitalism, an entire social system. Capitalism's first form was that of a commercial revolution, marking a qualitative increase in trade. Though referred to as a revolution, it was the result of a long, slow process of accumulating exchanges, which had been spurred by the Age of Discovery. By the seventeenth century, the commercial process posed problems of balance of trade, currency transactions, and banking that brought forth mercantilist tracts and writings on "political arithmetic" giving guidance in dealing with such practical issues.

As for general theory at the time, there was only a vague, diffused set of notions that has come to be called mercantilism. Such theory was oriented to the state and its needs, and not to a market economy. The new, emerging reality, however, demanded a new kind of understanding of the phenomena confronting the West. The need was for a rational system offering scientific rather than merely philosophical comprehension.

The breakthrough seems to have come with François Quesnay and the Physiocrats, or *Economistes*, as they were called.[9] Quesnay was an eighteenth-century French physician familiar with William Harvey's treatise on the circulation of the blood, which probably served as his economic model. In *Tableau économique* (1758), Quesnay postulated an annual circulation of wealth in a closed nation-state. In his scheme, wealth was produced solely by nature, from the land, and transformed in shape by manufacturers, whose products were then circulated by merchants. Quesnay's economy was basically an agricultural one, not a commercial one. The Physiocratic contribution to commercial economics was an advocacy of "Laissez-faire, laissez-passer," that is, let goods circulate free of such artificial restraints as tariffs and tolls, because the manufactured goods added nothing of value to what had been produced by nature. Indeed, the term *physiocracy* means "rule by nature" (*physis*, as in *physics*).

The real breakthrough to a science of economics was made by Adam Smith. Although he wished to dedicate his *Wealth of Nations* to Quesnay (who died before its appearance), Smith went well beyond his precursor. It was Smith who fully recognized the emergent commercial revolution (and groped toward the coming industrial one), with all its accoutrements of markets, labor, and capital.

What is more, Smith linked the notions of free markets, free labor, and so forth, with his work as a moral philosopher. The result is one of the most blinding revelations in the social sciences. As another scholar has put it, "The concept of a free market's capacity to harness private self-interest in the service of the public welfare and to reconcile the profit-maximizing behavior of individuals with the public good without any centralized plan or control is surely one of the most intellectually breathtaking and provocative ideas ever conceived."[10]

Smith's theory of how the pursuit of self-interest was led to serve the general good comes to us trailing religious clouds. In places he speaks of Providence rather than the "invisible hand" as the guiding force. In fact, he was recasting the religious debate over theodicy — how to justify evil in a universe ruled by a God who is all good — in secular terms. In the end, Providence-cum-invisible hand resolves into the laws of the market — a science of economics.

Smith was not alone in his efforts to establish economics; he had predecessors as well as contemporary coworkers. Nevertheless, it is in Smith's writings that the recognition of the emergent phenomena of the market and the concepts and theories with which to explain them takes most forceful shape. Since he established the categories by which we still operate, let us deal in a bit more detail with his concepts.

Most of human existence has been in terms of basic subsistence: humans engaged in hunting-gathering and then agriculture for more or less immediate use. Yet even from the beginning, simple exchange took place. Such exchanges are rooted in the nature of the species. As Karl Marx notes, "The fact that the need of the one individual can be satisfied by the product of the other and vice-versa, and that the one is able to produce the object for the other's need, and that each confronts the other as possessor of the object of the other's need, shows that as a *human being* each transcends his own particular needs, etc., that they are behaving toward each other as men, that their common species being is known by all. This is unique. Elephants do not produce for tigers, or animals for other animals."[11]

Money, a relatively recent invention, facilitates the exchange of objects. When money creates capital, abstraction upon abstraction takes place, with enormous explosive power. Coupled by Smith to an equally abstract notion of free and marketable labor, capital becomes capitalism. Marx is worth quoting on the importance in this case of making labor an abstract commodity: "It was an immense advance when Adam Smith discarded any definiteness of the wealth-producing activity — for him it was labour as such, neither manufacturing, nor mercantile, nor agricultural labour, but all types of labour."[12]

All parts of the economic system, and the theory, are commodified — land, labor wages, human beings (as Smith admitted, "Man becomes a commodity just like any other commodity") — and thus conceptually homogenized and translatable. What is more, all of these factors became quantifiable. They can be reduced to monetary terms and made calculable. As Edmund Burke lamented in his *Reflections on the Revolution in France* (1790), his was an age of "sophisters, economists, and calculators."

One more part of the system remains here to be identified: the notion of economic man — a rational, calculating, maximizing, completely informed, profit-seeking animal. This definition takes on prescriptive force. Only if men behave as indicated will the economic system function as a market one and will economic theory be possible. Theory then becomes prescription, which is essential for the theory to be true.

Capitalist societies operate so as to turn prescriptions into self-fulfilling prophecies, rewarding those who conform and punishing those who do not. There is, moreover, an impetus for the system to be extended to other societies and for their institutions to be remade in the capitalist image, so that a world market can exist. Thus, for example, in India, as Louis Dumont summarizes, "the British domination emancipated wealth in moveables and chattels by substituting for a political regime of the traditional type, a modern type of regime, one of whose fundamental tasks was to guarantee the security of property, a regime which, compared to the previous one, abdicated part of its power in favour of wealth. The transformation of land into a marketable commodity is only a part of this change." [13]

This quotation reminds us that a noneconomic feature of economics is its assumption, first, of a nation-state framework — wealth is measured in gross *national* product — and the necessity of certain minimum structures of law and order, guaranteeing property and especially capital. Within this framework, often taken for granted, the emerging features of capitalism could manifest themselves, to become the subject matter of classical economics.

A reading of *The Wealth of Nations* makes clear one other essential part of Smith's analysis of capitalist society. Much of the book is devoted to a historical treatment, not just an abstract one, of capitalism. We are given insight into how it and its constituent elements emerged over time. Many pages are devoted to the history of money.

Most important, Smith speaks of an unintended historical revolution in which the feudal order is replaced by a bourgeois society. The aristocrats gave up their power, based on the ability to raise armies on their land, for the baubles of luxury; and they did this by abandoning agricultural subsistence and producing for the market and thus for ready money. The "pedlars," as Smith calls them, who were selling the luxury items, only intended to make a profit. The unintended consequence, however, was not just economic; it was also the overthrow of the existing social and political order. In Smith's world, we are dealing not just with economics but with political economy.[14]

Most of Smith's fellow classical economists, such as James Mill, John Ramsay McCullough, and David Ricardo, tended to drop the historical element from their account of the capitalist system. Both Mill and Ricardo wrote about economics as a deductive science. It was left most prominently to Karl Marx in the mid-nineteenth century to reintroduce the historical note.

In the time between Smith and Marx an industrial revolution occurred. It is difficult to overemphasize how important a transformation this was, and has been, in setting the context for the existence of the human species. Many thoughtful scholars have dealt with this transformation in detail elsewhere. Here, we need only remind ourselves that an industrial form of capitalism succeeded to a commercial form.

Even in its industrial form, capitalism fell under the theoretical empire established by Smith. The great intellectual breakthrough, though requiring many modifications, had already been made. Smith's economics could contain the leading features of the industrial revolution and its subsets: an agricultural revolution, which involved in England, partly as a result of enclosures, the pushing and pulling of laborers off the land and into the cities, and the speeded-up conversion of farm production to market purposes; a demographic revolution, which drastically and rapidly increased both the labor force and the consuming public; a commercial and banking revolution, which furthered capital accumulation and investment; a transportation revolution, whose canals and turnpikes and, eventually, railroads, widened the market in an unprecedented fashion; and a technological revolution, unforeseen by Smith, involving the great mechanism of the factory and using innumerable inventions, like spinning and weaving devices, the steam engine, new iron-puddling processes, which yet fitted nicely into his analysis of the division of labor and the extraordinary increase in production that followed from the union of human and machine.

These developments in actual life and economic theory went hand in hand, ultimately leading in the industrializing countries to an increased standard of living for many. But the resultant unprecedented change in human relations (the agricultural revolution twelve thousand years previously, when humans shift from hunting and gathering, was much more gradual), also placed extraordinary stress on the social system by introducing such phenomena as busts and booms, labor strife, and urban distress of a new intensity. One consequence, following close on the heels of capitalism, was the arrival of socialism as both a body of thought and a movement. In reaction to the reliance on individualism and self-interest in capitalism, socialism put the accent on society and altruism.

In this changed and changing context Marx succeeded Smith, continuing the latter's historical approach. Where Smith's contemporaries generally saw the economic laws of capitalism as eternal and universal, Marx maintained their transitory and limited nature: they were the product of, and the explanation for, a specific historical period. In contrast, inspired by Hegel, Marx placed classical economics in a framework of change, emphasizing how its birth was recent and how its own mechanisms must produce its downfall *and* its successor, communism.

Marx's emphasis on capitalism as a system, historically created, did not find followers among the neoclassical economists who came to dominate the profession in the late nineteenth and early twentieth centuries. Marginal utility theorists, such as William Stanley Jevons, removed the political from political economy and referred to their field simply as economics.[15] The capitalist order was perceived as largely static, existing autonomously within itself, though marked by rapid internal, nonsystemic changes.

Most of economics has tried to free itself from the historical conditions that produced it and still surround it. Economists today, leaving behind Smith's concern with history, strive to perfect the mathematical models that are possible because of the monetarized market and the calculable relations that he pioneered in analyzing and which were produced by the commercial and industrial changes. As a consequence, within neoclassical economics, development and labor economics are treated as stepchildren. The situation is well described in the following quotation: "Mainstream economics persists with the Newtonian prejudice that a complex phenomenon is best understood by reducing it to individuated and isolated causal agents, and studying the interaction of these parts. The problem is that this approach cannot cope with systemic change: it cannot see the wood for the trees. Economists are thus reluctant to theorize about the dynamics of change in economic and social systems over long periods of time and through major structural changes. Modern mainstream economics has dodged the questions that Adam Smith, Marx, Herbert Spencer, Hayek and Joseph Schumpeter embraced as primary objects of study."[16]

We could go on—how, for example, omit mention of Keynes?—but our purpose is served with these hints about the background of the current field of economics. Now let us evaluate its claims to scientific knowledge. Our concern is with the general question of its status as a social science. What do we know in its field of study, and how scientific is that knowledge? In fact, we know very little that is even approximately certain about economies in the process of development or structural change. Rejecting history, neoclassical economics is nevertheless dependent on history to decide between its models, for it has no controlled experiments of its own. Even within an existing capitalist economy, neoclassical economics is not strong on predictions, one of the hallmarks of the physical sciences. We could continue to list the glaring deficiencies of the dominant school of contemporary economics. We could emphasize its inability to bring its models to bear on reality, and vice versa, its lack of experimental possibilities, its unadmitted ideological nature, smugly justifying the existing capitalist system, its jettisoning of moral concerns other than that found in the invisible hand, its blindness to the historical dimensions of its ordering theories. Instead, let us focus on the achievements of economics.

Economics embodies the awareness of the new world created by the commercial and industrial revolutions. It permits at least a partial understanding of a capitalist society because it has conceptualized the major categories — capital, free labor, the market — that can give us that understanding. In its technical, mathematical aspects — give the devil his due — it does offer partial and limited intellectual control, sporadically realized in practice, over such phenomena as the business cycle, inflation, and currency transactions.

Seen from this perspective, the theoretization encountered in economics is a scientific achievement, not found in previous times or places. True, the science is girded round with prescriptive necessities, plagued by enormous uncertainties, and dreadfully lacking in predictive abilities (especially in regard to historical change). In the long run, economics can be said to fail badly in its neglect of an evolutionary perspective and its disregard for the nature of emergent phenomena. Yet, in its hobbling way, it is cumulative and subject to reasoned argument, taking new evidence into account and giving it continued theoretical expression. In this sense, economics can be said to strive after and partly achieve a suitable social scientific method — no mean feat. In sum, economics, as the conceptualization of the "free" market (and perhaps also of state economies) marks a great advance in human understanding and achieves this understanding in an approximately scientific manner.

Sociology

Not surprisingly, the origins of sociology were multiple. It is useful to think of sociology as having had two accouchements. The first birth was in the first half of the nineteenth century, presided over by Auguste Comte, and the second, related birth was in the second half of the same century, at the hands of a number of classical sociologists. Of course, to complicate matters, the births were intertwined, like Siamese twins. We must be careful, in short, that our metaphors and analogies do not strangle us or the infant discipline in their toils. I have already had something to say about Comte, in the chapter on positivism. It was he who first gave currency to the term *sociology,* meaning the final body of scientific knowledge that would come into existence after astronomy, physics, chemistry, and biology and that would serve as the capstone of human efforts to understand and to control human life.

All of the previous sciences abide by the law of the three stages, first envisioned by Turgot: theological, metaphysical, and positive. Astronomy, the first science to be developed, had begun by explaining planetary phenomena in terms of gods, then in terms of nature, and finally in terms of pure relations — in this case, mathematical ones. To explain social phenomena, the new science of sociology would also have to make its way to a positive form. It would have to do so, however, by developing its own proper methods, for mathematics, as Comte pointed out, was not suited to its materials.

Much of Comte's inspiration came from the history of science, with its record of achievement in various fields. Another large part of Comte's inspiration came from the events of his time. It was the French Revolution that demonstrated conclusively that the social order was not ordained by God but, rather, was made by individuals. Here, in the events of 1789, the lesson finally sank in: society, or so many believed, was a construction, or contrivance, by humans that could be more or less rationally ordered. Unfortunately, as the revolution rolled on, turning into a Reign of Terror, it became clear that the faith in reason was hubristic and that the forces of unreason had gained the upper hand.

Comte saw his task as redeeming reason from this disturbing encounter with reality. The problem with the French Revolution was that its adherents lacked true knowledge — scientific understanding — of social matters. Positive sociology would remedy this situation. It would not only be intellectually correct but would, as a result, bring order out of social chaos.

This first birth of sociology gave rise to a future brood of positive sociologists. For them, as we have already seen, positive, certain knowledge was obtainable by sociology and could serve as a secure basis for social action.

Another major upheaval besides the events of 1789 stood at the first birth of sociology: the industrial revolution. It must now be spoken of in regard to Comte's new discipline, for, like the French Revolution, the industrial revolution underlined humanity's shaping force in bringing about and then displacing various forms of society. Tools, as well as ideas, constructed and deconstructed the different stages of social organization. Industrial capital as a force displaced God in the creation of social systems.

It is, in my view, primarily economics that gave rise to sociology. Or, rather, it was the gap left unfilled by economics into which sociology stepped. Eighteenth and early nineteenth-century economics, with its cyclopean fixation on the individual and its "invisible" concern with the social whole, provoked the rise of socialism both as a theory and as a social-political movement at the same time as it called forth the establishment of a new discipline in the social sciences: sociology. Although socialism and sociology often overlapped, the latter makes a separate and distinctive claim to scientific knowledge. It aspires to be a "new science," unknown to earlier times and societies.

Sociology arose, in large part, from what were perceived by many, whom I call lamenters, to be the corroding features of industrialization and the additional deficiencies of economic science.[17] Starting in the eighteenth century and strengthening in the early nineteenth, a growing chorus of voices deplored what they saw as the breakdown of all connections except one: the cash nexus, or the purely market and money relations of humans to one another. This humanist outcry at the lamentable effects of economics helped give rise to the establishment of the social science that came to be called sociology.

One effect of economics, for example, was to create class as a form of social stratification. As I have stressed, essential to an understanding of the human species is the nongenetic ascription of rank and position, a characteristic linked to the heterogeneity of human beings. This, in turn, links to the possibilities of the division of labor, so critical for Smith's conceptualization of economics. Although in the *Wealth of Nations* Smith speaks of "ranks and orders" of society, not "class," his economic analysis of the factors of production (land, labor, and capital) and their rewards (rent, wages, and profit) led inexorably to the new social stratification.

Marx and Engels saw this clearly. When land is used to produce capital rather than fighting men and exists in the service of capital rather than subsistence, the two social ranks—the aristocrats with their attendant subordinates, and the merchants and manufacturers—become collapsed into one. A single capitalist class is created that stands opposite a newly created working class, produced by the textile and other factories as ceaselessly as they produced fabrics and other mass-produced products.[18] Class, an economic-based classification, replaces rank, order, estate, and the myriad other status and legal classifications as the dominating principle of stratification in industrial society.

In more rigid moments (which were many, for revolutionary desires often triumphed over intellectual discriminations and hermeneutic insights), Marx and Engels claimed that their sociology was as scientific as Comte had hoped sociology could be (although Marx despised his French contemporary, even while borrowing from him). Though not using the term *sociology* for their results, they pioneered in the introduction of the concepts of class, ideology, sociology of knowledge, and many other elements of modern sociological analysis. Combining economics and sociology in a unique blend, with history as the basic ingredient, Marx and Engels presented a powerful analysis of capitalism that, it hardly needs saying, has had incalculable effects. In this aspect of their thinking, they, like all good positivists, believed that with science in hand one could act with certainty in human affairs.

The aspirations of Marx and Engels were Promethean. There was, and is, another version of social knowledge and its power to remold society. This version was both less positive and less sweeping than the Comtean and Marxist sociology in its expectations. I refer to social science, or social studies. This, too, forms a part of sociology and, in fact, may be seen as furnishing its bread and butter. This version of sociology arose primarily as a reform movement whose nonrevolutionary intent was to ameliorate the worst features of industrialized society. Abjuring grand theory, it sought, and seeks, by practical means to deal with contemporary social problems, whether they be poor laws, housing conditions, juvenile delinquency, or care of the elderly.

Such problems, especially welfare issues, sit at the heart of social science as practiced by Beatrice Webb and others like her at the beginning of the twentieth century. By amassing data through various techniques, with statistics being a prime instance, sociologists try to accumulate enough empirical findings to provide a solid, or more solid, base on which to recommend social solutions. Whatever theory is used tends to be low-level and as much political, or policy-oriented, as social. Whatever its deficiencies as science, however, such research claims to provide tangible guidance for socially worthwhile projects and, because numbers seem to provide hard data, inspires confident policy making. (For more on statistics, the preferred form of hard data, see the Appendix).[19]

So far, we have been discussing what I have called the first birth of sociology and its development. The second birth, more hermeneutical in its inspiration, occurred in the second half of the nineteenth century. Classical sociologists like Tönnies, Simmel, Durkheim, and Weber, along with Marx and Engels, provided the leading ideas, concepts, and theories by which today we try to understand the nature of modern society and its connections.

Like the positivist sociologists, as I am calling them, the more hermeneutical or historical sociologists are aware that society is a human construction, subject to continued change, and not a God-given, eternal entity. Their emphasis is on history and on the application of an interpretive method to its data. Their concept of society also embraces the specific recognition that capitalism is a form of social-economic organization in conflict with past forms (feudal, aristocratic) and future ones (socialistic, communistic). It is these particular circumstances — capitalism and its early and late vicissitudes — that surround the central problem: how to deal with the perennially conflicting claims of altruism and self-interest and in sociological rather than philosophical or literary terms.

Let me illustrate the focus on capitalism, historical change, and hermeneutics. Tönnies distinguishes between present capitalist society (*Gesellschaft*), afflicted by rootlessness and constant change, and precapitalist community (*Gemeinschaft*), solidly based on eternal values of land and blood. Simmel almost novelistically treats of capitalist society, in fairly approving terms, as characterized by money relations, changing fashions, fragmentation (and yet with certain constants of superordination and subordination), the role of "strangers," and so forth. And Durkheim writes favorably in his earlier work of the new form of connection — modern society based on the division of labor — while analyzing its potential for anomie and turns in his later work to a study of the elementary forms of the religious life, seeking to trace the initial forms of human connection and subsequent construction of society that serve as the basis for any knowledge, religious or scientific.

Last there is Max Weber, the polyhistor, who offers a monumental comparative, historical analysis of modern society and its antecedents. With Weber, the hermeneutic aspect of classical sociology presents itself dramatically, especially in his handling of the idea of Verstehen. The very title *The Protestant Ethic and the Spirit of Capitalism* places us in a hermeneutical world that occupies a different space from that, say, of Comte, with his emphasis on raw positivism, or Marx, with his similar emphasis coupled to materialism (though, admittedly, Marx also relies heavily on the interpretive method). In the end, Weber's ambition — and achievement — is to combine a maximum of positivism with hermeneutics.[20]

Out of the struggle for the soul of sociology has come a claim to a particular form of scientific knowledge, as well as the leading terms with which we today discourse on the nature of society, especially modern capitalist society. Gemeinschaft and Gesellschaft, alienation and anomie, class and status (Weber's expanded notion of stratification), communism and bureaucracy, ideology and charisma (the latter, again, Weber's contribution), superordination and subordination, exchange and division of labor, the organic and the mechanical, the sacred and the profane — these are the modern words and concepts by which we try to deal with the issues of individualism and community.

How valid is such knowledge as science? For reasons intrinsic to the phenomena being studied, the concepts and theories of the great classical sociologists can hardly be defended in terms of positivist aspiration for verification, repeatability, predictability, and so forth. If, however, by science we mean greater systematic understanding, based on empirical observations, of a phenomenon, then I would argue that the combined work of Engels and Marx, Tönnies, Simmel, Durkheim, Weber, and others like them has made a definite contribution to the science of sociology. Their work helps us to organize our understanding of society as a peculiarly human creation, providing a relatively new way of structuring such understanding. Further, their ideas allow us to bring intellectual classification to what otherwise seems a chaos of social experience and to frame that experience in ways that accord with our desire for the rational ordering of empirical data, that is, with our need for scientific, as opposed to religious or literary, understanding.[21]

How cumulative is such knowledge? The emergent nature of the phenomena under consideration imposes a severe burden on any simple effort at accumulation. Class, for example, is no longer what it was in Marx's time or through much of the twentieth century; its use as an analytic tool in regard to capitalist society is much dulled. Nevertheless, there is a great deal to build on in the work of the classical sociologists. We can retain belief in the scientific nature of sociology, with its aim of giving theoretical expression to new empirical data as they manifest themselves historically. Such an effort requires a keen awareness that sociology not only studies human meaning and morality but in the process necessarily helps to shape them. Edward Ross's injunction that the sociologist by his science controls society has a grain of truth, but it is a grain that must be milled through the fine sieve of hermeneutics.

Sociology, I would argue, viewed as I have suggested, has major achievements to its credit, achievements that can shelter in the penumbra of one version of science. But sociology needs to revive the grand theoretical aspirations of its early founders. It needs to take account of the historically changing phenomena of modern (and supposedly postmodern) society. This means that it must tackle once again the large-scale problem of profound and sweeping social transformation. (This still leaves room — for there is no single "right" way of doing sociology — for practical social work, which, however, has only limited defensible pretensions to theory.) As in its earlier incarnation, sociology must respond to the phenomena that form the substance of its companion, economics.

Herein lies a problem. Neoclassical economists, having abandoned the broad, synoptic vision of an Adam Smith, have little to say about the changing nature of capitalist society. Economics itself, then, must produce a revivified form of development economies (aided in the process by a developing sociology; images of feedback processes come to mind). With the new empirical data coming from developing economies and aided by whatever economic theories accompany them, modern sociology can once again aspire to the grand theoretical sweep, though in an updated form, of its classical founders.

This can happen only if the lens of evolution and emergent phenomena is brought sharply to bear on the problem of constructing a science of sociology. The other precondition is fecundation from a modified science of economics. As in the natural sciences, much of the development in any one field is predicated on advances in an adjacent field. Questions about the scientific validity of sociology and economics and their possible future states can be answered only by taking into account how intertwined the two efforts have been in the past and are now in the present. Economics and sociology are, in fact, one science that has been split in two, mirroring perhaps the double nature — social and unsocial, to use again Kant's formulation — of humankind.

Anthropology

If, as I have been arguing, a sustained discipline of sociology arose as a response to economics and its omissions, in its own beginnings sociology was a very diffused subject, with no clear lines separating it from the related discipline on its other side — anthropology. This was certainly true in the seventeenth and eighteenth centuries, before the disciplines as such were recognized. It remained true, institutionally, more or less until the middle of the nineteenth century and the establishment, for example, of the Société d'Anthropologie de Paris and the Anthropology Society of London. It continued for a while to be true intellectually as well, for reasons hinted at by J. W. Burrow when he remarks, "It is often difficult in the nineteenth century to distinguish between sociology and the philosophy of history and, if the central preoccupation of sociology is with discovering laws of social development, this is not surprising."[22] This effort at discovery motivated much of anthropology, too, with a figure such as Sir Henry Maine playing a major role in the development of both classical sociology and early social anthropology.

By the second half of the nineteenth century, however, the disciplinary line had been sharpened, and anthropology claimed its own terrain (as did sociology). While the two might be mixed, as in Durkheim's *Elementary Forms of Religious Life* (1912), anthropology eventually and carefully proclaimed its autonomy and boasted of a unique knowledge of humankind. In what follows, we shall try to understand the form that that unique knowledge took, and takes, for the form shifted over time, as might be expected. Here, however, I want to make one sweeping claim for anthropology. Because much of modern natural science, as we have noted, arose as an attempt to remove humans from their observations of nature, or what was outside the self, science aimed at objectivity, or, to put it another way, at the deanthropomorphizing of scientific method. The observer's desires and motives were not to be imputed to the subjects of observation. Anthropology, in its early phases, struggled to achieve this same viewpoint. Starting in the late nineteenth century, it became clear that this was not only an impossible aim but a mistaken one. Anthropomorphizing, in fact, was the very core of anthropology, for that is what its observed subjects did. To understand these subjects, then, the anthropologist had to inject himself or herself into these other humans, in order to see how they, in turn, injected themselves into nature and society. To achieve scientific understanding of the Other — one of the great achievements of anthropology — meant to embrace a special form of anthropomorphism. Paradoxically, the effort to distance oneself from the phenomena, to remove one's personal desires and motives, ended up by placing those desires and motives at the center of the stage.

Human interest in other humans is not new. In this sense, Herodotus's *History* can be classified as a work of anthropology. But the sustained effort to foster that interest in the form of modern anthropology arose to a large extent out of the travels and explorations of Westerners, starting in the fifteenth century.[23] We have already noted the importance of the Age of Discovery for the emergence of human science. By the Enlightenment, under the further inspiration of advances in natural science, the effort at differentiated human sciences surged forward. It was supported and given additional momentum by the European voyages to the South Seas, which now supplement those to the west.

Imperialism was undeniably the main motive force behind the voyages and exploration. But another force was present and is not to be underestimated: human curiosity. Often, this curiosity was placed in the service of philosophy and science and used for an inquiry into the human condition. Georg Forster, for example, writing in 1777 about Captain James Cook's voyage around the world, saw it as an effort to find out, through empirical research, the true foundations for human happiness and morality. Forster wished to surpass Descartes's doubting and his skepticism as to the possibility of a science of custom and morals by foreseeing a kind of protoscientific anthropology.[24]

Climate and custom were seen as the two shaping forces of culture. Comparative study was necessary to establish the varied effects and strength of the forces, and that required trips around the world. "Exhaustive and critical information and observations garnered from travel narratives supported the theoretical reform of knowledge and the indefinite and concrete documentation of a science of the species quite capable of refuting itself or being 'perfected.' "[25]

Much of early anthropology was concerned with bringing home truths (or fantasies) that would help "perfect" "civilized" human beings, often by holding aloft the virtues of a simple existence (as Diderot did in his "Supplement to Bougainville's Voyages"). Much of anthropology, however, was intended to help "perfect" the lesser "breeds," the "primitive" peoples encountered, by Christianizing and Europeanizing them. The travelers-cum-anthropologists were frequently missionaries.

There is, as Christopher Herbert explicates, a fundamental relation between "modern anthropological thought and evangelical theology."[26] What the Evangelicals saw in the Other was generally their own dark side, the devilish indulgence of naked passion, the secret lusts of Mr. Hyde. Speaking of the South Sea voyages, Bernard Smith adds that "the incursion of evangelical Christianity into the Pacific mission field . . . did much to present the unredeemed savage as an object of pity or dislike."[27] One motive, pity, tended to inspire the spiritual extermination of the savage through conversion; the other motive, dislike, often led to another sort of extermination.

In any event, travel accounts proliferated, whether by explorers like Cook, scientists like Alexander Humboldt and later Darwin, or the innumerable missionaries and traders. The problem of verification loomed large from the very beginning. Were the accounts fictions, fantasies, or sober documents of facts?

Jonathan Swift places the controversies squarely in front of us, albeit in satirical form. It is often forgotten that what is called *Gulliver's Travels* was originally entitled *Travels into Several Remote Nations of the World,* reputedly by Lemuel Gulliver. Swift is at great pains to establish the veracity of his account. At one point, he has his traveler, Gulliver, explain why he hesitated to put pen to paper. "I thought," the presumed author declares, "we were already overstocked with books of travels; that nothing could now pass which was not extraordinary; wherein I doubted some authors less consulted truth than their own vanity, or interest, or the diversion of ignorant readers." As Gulliver explains, "My story could contain little besides common events, without those ornamental descriptions of strange plants, trees, birds, and other animals, or of the barbarous customs and idolatry of savage people, with which most writers abound." [28]

It was the naturalists on board during Captain Cook's voyages of the 1770s who would be interested in the strange plants deplored by Swift. Even earlier, Linnaeus himself had made a voyage to Lapland to report on the plants and presumably barbarous customs of the native habitat; and his disciples had spread out over the known and unknown world to implement his biological and anthropological ambitions. Swift's satirization of such efforts to come, as well as of the ongoing inquiries of the Royal Society, shows the way anthropology and natural science were linked in the beginning.

Traders, missionaries, military men, explorers, naturalists, and even humanists were there at the creation of anthropology. As their varied and problematic accounts proliferated, the initial attempt to construct a science of anthropology came increasingly to be focused in one direction: toward seeing its subject, primitive humans, as an object of natural history, along with flora and fauna. Not surprisingly, Buffon entitled his book *The Natural History of Man.* An attempt was made to objectify the Other, which could entail placing "savages" on exhibition, alongside machines, and displaying their artifacts in museums, as if they were fossils.

This perspective prepared the way for the dominant attitude of nineteenth-century anthropology: the evolutionary. Even before Darwin, conjectural historians like Turgot, sociologists like Adam Ferguson, and philosophical anthropologists like Johann Gottfried Herder were postulating a developmental sequence in which humans moved from hunting through nomadism to agriculture and eventually to what we know as civilization. In the sequence, "primitive" humans were the lowest form of humanity, lingering on in faraway places, occasionally appearing in existing civilizations in the form of the Wild Man or, as vaguely intuited, for example, by Diderot, persisting beneath the veneer of civilized humans.

Evolutionary theory, therefore, seemed to confirm and give greater scientific order to this perspective. Darwin himself, starting out as a voyager and writing an account of his travels on the *Beagle,* displayed the typical attitude to the primitive in his horrified comments on the miserable Fuegians. *The Descent of Man* was really misnamed, for Darwinian evolution showed the ascent of man (as Jacob Bronowski's BBC series demonstrated) from a primitive state of culture.

In this context, anthropological theory became evolution applied to humans (in fact, this connection was intimated even earlier, in the eighteenth century, by Lord Monboddo). Great emphasis was placed on physical attributes.

Anatomy was central. We have already noted the signal service of anatomy in making humans a subject of scientific research. Now anatomy became the basis of much of anthropology. Prehistory, based on human fossils, came into its own. Skulls of living people were measured, colors discriminated, hair types identified, and eye shapes distinguished. Racism was rarely far removed from the early inquiries. Anthropology in this form was physical anthropology.

Yet the travel accounts had described primitive social customs and structures, not just physiques. Anthropology had to shape itself about these facts as well, and the early ethnologists cast them in the framework of social evolution. "Armchair" anthropology was a fitting counterpart to this theory-cum-ideology, whose practitioners took it "for granted that science could make systematic use of data collected by persons 'of good intelligence and fair accomplishment' whose life experience carried them to far corners of the world."[29] Given this assumption, the dominant methodological problem was verisimilitude: how to sift truthful textual accounts from false ones. And all this could be done at a distance—the second form of distancing oneself from the Other—by applying logic and reason while at home, without any need for the anthropologist to visit and live with the natives.

The great transformation of anthropology began to occur in the later decades of the nineteenth century. It was a hermeneutic turn, circling about the concept of culture. The title of E. B. Tylor's book *Primitive Culture* (1871) encapsulates the shift. Indeed, the notion of culture still stands at the heart of anthropology, setting it off from sociology, which tends to revolve around society (although the circles have continuously intersected).

The word *culture* is neither new nor restricted to anthropological usage. A moment's reflection on the "high culture" of literature puts us straight. As Raymond Williams reminds us, "The idea of culture rests on a metaphor: the tending of natural growth."[30] It is associated with the notion of cultivation, as in agriculture. Cultivation can be extended to humans, as in the German notion of Bildung; one becomes a cultivated person and only thus "human."

Culture is viewed as organic, hence basically unplannable. We can immediately see the presumed opposition to society which is tinged with the modern, commercial, and industrial (compare Gemeinschaft and Gesellschaft). In this view, society can be shaped and reshaped, perhaps even consciously, whereas culture is deep rooted and more or less unconscious, fundamental, and enduring. In eighteenth- and nineteenth-century Germany, *Kultur* was generally placed in opposition to *Zivilization.*

Echoes of these matters vibrate in modern anthropology. One of its founders, Franz Boas, appears to have been influenced in his conception of culture by the German historicist and Geisteswissenschaften tradition. He argued that a museum exhibit of, say, primitive musical instruments ought not to be organized as leading in an evolutionary line from the simple to the more complex; rather, all the instruments should be restored to their context—the whole life of the people who played and listened to them. That whole life was culture.

No single theorist can be given the credit for the idea of culture. It became the core of anthropology—its great conceptual contribution—as the result of many thinkers, a diffuse reorganization of sensibilities, and much thrashing about. One consequence is that different definitions abound and proliferate. The anthropologists A. L. Kroeber and Clyde Kluckhohn have indicated more than one hundred meanings assigned to the term in their field.[31]

Tylor defines culture as "that complex whole which includes knowledge, belief, art, morals, law, custom, and any other capabilities and habits acquired by man as a member of society."[32] Geertz, a more modern anthropologist, places the emphasis on culture as a symbol system. For him, "culture patterns—religious, philosophical, aesthetic, scientific, ideological—are 'programs'; they provide a template or blueprint for the organization of social and psychological processes."[33] In a later publication, Geertz declares that "culture is not a power, something to which social events, behaviors, institutions, or processes can be causally attributed; it is a context, something within which they can be intelligibly—that is, thickly—described."[34] In almost all definitions, the key elements are the complex whole, which can, however, be constructed as a hermeneutic circle, and the context, which, while shaping us, is made up of symbols that can be understood semiotically. Such postulates leave us with culture as not just the grand conceptual achievement of anthropology but also the source, methodologically, of its, and our, profound disquietudes and difficulties.

What are some of the difficulties occasioned by the concept of culture? A fundamental one is that it tends to obscure the heterogeneous factors in all human society. Although it offers the inestimable virtue of homogenizing the heterogeneity of life into a whole—and does so in practice by making members of a society conform to certain dominant values—homogenization is achieved at the price of setting up a mythical whole. If the notion of a whole is understood to be a heuristic device, its dangers are limited. When it is taken as a reality, dangers abound.

The subject of anthropology, at least as practiced in the past, has been primitive tribes. These may range, ideally, from groups of fifty to groups of a few thousand (and anthropology, true to its imperialist origins, has tended to spread "imperialistically" to much larger groupings). The whole society can be viewed, presumably by one person, and its values and behavior carefully observed and recorded. The temptation to see a fixed and unchanging whole is sirenic.

Seeing a fixed whole leads to a related danger: seeing them as without history, as more or less changeless. Historical anthropology has only recently come into its own.[35] Many earlier anthropologists, fiercely in love with "their" tribe, resisted passionately the realities of ever-moving history and tried desperately to protect their primitives from the forces of modern society.

The primitive, unchanging, and presumably harmonious whole was frequently used as a stick with which to beat modern society. Margaret Mead's idyllic and, in places, false picture of *Growing Up in Samoa* is a case in point. As Eric Wolf perceptively remarks, "Without imperialism there would be no anthropologists, but there would also be no Dené, Baluba, or Malay fishermen to be studied. The tacit anthropological supposition that people like these are people without history amounts to the erasure of 500 years of confrontation, killing, resurrection, and accommodation. If sociology operates with its mythology of Gemeinschaft and Gesellschaft, anthropology all too frequently operates with its mythology of the pristine primitive. Both perpetuate fictions that deny the facts of ongoing relationships and involvements."[36]

Today, this danger is fading as anthropologists are running out of tribes — or else seeking new "tribes," such as laboratory scientists, where the concept of the whole is more difficult to apply. Even so, the old attitudes linger on. The sense of alienation from contemporary modern society runs strong in anthropology. In its classical form, there is an irony or contradiction that permeates it and has often taken on disguised form in the present. Thus, to take up a point made earlier, the very science that is intended to give us greater self-and social knowledge, by making us comprehend the Other, is denied to the native, who ideally will remain undisturbed by the "white man" and his modernity. There is an asymmetry between the assumed holism of primitive culture, from which comparisons are drawn, and the fragmentation of modern society, with the result unfavorable to the latter.

Perhaps one example may illustrate the disparity between the primitive and the modern and the way knowledge may be time and culture specific. Absolutely central to the modern anthropological study of the tribe has been the concept of kinship. It seems to be the rare study that has not spoken of consanguinity and provided elaborate tables tracing the links that are so vital to the members of the tribe. This is entirely proper, for kinship is, as Victor Turner says, one of those "individual and group structures, carried in people's heads and nervous systems, [which] have a steering function in the endless succession of social events, imposing on them the degree of order they possess."[37] The problem is that kinship, though still existent in a much diminished form, is no longer central to modern society. The virtues of primitive societies, so bound up with kinship systems, are not particularly relevant to today's systems. What is worse from the point of view of the human sciences is that a static, nonhistorical version of anthropology, here, emphasizing kinship, has little to say about the emergent phenomena of modernity.

Yet anthropology does not have to be this way and, indeed, has been changing. A more dynamic definition of culture, allowing for a more sophisticated analysis of historical process, can help with the problem. Terence Turner's view of culture as "a flexible capacity for collective adaptation and self-creation" points us in the right direction.[38] It pushes anthropology, and us, into a truly contemporary and yet imaginative encounter with the Other.

That encounter should ideally be placed in a broader context: the search for human universals. In a sense, this inquiry was the original inspiration for anthropology. It was to be an inquiry, scientific as opposed to mythical, into humanity's origins and subsequent development. Deeply Eurocentric at first, badly flawed by simpleminded evolutionary convictions, it nevertheless was a worthy attempt to deal with the nature/nurture controversy by comparative means. What, if any, were the limits to human plasticity? How much of human development could be explained by climate and how much by custom? Before sociobiology and the debates over genetic determinism, anthropology wrestled with similar large-scale, philosophical questions.

There were dangers. Facile and sweeping generalizations came readily to hand. Parochial practices could be seen as fundamental attributes of humanity. Although the effort at human universals is necessary to keep anthropology itself from falling into a parochial corner of the human sciences, "an anthropology confined within the boundaries of a special branch of science has no future." Continuing, Axel Honneth and Hans Joas assert that "anthropology must not be understood as the theory of constants of human culture persisting through history, or of an inalienable substance of human nature, but rather *as an inquiry into the unchanging preconditions of human changeableness*" (my italics).[39]

Indeed, it is exactly those preconditions that allow us to enter empathically into an understanding of the Other. We can understand a Balinese cockfight, to take a classic Geertzian example, in all its strangeness and alterity because we have similar fighting games. Some combination of Honneth and Joas's unchanging preconditions and Turner's flexible capacity for collective adaptation, then, must stand at the center of the anthropological inquiry into the ways, synchronic and diachronic, of the human species.

Once culture, however defined and conceived, is placed at the core of anthropological inquiry, the hermeneutical increasingly makes its presence felt. As Evans-Pritchard noted, anthropology was about observed facts. But what if the facts were social relationships? These depend on the meanings brought to them by the participants, and such meanings cannot be observed in the usual positivist fashion. The observer must enter into the minds of the participants and generally this means into their linguistic productions. In turn, this means taking seriously the symbol-making nature of humans.

Remaining in the armchair obviously wouldn't do. Anthropologists had to enter the field. They themselves had to become travelers in order to enter directly into the cultural world of their subjects of study. Yet while subjecting themselves to the Other, they had still to maintain their stance as scientific investigators.

The strain is tremendous. We see and feel it keenly in Bronislaw Malinowski, the great exponent of anthropological fieldwork. The positive and the hermeneutical contest mightily for dominion over his mind. Himself an alien in British society, Malinowski was prepared to wrestle with the problem in a more limited, "primitive" setting. Although he had no doubt that, done correctly, anthropological investigation was scientific, he was aware of how tortured its methodology might be. Its "sources are no doubt easily accessible," but they are "also supremely elusive and complex; they are not embodied in fixed, material documents, but in the behaviour and in the memory of living men." [40]

The methodological problems are manifold. Anthropologists must observe external facts carefully. They must interview their subjects as to their meaning. Can the natives be trusted? Can differing interpretations, if they arise, be integrated successfully? What of an anthropologist's own self? How neutral is the observer, how able to recognize personal projections and prejudices? Then, at the end of surmounting all these difficulties, anthropologists must write up their materials. How convey the months of fieldwork accurately in the pages of a monograph? How overcome the pitfalls of literary expression?

Malinowski was very aware of these problems. At one place he writes, "I consider that only such ethnographic sources are of unquestionable scientific value, in which we can clearly draw the line between, on the one hand, the results of direct observation and of native statements and interpretations, and on the other, the inferences of the author, based on his common sense and psychological insight." [41] Alas, what Malinowski considered unquestionable has been questioned, and his whole approach, typical of anthropology, made problematic. A critical hermeneutics has been applied to his hermeneutics of the primitive.

He himself supplied the starting point. As he announced, "an adequate ethnographic description must reproduce in miniature the gradual, lengthy, and painful process of field-work." [42] The natural way to do this was by writing a narrative account. It would, as much as possible, enable the reader to relive the experiences of the ethnographer in the field and thus judge the evidential base from which the ethnographer made theoretical excursions. In his book about the Argonauts of the western Pacific, Malinowski invites each reader to become an argonaut of a sort, venturing into this new physical and cultural world. "Imagine yourself suddenly set down, surrounded by all your gear, alone on a tropical beach close to a native village." [43]

Such is Malinowski's, and anthropology's, version of Freud's "Just So" stories. What means of verification do we have? Can we recheck the interviews? Can we go back and observe the same artifacts and, even more important, reported social relationships? As Geertz points out, commenting on both Malinowski and Evans-Pritchard, "Even if, as is now increasingly the case, others are working in the same area or on the same group, so that at least some general checking is possible, it is very difficult to disprove what someone not transparently uninformed has said. One can go look at Azande again, but if the complex theory of passion, knowledge, and causation that Evans-Pritchard said he discovered there isn't found, we are more likely to doubt our own powers than we are to doubt his — or perhaps simply to conclude that the Zande are no longer themselves." [44]

We are, in short, very much dependent on the original ethnographers and their write-up of their experiences. This dependence has preoccupied many contemporary anthropologists and caused many critics to almost erase the boundaries between anthropology and literature and thus deflate the pretension of the former to be a social science. [45] Suffice it to say, for our purposes, that hermeneutics appears to have conquered present-day anthropology as effectively as evolutionism colonized nineteenth-century anthropology.

Even this brief discussion of the births of anthropology allows us to draw a few conclusions about the accomplishments of this discipline, however controversial they may be. The original positivist and evolutionary ambitions, though discredited and superseded, serve nevertheless as a reminder that anthropology can and must pay heed to the historical and emergent nature of the human phenomena it studies. Anthropology *does* contribute to our attempt to comprehend human universals. And it does so in a disciplined fashion that raises the inquiry from the level of the earlier, often mythologized travel accounts to social science.

We are now possessed of an incredibly wider and more securely based knowledge than before of the manifold forms that human society and culture have taken in the past and take in the present. Modern anthropology has gone beyond the Eurocentric evolutionary perspective of the late nineteenth century and can now entertain a revised evolutionary tale in which humans figure as symbolic animals of a special sort.[46] This awareness has become part of our historical consciousness. Although the expanded knowledge of what it is to be human has been based mainly on the study of primitive societies, its implications for any society are evident. Our other selves are all around us, still living and with us, so to speak, in our now heightened awareness.

We are aware of a cultural web, in which all peoples must live. The perception is true to our nature as a symbol-making animal; if nothing else, anthropology has sought to explore more scientifically the manifestations of the human symbol-making activity in "other places," both for their intrinsic interest and for the sake of increasing awareness of our own semiotic web-making.

Another of the great achievements of anthropology concerns the Other, a subject that is a bit like a Möbius strip, curling in upon itself. Anthropology especially, though along with other sciences, makes clear to us how essential it is for human groups to define themselves by setting up boundaries, by establishing the limits of acceptable thinking and behavior, and by classifying group, self, and Other. Even within the group, classification and differentiation exist in the form of stratification.

Yet anthropology also takes upon itself the task of making that Other a part of ourselves, of bringing it within reach of our comprehension of what it is to be human. Anthropology shares this task with literature, and literary critics like Mikhail Bakhtin voice the same insight, summarized thus: "As I learn to see myself 'through the eyes of the other,' as well as through my own, I am granted a more complete perception of myself than I could gain alone."[47] Such overlaps cause some to equate literature and anthropology.

One difference should not be overlooked. In dealing with the Other, anthropology strives for a distancing, which, although it also involves a rapprochement with the self, is more overtly scientific than the mere literary gaze.[48] In this it is closer to psychoanalysis than to much of literature. The distance sought is an empathic one that combines detachment and an integration with the Other.

Anthropology, if it is anything, is the attempt to realize in regard to culture a scientific concern with perspective. Where the physical sciences emerged in large part from an awareness that perspective dictated our observation and conceptualization of phenomena, as in Copernican astronomy, anthropology has had to wrestle with the awareness that perspective, in the shape of the Other, is itself the phenomenon to be studied. Anthropology allows us to adopt the perspective of the Other and thus to change our own perspective on what is possible and proper for a human being to be and to achieve.

Whatever the methodological problems, and they are many, anthropology seeks to take the scientific perspective on humanity's evolutionary development, emphasizing the cultural more than the physical human being. While attention to the evolutionary side of that development has recently been overlooked or dismissed, along with concern for human universals, the necessary correctives — involving attention to culture and the Other — to the earlier one-sidedness amount to major achievements. The balance between positive aspirations and hermeneutic intentions is precarious and necessarily shifts over time. Such is the nature of accumulated knowledge in the social science that is called anthropology.

Political Science

Political science is, by name, the most ambitious of the social sciences and, in fact, the least scientific of them, for it relates more to values and the actions to which they should lead than it does to empirical data and the theories to which they might give rise. Political science is the Western "science" that differs least as a form of knowledge from what is found in other parts of the world. Robert Wokler comments that political science "may in fact be the oldest of all the human sciences," and he wonders "whether its apparent lack of progress since antiquity just indicates that politics is unamenable to scientific study." [49]

There are major Chinese, Indian, and other philosophers who early on offer valuable reflections on politics, but the roots of political science as a science lie in Greek thought. Plato, it is true, offers a philosophy of politics, one that is more idealistic than realistic. Aristotle, however, conceptualized politics scientifically, although he thought of it as the continuation of ethics. For both thinkers, politics referred to the polis, the city-state. The citizens — meaning free men, for women and slaves did not count — were to act politically to nurture their character and to give glory to their city. Politics meant pedagogy. But as Habermas sums it up, even "Aristotle emphasizes that politics, and practical philosophy in general, cannot be compared in its claim to knowledge with rigorous science or with apodictic *episteme*." [50]

Most political systems in the fifth century were not city-states. They were kingdoms or empires or tribal entities or what have you. Whereas Plato and Aristotle had a theory of cyclical forms — despotism to monarchy to aristocracy to democracy to despotism — it applied only to city-states. For other political systems, we would do as well or better to turn to non-Greek philosophers, like Confucius and his Chinese disciples, or to Indian mythology and philosophy.

In the West, feudalism brought political forms unknown to the Greek and Roman sages. Christianity became the necessary frame for any speculation about the nature of politics. There was little that could be called political science. Morality was at the core of political thought, as evinced, in the late Middle Ages in the "mirror of princes" literature. Here, the prince, unanalyzed, was exhorted to rule wisely, guided by his moral image in the "mirror," that is, by the book he was reading—which was written by the philosopher or theologian making the exhortation.

The same Renaissance that produced Copernicus also gave rise to Machiavelli. It is he who instigated a revolution in political thought. He consciously sought to open a "new route" by which, putting aside morality, he wished to look at what men do, not at what it is said they should do.[51] In this perspectival shift, the Florentine was claiming to turn the consideration of politics into an empirical science and away from moral philosophy. As with the Greeks, he has city-states most in mind, but now with their new, self-made princes.

Machiavelli insisted, as so many others did, that human nature was everywhere and at all times the same. Because one could assume constant motives and behavior on the part of ruled and ruler alike, material from ancient history served as well as present data. Machiavelli saw no reason for differing comparisons such as would be made by later anthropologists.

In the end, although Machiavelli claimed that his political science was founded on empiricism, his findings were rather airy generalizations. This rationalism on the part of Machiavelli, not his emphasis in principle on the empirical, is what prevailed among his followers and served as the basis for a presumedly scientific politics.

Hobbes and Locke are the great thinkers in the next part of this historical reconstruction. With the English Civil War much in his mind, Hobbes sought to achieve stability through an apodictic political science. He viewed the state as Leviathan, an artificially constructed entity, whose principles could be discovered rationally and demonstrated conclusively, like the theorems of geometry.

The organizing concept was the social contract. Men, as rational beings, entered into an agreement to establish a state, whose nature Hobbes rationally demonstrated. That state was authoritarian. Locke, who agreed with Hobbes in discarding the theological and traditional underpinnings of the state, opted for a more democratic outcome. His social contract was an equally rational decision, but it provided safeguards against despotic rule. Whatever their differences, Hobbes and Locke together laid down the basic ideas of the modern capitalist political ideology.[52]

Ideology, however, is not science. Nor is philosophy science. Machiavelli's call for attention to "what is" turned out not to be a call to the empirical in a meaningful way. The ideological triumphed in the guise of rationalism. Political science remained unscientific in any real sense.

In the eighteenth century political science divided mainly into a study oriented to classifying the forms of government, along with any constitutions they might have, and a philosophy, which still emphasized the political values that should animate the forms of government. The divided political science continued to offer much wisdom but little that could be called scientific. Its claim, of course, was otherwise.

The state of affairs is illuminated by one of the wisest writers on the subject of government, James Madison. Addressing, in *The Federalist Papers*, the difficulties of distinguishing among the faculties of the human mind and discriminating between vegetative and animal nature (quite a starting point), Madison proceeds to the divisions and balances among governments. He writes,

> When we pass from the works of nature . . . to the institutions of man, in which the obscurity arises as well from the object itself as from the organ by which it is contemplated, we must perceive the necessity of moderating still further our expectations and hopes from the efforts of human sagacity. Experience has instructed us that no skill in the science of government has yet been able to discriminate and define, with sufficient certainty, its three great provinces — the legislative, executive, and judiciary; or even the privileges and powers of the different legislative branches. . . . Besides the obscurity arising from the complexity of objects and the imperfection of the human faculties, the medium through which the conceptions of men are conveyed to each other adds a fresh embarrassment.[53]

Hermeneutics was far from Madison's mind, but he intuited its power in political life. Of course, he had contemporaries who believed otherwise: that a positive science of politics was possible. Condorcet thought that he could construct a "social art" that would give a significant amount of certainty to the political state. Jeremy Bentham believed that a Utilitarian felicific calculus could be worked out and the ambiguity of words removed, leaving scientific principles by which a legislator could rule. A close look at their work, however, shows how much more philosophical they were than scientific.

Montesquieu was perhaps "the first and most important author" to seek a new political understanding in the context of the unprecedented change from a feudal to a modern society occurring in eighteenth-century Europe. His work has been well described as exploiting "new opportunities for the comparison of law and government across space as well as time" and was "broadly cultural and comparative, a veritable museum of *moeurs et manières*, encompassing every clime and polity."[54] His *Spirit of the Laws* was an attempt to introduce a more scientific attitude by classifying governments into three kinds and seeking the general spirit that animated each of them. The doctrine of the separation and division of powers that Montesquieu elaborated is an enduring contribution to political philosophy, bordering significantly on an emergent scientific conception of politics (doing for it what Smith's division of labor was to do for economics, and in spite of Madison's doubts). But Montesquieu's scientific aspirations were followed up especially in sociology, where he was acknowledged by both Comte and Durkheim; in political science Montesquieu's contribution, like his successors', was mainly philosophical.

So, too, was the contribution of Jean-Jacques Rousseau. His corpus goes well beyond the political. The totality, however, informs the political and, together with the specific political writings, places him at the center of revolutionary and postrevolutionary political philosophy. It is Rousseau who takes the social contract theory on its most democratic tangent but also implies its possible despotic outcome. More generally, his observations on political man still shape much present-day thinking. Yet, in terms of political science as we are trying to define it, Rousseau offers only a fitful inspiration at best.

One important thinker has been left out in this far too brief account of the history of political science: Karl Marx. He pioneered in the effort to combine theory and praxis. His thinking has had a profound impact on the actual political world. He certainly claimed, in most of his moods, that he was offering a science. But few today would subscribe to that claim. Marx was a great political philosopher, a subtle hermeneutist, a historian of capitalism, a classical economist, a pioneer in sociological theory, and something of an actual revolutionary. It is difficult to think of him as a political scientist.[55]

In fact, Marx can be said to have abolished politics, and thus political science, in his work. The state, as Engels said, would "wither away." Only a form of administration would remain to keep things running. After all, in their view, the political was an expression of class rule and would be totally unnecessary in a communist society. Having grasped that the emergence of capitalism had transformed political science, Marx drew the further, unwarranted conclusion that economics would do away with politics. Political science, for him, became a non sequitur.

Even the quick runthrough in which we have been engaged should allow us to make a number of significant points. Until the mid-twentieth century, political science as such consisted mainly of political philosophy, on one side, and constitutional classification and analysis, on the other. Although there were frequent cross-references, there was almost no true cross-fertilization. In the twentieth century, an attempt to model political science on the order of the behavioral sciences flourished for a few decades, but with little in the way of solid accomplishment. This initiative has now trailed off into the creation of rational choice models (whose inspiration comes from economics), which, promising much, have delivered little, at least so far. On another side, political science has mainly thrown its lot in with history, with an emphasis on comparative studies.

In the end, then, it is hard to discern in a history of the discipline even the outlines of a science. The data shift so erratically that verification of findings in the field is almost impossible. Experimental proof is more or less nonexistent. Many political scientists think of history as their experimental laboratory, but this is mainly a figure of speech. For one, history is not a controlled experiment. As Malthus reminds us, "An experiment with the human race is not like an experiment upon inanimate objects. The bursting of a flower may be a trifle. Another will soon succeed it. But the bursting of the bonds of society is such a separation of parts as cannot take place without giving the most acute pain to thousands." [56] For another, ephemeral emergence plays such a powerful role in political life as to more or less preclude meaningful generalizations.

What political scientists do have is comparison, which may be synchronic, comparing coexisting political systems, or diachronic, comparing different systems over time. Such comparisons can be enormously interesting and even revealing, but they are not particularly so in the manner of what I have been calling science. The masses, for example, have emerged in modern times as an active, apparently continuing political force. Comparing their manifestations in Roman times and in the present, or in the United States and in the Soviet Union, may offer an important form of comprehension. But such comprehension does not lend itself easily to the more precise terms of scientific method.

Similarly, past and present political science has had much of value to say about the great man in history; current political psychology makes the discussions more up-to-date and informed. Discussions of democracy have reached new heights as the concept is brought into contact with such notions as civil society, which can be thought of as an emergent development. (In this regard, economics, with its creation of a private market sector of society, can enter into collaboration with political science.) So, too, a heightened awareness of human heterogeneity suggests the argument that of all kinds of government, a liberal, democratic government allows the most humans the greatest opportunity to live their own way of life.

None of this, while often of great value, has much to do with politics as a science. On the great issues of political life—war, leadership, security—we have no real predictive powers, nor means of experimenting with and verifying analyses, nor profound new insights going beyond those found in ancient times or non-Western societies.

To take one example, war: Although the details may be dated, we have little, if anything, today that can surpass Jonathan Swift's analysis in *Gulliver's Travels*. When asked by the Houyhnhnms "what were the usual causes or motives that made one country go to war with another," Gulliver answered that "they were innumerable, but I should only mention a few of the chief. Sometimes the ambition of princes, who never think they have land or people enough to govern; sometimes the corruptions of ministers, who engage their master in a war in order to stifle or divert the clamour of the subjects against their evil administration. Difference in opinions hath cost many millions of lives: for instance, whether flesh be bread, or bread be flesh." [57] Given recent events in Bosnia, Somalia, Haiti and the continuing pertinence of nationalism, racism, and ethnicity, among other political concepts, can we do better today?

I am forced to conclude that the main modern accomplishment of political science has been to further the secularization found in the great Greek, Chinese, and other early thinkers. Secularization has been a prerequisite for developments in the social sciences. Unfortunately, in political science, it has not led very far.

A remark by Charles Lindblom and David Cohen hints why this is so. "Practitioners of PSI [professional social inquiry] know that some kinds of issues have to be settled by the interactions called 'politics' rather than by analysis." [58] Politics, by nature, is characterized by rapid and unpredictable change, the intermingling of interests and ideologies, aspirations to both equality and tyranny, loves and hates, values and power drives.[59] We must try to make sense out of the bewildering kaleidoscopic picture. Political philosophy (sometimes called political theory) is essential as an irreplaceable guide to values. Detailed analyses of such transient local phenomena as legislatures, laws, and leaders enter into the very politics they seek to study. But we must not mistake such efforts for lasting or cumulative science.

Still, we do have to recognize the aspiration to science. Desperately we seek understanding and control in this vital area of our lives. In the past, how did the Politburo behave? Today, how will matters in post–Soviet Russia unfold? Can the U.S. Congress cope with budget deficits and the problems of the welfare state? And on and on. We read the tea leaves of scholarship, the pious pronouncements of experts and policy wonks, but — are we reading anything that smacks of truly rewarding scientific method?

Little among the emergent phenomena — perhaps civil society is the exception — is politically new in a fundamental theoretical sense. Democracy, for example, has conceptually remained the same, although it has taken on dimensions and exhibited relations (with, say, the media) unknown in previous times. So we have a more complicated understanding of the manifestations of democracy, but we do not have a more insightful theoretical knowledge about democracy. What has remained the same is the fact that humans must and do act politically, but in so unpredictable and random a way—who would have predicted a revolution in 1789 or the revolutions in eastern Europe in 1989?— as to preclude positivist scientific knowledge. All we do have, although it is of major importance, is the sort of science embodied in historical analysis and reconstruction.[60]

If political science is not scientific, we do best to acknowledge that fact and then to prize the kind of wisdom and understanding that practitioners of political science, in spite of its misnomer, can ultimately bring to us. In accepting this kind of understanding, we should also realize that this form of Western social science has no superiority over the political thinking of other societies, unless in terms of certain values, such as human rights (and this already expresses a "value" position). A universality of scientific thought is not to be encountered in this area, although political action in the form of physical, economic, or cultural conquest may lead to more sharing of values and institutions.

History

History, it is frequently claimed, straddles art and science, with a foot in both the humanities and the social sciences. This is a judgment with which I am in accord. At one extreme, however, are those who resolutely restrict history to an art form, according it little or no claim to more certain knowledge. By denying the possibility of scientific history, such commentators see little or no distinction between fiction and history (with its claims to truth), nor any significant difference between history and myth or literature.[61] Yet the history of history, its historiography, is in large part the story of history's transcendence of myth and its reach to scientific status. The problem, as with all the social sciences, is what science means in this context.

It is probably in China that the memory of the past is first systematically put down in written form rather than entrusted to oral transmission. The writings are chronicles, generally records of dynasties. History as an inquiry into the authenticity of past events and their secular causes, not just a record of those events, arises in fifth-century Greece.[62]

Herodotus and Thucydides are the well-known names, but they did not come out of nowhere. They emerged from the Greek efforts at ethnography, geography, chronography, genealogy, and so forth, all from the perspective of the polis (even when they aimed at universal history, which meant that of their immediate neighbors).[63] More generally, the historians were part of that extraordinary outburst of philosophical thought in the fifth century B.C. that saw Greeks groping toward the notion of science. Inquiries into physics were matched by inquiries—as I have noted, the Greek word for *history* means exactly that—into human affairs. The conceptual leap, though executed in slow motion, is of incalculable importance in the history of the human species.

In the opening sentence of the *History of the Persian Wars*, Herodotus states his intentions: to preserve the renown of both Greeks and non-Greeks and to explain the cause of the fighting between them. His stories are comparable to those told of the gods and epic heroes, but now the achievements are human ones, to be explained in human terms. Although supernatural manifestations are not entirely removed, "human, rational causes," as one author tells us, "are to be found whenever possible." "Herodotus' *History* stands somewhere between the narrative fiction (or legend) of the Homeric poems and the totally non-poetic productions of certain of his successors."[64]

Herodotus made the major breakthrough. No Persian Herodotus gives the other side of the story. Instead, the writing of history went forward in the work of Thucydides. It is he who sharpened the scientific edge of the emerging field. More severe than Herodotus in his rejection of mythical accounts and more cutting in his rational evaluation of his sources and explanatory devices, Thucydides, the author of the *History of the Peloponnesian Wars*, a relatively parochial event (though called "the greatest in history"), imparts forcefully to history its first claims to universality and scientific method.

The Greek accomplishment was to substitute logos for mythos (or, more accurately, to add one to the other), in the realm of human affairs, as well as in the realm of nature. Often falling short of aspirations, Greek history nevertheless showed the way to a focus on the events of humans, not gods, while subjecting the invoked evidence and causal explanations to the test of rationality. Carried on a few centuries later by Roman historians, this procedure became a priceless heritage of the West.

The decline of Rome, which had absorbed Greek culture earlier, and the emergence in the West of feudalism, accompanied by a lapse from high culture and a lack of aspirations to secular knowledge, more or less temporally ended the experiment with history as scientific inquiry. Under the domination of Christianity, the history that counted was sacred history, the text that mattered was the Bible, and the only secular form that survived was the chronicle.

The Renaissance meant the rebirth of classical knowledge (though originally in the service of Christianity) and thus of the impulse toward history.

Again, the first setting was the city-state, this time Italian. It should come as no surprise that Machiavelli wrote a *History of Florence* as well as *The Prince;* and his friend Francesco Guicciardini correspondingly wrote both on Florentine history and on the politics of the time.

In the seventeenth century, the time of the scientific revolution, the battle for the soul of history in the West was joined. As in Greek times, the effort at historical science occurred in the context of the effort at natural science. All too often it is overlooked that the battle of sacred and profane history took place as part of a single overall argument about the history of the earth, its time span, and the reason for its present shape—long, slow processes or violent catastrophes. Fossils stood at the center of the debate. And the future science of geology, like history, was being forged on the anvil of interpretive debate. Was time infinite? Were rocks, plants, animals, and humans produced by operating laws, or was there a unique beginning and a once-and-for-all creation of artifacts and creatures, with the world a more or less static place?

In the eighteenth century Buffon spoke of a "dark abyss" of time. Was time interpreted aright by the Bible, with its seven fixed days of creation and six thousand years of history? Were the Hebrews the first people? What of the Chinese, the Chaldeans, and others whose historical existence seemed to predate that of God's chosen people? Were the fossils being discovered God's signs, shifted from one spot to another by his hand, or evidence of fundamental mutability?

Only by crisscrossing from field to field, moving between arguments over scriptural and secular history and inquiries into geology and language, might the answers be found. As Paolo Rossi says, "This was not a matter of unconnected processes . . . but of paths that interwove at more than one point: this happened not only because the men of those times spoke of monuments when they spoke of the history of the earth and of documents when they were considering fossils, but because during those years natural history came to be the history of nature, and the awareness of a new relationship between natural history and human history was born."[65]

At these crossroads, modern secular history found its directions. In sacred histories of humans and the earth, any newly discovered data had to be reconciled with the assertions of the Bible. Here, the answers were already known and could only be further supported. The great tectonic shift of mentality occurred when this procedure was reversed and the data were allowed to call into question the assertions of scriptural text. This was truly the Copernican revolution in history. Critical history, with Pierre Bayle and his *Historical and Critical Dictionary* (1697) in the forefront, gradually emerged, vindicated.

History became professionalized and enlarged in the course of the eighteenth century. In the German universities, such as Göttingen, seminars were set up to give training in archival research, and high standards of evidential verification were held aloft. The footnote became a new tool in history. Histories, like those written by Voltaire and Herder, encompassed cultural and social as well as political activities.

In the nineteenth century, to continue our compressed and corseted historical account, we notice the fulfillment of these two tendencies — to scholarly training and to broader coverage — in Ranke and in historicists generally. With Ranke, the historical seminar conquered the universities, and the footnote became its triumphant emblem. His "scientific" history was based on archival research, which, with all its drawbacks, has served as the Antaeus-like foundation for most modern historical investigations.

With historicism, called by one historian, Friedrich Meinecke, the "greatest spiritual revolution of the Western world," the holistic and hermeneutical nature of history was sweepingly asserted.[66] Here, *Anschauung*, or "the intuitive glance," became the favored approach. Preferably darting forth after Rankean archival work, Anschauung was the means of bringing together the scattered pieces of the whole that was history. Historians apparently abandoned perspective, or at least linear perspective, with its emphasis on a chronological beginning, middle, and end. The substitute was a holism akin to the anthropologist's approach to small societies, but now applied to complex cultures.[67]

In any event, subsequent history has built outward in all of the directions mentioned above. Professionalization and specialization have grown apace, with institutional manifestations at all levels. Family history, psychohistory, quantitative history, prosopography — all have arisen as subfields. Given its hermeneutic nature, much of contemporary history has also taken a linguistic turn, with much attention to language and localized discourse.[68] In another direction, universality, implicitly opposed to historicism, has been claimed by Fernand Braudel and the Annales school, but a universality based on the minute study of local history and on a priority accorded to the material aspects of human doings. The historical sense, in whatever manifestation, has become a sort of second nature for modern humans, even as they seek to ignore or escape its presence by moving quickly into the future.

What is the nature of history? How valid is its claim to be scientific? What sort of knowledge does it bring us? Having secured some idea of the development of the discipline, we can step back and try to secure some idea of its achievements and lack of same.

Let us consider a few general points before proceeding to more particular ones. The first is that, in principle, history, like anthropology, is holistic in ambition. It is fully aware that everything is connected to everything else—that history itself contains, and must employ in its own explanations, all the social sciences—and that to explain any one thing means placing it in a larger context. In this manner, history necessarily embraces the notion of a hermeneutic circle.

One result is that boundaries—which can contain a problem in a manageable way—are difficult to erect in history; thus its "scientific" findings are always porous. Unlike anthropology, however, this boundary problem relieves history of pretending to give detailed holistic accounts of societies (except, perhaps, in the form of philosophy of history, hardly the same as historical research), for, instead of dealing with fifty to five hundred people, it is dealing with one million or one billion or now almost six billion, and the number is changeable over time.

Another point is that history, aside from establishing facts, can emerge from its inquiries with few large-scale positivistic historical statements—generalizations or laws or predictions—although its aspirations in this area are sometimes high. Even more important in this regard is that, like political science, history deals to a large extent with human actions that are heterogeneous, seemingly random, and often turned to unintended consequences, where prediction, verification, experimentation, and even observation are precarious, if not nonexistent. Nevertheless, paradoxically, scientific history is both possible and available in the sense that the practice of history accords with a scientific method that embraces both empiricism and rationality. As the history of history shows, the move toward secular, rational history is one of the great transformations in human consciousness.

The sign under which modern history marches toward scientific method is the lowly footnote. It appears to have emerged in the seventeenth century "as part of an effort to counter scepticism about the possibility of attaining knowledge about the past." The invention of the footnote "made it possible to combine a high literary narrative with erudite investigations. . . . Once the historian writes with footnotes, historical narrative becomes a distinctly modern, double form. . . . In documenting the thought and research that underpin the narrative above them, they prove that it is a historically contingent product, dependent on the particular forms of research, opportunities, and states of particular questions that existed when the historian went to work." [69]

In the footnote, historians implicitly acknowledge the uncertainty of their work, its contingent and open-ended quality. By contrast, traditional historians, as Anthony Grafton reminds us, "claimed universality; their examples of good and evil, prudent and imprudent action, offered their readers instruction valid in all times and places." In this they resembled the mythologizers, who also thought "out of time," and even a "realistic" Machiavelli who inhabited the same semi-mythical time zone in his universal examples. The footnoting historians are content to plod along in their time-bound research.

By using footnotes, historians are able to distance themselves from their narrative and to call the narrative at least partly into doubt. Honest scholars are forced to point out research contrary to their own, interpretations of the same material that reach different conclusions, and omissions in the evidence that they are presenting. Where myths and sacred texts proclaim undebatable truths, scientific historians, especially through the use of footnotes, invite readers to a rational inquiry; they invite them to test the evidence and the inferences by which the inquiry proceeds.

According to Grafton, the result is a "double narrative . . . the narrative in which a text states the final results, while a commentary describes the journey necessary to reach them." [70] Such footnoted history, I would argue, is a scholarly version of the accounts of travel and adventure discussed earlier. Readers are invited through the footnote to be on the team engaged in historical exploration, making its way through dusty archives and then the uncharted realms of imaginative reconstruction.

Sometimes the result of annotation is more pedanticism than intellectual adventure. We are rightly suspicious of the longest footnote in historical literature, which occurs in a nineteenth-century work of antiquarian scholarship: volume III, part 2, pages 157–322, of John Hodgson's *History of Northumberland*. Here we seem to be in the presence of a Swiftian or Baylesian joke (some of Bayle's footnotes do go on for pages). Such mockery, however, would merely be the tribute paid to the virtue of the historian's attempt to practice scientific history.

The serious footnote is one way, and a most important one, of marking the boundary between fact and fiction. As we noted earlier, it has become fashionable recently to call into question the distinction between history and fiction. Both, we are told, are mere narratives, works of imagination and interpretation. In a twisted comment, Oliver Stone, the director of a movie on the assassination of JFK that plays fast and loose with the facts, declares that "it is a restriction and it is a form of censorship to demand of history a fact-only basis, because history is subject to interpretation and re-interpretation. And the facts are often in dispute." [71] Similar statements — for example, about the Holocaust not having occurred because the events can be disputed (what can't?) — exhibit the same gross disregard of what is involved in scientific history.

One cannot so easily dismiss reputable historians (and one must be careful not to tar them with the same brush that was applied to Stone) who seem disposed to erase the boundaries between history and fiction. Often brilliant narrators versed in the use of footnotes, such historians appear to confuse uncertainty with relativism. Because history and fiction use the same form, narrative (analysis is dismissed in their accounts), the two contents must, it is argued, be equally fictitious. Or, in another version of the argument, because history uses narrative, it is simply a form of dramatic emplotment.[72]

In most cases, the historian either wishes or is forced to cast the research results in a narrative. The narrative is a form of causal explanation. It says that something happened because . . . , and the story unfolds. As with Freud and psychoanalysis, the narrative—the "Just So" story—details the connections that make for scientific explanation of a particular sort. In this sense, historians do construct a plot, with an artificial beginning, middle, and end.[73] Because it is a historical plot, however, it is constrained by the facts and generally held close to the ground by the footnote.

As for the plot itself, out of the whole that is history a section must be delimited. A beginning must be made, and an end reached. The problem of selection is key, for out of the welter of facts the historian seizes on some and presents some of these as more important than others. He or she tries to make sense out of the facts, to offer an interpretation of what they mean, how they add up. Another historian may make different choices.

How can a reader decide who is right? In talking of selection, we are also talking about perspective. From what point of time are we making our judgment as to what really happened and what is important? In a static or even cyclical view of time, the meaning of an event is already fixed, for no new perspective can ever arise. In the linear, changing view of time embraced by modern history, we do not know the meaning of an event once and for all. The consequences, for example, of World War I may be seen one way in 1920, another in 1939, and yet another today, for the context in which we seek to understand them has necessarily been changing. There is, in short, no final history and cannot be.[74]

Even at a specific time there can be no one perspective, given the heterogeneity of the human species. In spite of efforts at the sociology of knowledge, we have not been able to discern an impartial spectator, no Martian who stands above all events while empathically understanding all participants in them. Every historical interpretation is from a given perspective, and others are possible.

Thus, it must be admitted, history often takes the form of ideology, a justification of domination—the victors usually write the histories. Censorship and hegemony mute the voices of offstream and dissident historians. In other words, the pursuit of scientific history is freighted with external obstacles as well as inherent problems. Yet scientific history as such is an inquiry that, like any other effort at science, represents one of humanity's best hopes for acquiring "true" knowledge about the world. Accepting its uncertain and problematic nature does not mean rejecting its methodological claims.

So far, only narrative has been cited as a form of historical inquiry and presentation. In fact, much of modern history is devoted to analysis. Instead of narrating the course of the Russian Revolution, the historian might inquire into agrarian conditions on the eve of 1917, or do a prosopographic study of the Bolshevik leadership, or examine the military structure of the tsarist regime. Along the same lines, a comparative study of the Russian and French Revolutions might be undertaken. In such work, overt attention will, or should, be given to the concepts, theories, and techniques—quantitative methods might be one—being employed.

Although the results of analytic history are often dull or too technical for the average reader, analysis serves as an indispensable part of historical methodology. Analysis often entails borrowing from the other social sciences. In a history of the industrial revolution, it will be useful to bring economic theories to bear. The same is true in regard to sociological theory, concerning the transformation to industrial society. And so on.

Narrative history generally incorporates analytic studies, and analytic history presupposes a narrative structure. In both narrative and analysis, the role of unintended consequences looms large. Given the multicausal nature of historical subject matter, most, if not all, of what happens in history is not the result of a single conscious intention but emerges out of the interplay of many intentions to an end not envisioned by any of the actors.

Often we have no way of comprehending past events or ideas except through their historical reconstruction. Debates between and among medieval and humanist thinkers cannot be understood—literally, do not make any sense—except in terms of their historical development. Why they debate a particular interpretation to the point of being willing to kill one another is not a matter of logic but of inherited positions.[75] Similarly, only a historical reconstruction helps make sense of why men in Western society wear ties—a hangover presumably from knights tying their ladies' scarf around their neck as they jousted in tournaments—or why Serbs and Bosnians kill one another, which is partly because of events at the fourteenth-century battle of Kosovo.

The knowledge that history gives is based on memories, which are generally embodied in texts that fetch up, often accidentally, in archives and are then subject to hermeneutical inquiry.[76] Whether history is analysis or narrative, selection and perspective, uncertainty and interpretive reconstruction surround the attempt to make it scientific. In the end, given the nature of history and the phenomena with which it must deal, what emerges is not so much a discipline modeled on positivistic science (although I have emphasized the scientific method that functions as a part of such science) as a deepening of interpretation.

This point can be illustrated only by immersion in historical debates and literature; by showing that historians differing with one another, even drastically, are advancing the state of knowledge by giving it greater depth. One historian may argue that John Locke was an apologist for a mercantile and manufacturing bourgeoisie, another that he was a theorist for agrarian capitalists.[77] But both are right, for Locke was himself ambivalent and ambiguous, a man whose work changes in its significance as further events transpire and additional contextual materials emerge.

As long as the historian is true to the facts—though they may be different facts from those of another historian—and demonstrates the procedures followed, remaining open to criticism, that historian is engaged in scientific history as we have defined it. Accumulation takes place when yet another historian seeks to reconcile the evidence and inferences in a more extended treatment, a larger synthesis—a deeper interpretation.

Accumulation and interpretation constitute no mean achievement. The extent and the depth of our present historical knowledge, in spite of all its uncertainty and lacunae, is almost overwhelming. As shown earlier with prehistory and the advances in its practice, we now know vastly more about humanity's history during the past few thousand years than we did even a few decades ago. Qualitative gains in detailed knowledge stand next to the other transformations brought about by the practice of history. Today, historical thinking comprises the effort to situate humans, not gods, in a world of actual happenings, established by both empirical and rational means and constantly subject to new interpretations. In history, secular humans seriously try to understand their continuing evolution, which they now experience as a cultural more than as a physical "happening."

In Medias Res

At least one philosopher has declared that "the very idea of a science of man or society is untenable."[78] Others have emphasized the radical dissociation between the social and the natural sciences.[79] Still others blame positivism for distorting mainstream social science and attack the notion that the "whole 'development' [positivism] represents some sort of 'progressive' emancipation of the 'disciplines' from a 'pre-scientific' and confused past."[80]

I do not share the first two views and, while finding much that is compatible in the third, would rather retain the aspiration to some form of scientific method inhering in positivism while rejecting its linear simplicities. How I would accomplish this aim is contained in much of what I have written and will be spelled out more fully in the chapter that follows. The fact is that the acquisition and discovery of knowledge can take place in erratic ways, sometimes regressing as well as advancing, and is often highly dependent on both emergent phenomena and developments in other fields of science.

It must also be acknowledged that social (as well as natural) sciences are socially constructed. How could it be otherwise? As efforts to understand the social environment and the interactions of real human beings who use symbols and other forms of language, which are also social constructs, the social sciences necessarily are shaped by the conditions that they seek to understand. To understand the conditions, social scientists must strive for as much objectivity as possible—which is why objectivity is an ideal never fully realized. They must also become as conscious as possible about angles of perspective, including their own.

This task is enormously difficult. It requires looking at the lenses by which we look at phenomena while we use them, a kind of Escher-like recursive action. Let me make this more specific. A study of three empires, Ottoman, Mughal, and Safavid, in the premodern period, taking them as a region—a new conceptualization—may begin by bringing Western social science notions of class, power, and political structure to the subject. Gradually, it becomes clear that the notions obscure as much as they clarify. They are largely inappropriate, too Eurocentric, making it necessary to devise new notions, new concepts, that are both more useful and "truer" to the phenomena being studied.[81]

Hopefully, the result will not be muddle. The cultures and societies of the Other can bring more extensive and deeper concepts to the existing effort at social science. The Other can add perspectives perhaps undreamed of by the original social scientists (largely of the West). As I shall argue later, this possibility is partly the consequence of an emergent phenomenon, globalization, which is giving rise to further effort at social science and inspiring the explanations that social science can provide.

For explanations to be social science, they must employ the scientific method appropriate to the specific discipline. In my brief treatment of economics, sociology, anthropology, political science, and history, I have tried to give some idea of what in their efforts is scientific and to what degree. The most useful definition of *science* emerges out of the methods that workers in the disciplines develop to obtain as much knowledge of their chosen phenomena as they can. If this approach to the definition is accepted, people living today can claim to be witnesses to extraordinary strides forward in the scientific knowledge of the human species, especially the evolutionary and prehistoric states of the species. We know more that is accurate about the time and space in which humanity has existed than others before us — before the emergence of human science.

The conclusion is inescapable: Despite all the limitations of human science and the human sciences, their achievements are highly significant. This view must not be lost sight of nor obscured by otherwise legitimate criticisms of the fields under discussion.

6 / The Uncertain Sciences

*H*aving considered the achievements of the human sciences, let us now inquire into their future possibilities. What combination of positivist aspiration and hermeneutic intent can lie ahead? How can we reconcile past and present work in the natural sciences with that in the human sciences, for the two, as we have seen, are indissolubly tied together, especially, for our purposes, with the cord of evolutionary theory?

A foremost possibility is that the human sciences can develop only by humans becoming conscious of the ongoing achievements of the human sciences. By becoming conscious, I mean not only possessing ourselves of the existing state of knowledge but being willing to act on, and live in terms of, that knowledge. In short, we must become a community of human scientists.

Before I develop that idea, let us consider further such topics as unintended consequences, emergent phenomena, the mechanical nature of humans, the role of prescription, and the possibility of prediction. The overall argument is an extended and connected one and cannot be reduced to a formula or even a few sentences. It is what I might call an ecological argument, presented in historical and philosophical form.

Threads to Pull Together

In the short, sober book *Usable Knowledge* Charles Lindblom and David Cohen focus on the social sciences, arguing that what they call professional social inquiry (PSI) is of very limited value. It pretends to scientific authority but tends to be mainly of ideological significance and to be used to justify policies arrived at by other means. Whatever instrumental value PSI may have is largely drowned in a broader value context, about which it has little to say. If a practitioner of PSI is told "that he should take full employment to be an important value when he analyses and appraises fiscal policy," he has no way of knowing "how important it is relative to other relevant values like price stability or low taxes."[1]

For the authors, the only way of arbitrating among conflicting values and the people who hold them is through action. Social practitioners, misled by aspirations to model themselves on natural scientists, overlook "ordinary knowledge," or what one might call simple common sense. Common sense tells us that a result emerges from people's passionate and political actions, not from predictive scientific knowledge. Many, if not most, problems are too delicate and disturbing to the actors involved to resolve clearly and rationally, even if a solution is available. An example might be the allocation of resources, where the attempt to impose a rational plan often leads to a violent conflict of interests or to an unacceptable authoritarianism, or so it seems. The market may do a better job.

Thus, what solves social problems is social interaction—the thing itself—rather than a science that pretends to deal with social interaction. Actions, in turn, constitute a never-ending sequence of solutions that create new problems. In the late nineteenth-century United States, for example, progressives seeking to professionalize the police forces introduced pensions as a recruitment incentive, with the unexpected result that pensions became a crushing urban burden. Any solution to that problem would, in turn, create its own new problems.

Unintended Consequences

Lindblom and Cohen's analysis is realistic. They believe in knowledge, but "useful," as they define it, rather than "scientific" knowledge. Accepting their salutary cautions, let us try to go further. For one thing, they focus solely on social sciences and not on the human sciences. For another, the aspiration to scientific knowledge, as we have seen from Bacon's time on, does embody the desire to be useful.

There is a way to deal scientifically with the social interactions upon which Lindblom and Cohen focus and upon whose outcome they predicate their ordinary, useful knowledge: seeking to understand the nature of unintended consequences. With such understanding, we can attempt to exert control over social interactions, thus paradoxically introducing intention into what is otherwise unintended.

We find the phenomenon of unintended consequences everywhere in human history. The attempt to analyze it scientifically, however, is a recent effort. One pioneer was Adam Smith. In his hands, it has taken two forms. In one, it is given mathematical expression: the pursuit of self-interest leads to market transactions that can be analyzed, modeled, and given scientific shape (an economic social science). This form is too well known to need further comment.

The second form is less well known. It manifests itself, to take the example he gives, as a "silent and insensible" revolution, in which accumulated actions of pedlars pursuing a penny and nobles pursuing luxuries lead to the replacement of feudal society with a capitalistic one.[2] Here, Smith has given us a snapshot of an unintended consequence leading to a new form of society, the modern. Historians ever since have been trying to supply the details—a kind of detailed filming—of this enormous transformation.

As it happens, both the economist and the historian deal with the same sequence of unintended consequences—in this case, capitalism—with the former dealing with it in mainly synchronous terms and the latter in diachronic terms. Both aspire to some form of science. Indeed, the multitude of individual choices based on heterogeneous desires and motives do seem to shake down into identifiable unintended consequences, both economic and historic. We looked at the economic attempt with individual choices in the previous chapter. Let us now spend more time on the historical, for it is much broader, it is vaguer in appearance, and it is perhaps even more problematic.

Capitalism, our present example, emerged from a motley group of coincident developments. It was tied to the emergence of an absolutist nation-state, which provided one of the boundary conditions for capitalism, for that state could provide security for property with a legal system and a bureaucratic order. The absolutist state, in turn, was connected to changes in the religious order, with the state's supremacy made evident in the Westphalia settlement of 1648. Serfdom had to be abolished, and a market economy, based on free labor, allowed to expand. And on and on. All of these developments were unintended. Yet the origins and nature of capitalist society must be sought in such circumstances.

A brilliant description of these interlocked developments is given by Perry Anderson. Himself a Marxist, he provides correction and extension to Karl Marx's original effort to deal with the coming of capitalism as a matter of unintended consequences. Employing a comparative method, with contrasts between Western and non-Western developments, Anderson gives an enormous density—a deepening—to Marx's somewhat simplistic account of the disappearance of serfdom.

In addition to the factors mentioned above, Anderson pays attention to the cultural factors so often neglected by Marx. He notes, for example, that "the revival of Roman law . . . was accompanied or succeeded by the reappropriation of virtually the whole cultural inheritance of the classical world. The philosophical, historical, political and scientific thought of antiquity—not to speak of its literature or architecture—suddenly acquired a new potency and immediacy in the early modern epoch."[3] Such cultural developments are related to the rise of modern science and the Protestant Reformation, so central to Max Weber's analysis of capitalism.

Anderson reminds us too that "it was against this background that a cosmopolitan elite culture of court and salon spread across Europe." At this point, a move to Norbert Elias's history of the civilizing process in Europe, with its attention to manners, becomes apropos. In the second volume of *The Civilizing Process, State Formation and Civilization,* Elias explicitly draws some of the threads together. "It is more than a coincidence that in the same centuries in which the king or prince acquires absolutist status, the restraint and moderation of the affects . . . the 'civilizing' of behaviour, is noticeably increased . . . it emerged quite clearly how closely this change is linked to the formation of the hierarchical social order with the absolute ruler and, more broadly, his court at its head."[4]

To continue our compressed example, in the rise of capitalism the unintended absolutist state was indispensable. So, too, however, was the demise of that absolutist state, which also came about in unintended fashion. As Reinhold Koselleck tells us, it occurred because of the workings of the bourgeoisie and the emergence of civil society. "The Absolutist political system was vanquished by the indirect assault of a society which referred to a universal morality the State had to exclude, and through which—without apparently touching on the Absolutist system politically—[was] destroyed that very system from within. The concentration of power in the hands of the Absolute sovereign afforded political protection to a nascent society that Absolutism as a political system was no longer able to integrate. The State, as the temporally conditioned product of the religious wars whose formality had mediated the religious conflicts, had become the victim of its historic certainty."[5]

God and the State, it could be said, moved in mysterious ways. As we descend to specifics, we see that absolutist policies made about 400,000 émigrés pour out of France into northern Europe, with 80,000 of them going to

England, where many of them became ardent Whig propagandists, who subsequently flooded all of absolutist Europe with pro-constitutional arguments.
Out of such minor happenings, combined with innumerable others, came the
downfall of the French absolutist state in 1789.

I have focused here, for illustrative purposes, on the unintended emergence and then dissolution of the absolutist state. The same sort of analysis
can be used in regard to the unintended development of scientific knowledge
(or, indeed, any other cultural development). Hans Blumenberg has given us
a monumental explanation of how Copernicus's heliocentric theory emerged
out of thrusts and parries among intellectuals; as he concludes, there is "no
evidence for the assumption that the outcome of making history has some
resemblance to the ideas that people have while they are making it."[6] The accounts of the genesis of modern geological science given by Nicholas Rupke
and Martin Rudwick lead to the same conclusion.[7]Elias's summary is applicable to an entire range of human actions and interactions: "From the inter-
weaving of countless individual interests and intentions — whether tending in
the same direction or in divergent and hostile directions — something comes
into being that was planned and intended by none of these individuals, yet has
emerged nevertheless from their intentions and actions. And really this is the
whole secret of social figurations, their compelling dynamics, their structural
regularities, their process character and their development; this is the secret of
sociogenesis and of relational dynamics."[8]

Alas, the whole secret turns out to be perhaps even more complicated than
Elias had in mind. After all, we seek scientific knowledge about unintended
consequences partly so that we can make formal social decisions that allow us
to avoid them when undesired. This is certainly one intention of economics.
But can we even approximate such an intention in regard to large-scale historical movements? Or do no social regularities exist that can be extrapolated
from unintended consequence analysis? Can only broad directions of growth
be studied?[9]

Nevertheless, unintended consequences do lie at the heart of much social
science. They and their analyses are enormously important. They touch on the
problems of emergence, heterogeneity, stratified social interaction, and other
such matters that we have been stressing. In one social science, economics, an
effort can be made at curtaining off many of the conditions and axiomizing
and quantifying what remains; this is a useful, though limited, form of scientific knowledge.

In the less bounded, more extensive social sciences, such as sociology,
anthropology, political science, and history — especially the last — unintended

consequence analysis must be approached methodologically in a more ecumenical, less certain spirit. We are forced into a more narrative, less analytic mode of understanding. And such an approach immediately places us in the domain of interpretive science. The actors studied by history have intentions and motives; these are filled with meaning. To understand these meanings requires the form of interpretation that we have labeled "hermeneutics," with all the uncertainty attendant thereto. Such interpretation is apparently far removed from positivist aspirations to science.

Yet a combination of the two—the hermeneutical and the positive—is achievable and desirable. Max Weber pioneered in this attempt, and Jürgen Habermas continues with it philosophically. One commentator says of his work, "If social research is not to be restricted to explicating, reconstructing, and deconstructing meanings, we must somehow grasp the objective interconnections of social actions, the 'meanings' they have beyond those intended by actors or embedded in tradition. We must, in short, view culture in relation to the material conditions of life and their historic transformation." [10]

What we see, taking the lead especially of Weber, is that the hermeneutical approach is essential as we seek to understand the intentions that people have had as they acted in various economic, political, and other situations. With intentions understood, we can glimpse the way the interaction of heterogeneous intended actions produced an outcome—which we can analyze by at least semipositivist methods—unintended by any of the parties to the event. Such knowledge, in turn, has the potential for informing our future intentions by providing a foreknowledge of possible outcomes.

Knowledge and foreknowledge are at best asymptotic. Nevertheless, the hermeneutical and positivistic study of unintended consequences does give us scientific knowledge—at least the sort of science that is possible in the human sciences. While such knowledge does not match up to the positivist claim (now itself sharply qualified) to certain knowledge, with accompanying predictability and control, this form of studying unintended consequences does accord with the data and theory requirements of the phenomena studied. This match of methodology and subject gives us knowledge that, while uncertain, is scientific—it is soundly based on the evidence and legitimate inferences about humanity's path through time.

Accumulation of this sort deepens knowledge. It incorporates a way of seeing scientific connections among phenomena that would otherwise appear chaotic. It assuages our sense of what Adam Smith called wonder. And it does so in a scientific manner that we can find acceptable, even if it is on the uncertain end of the scientific spectrum.

Emergent Phenomena, Again

In considering the prospects of the human sciences, we need to look at the connection between unintended consequences and emergent phenomena. We can conceive of the latter as brought about by the former, which can be thought of as their temporal manifestation. In fact, emergent phenomena incorporate similar past developments, previous emergences, and thus can be seen as a form of cultural accumulation.

It is important to note what are not emergent phenomena, for example, a kitten being produced by a cat. Much misunderstanding is occasioned by vague use of the word *emerge* to mean "to rise from or to issue from." In contrast, the phrase *emergent phenomena* refers to a new entity whose whole could not have been predicted from an analysis of its parts.

I have already said a number of things about emergence. The twentieth-century biologist Ernst Mayr, among others, has written on the subject. He stresses the notion of the whole, with great insight. Earlier, in the late nineteenth century, the sociologist George Herbert Mead was employing the concept. He remarked, "We may assume that certain types of characters arise at certain stages in the course of development. . . . Water, for example, arises out of a combination of hydrogen and oxygen; it is something over and above the atoms that make it. . . . Anything that as a whole is more than the mere form of its parts has a nature that belongs to it that is not to be found in the elements out of which it is made." [11]

Mead's interest in emergence was connected with humanity's development of mind and consciousness, the emergence of social, economic, and political structures or the emergence of mental structures like those found in cultural configurations and sciences, especially the social sciences. These emergences have not been teleological, intended by some god or reason, but unintended consequences of the evolutionary process.

One example that we have dealt with is the emergence of capitalist industrial society. Could we have predicted it? Given that, à la Mead, it is a whole greater than its individual parts, the answer must be no. We could have looked at the rise of the absolutist state (as we did), the Renaissance recovery of classical wisdom and its effect on the Christian church, the gathering strength of the bourgeoisie under the state's protection, the increase in market exchanges, and a host of other factors — and still not guessed at the outcome: an industrial revolution.

Once the revolution had occurred, we could reconstruct its coming into being, put together the factors, the atoms and elements, that came together to create it, and then feel that we had achieved a scientific understanding of its origins. Could we now, as a result, predict accurately the next stage of development, perhaps the society that succeeds the industrial? The answer is no (in spite of the guarded optimism that I expressed earlier about unintended consequences becoming more intended). We were not even able to predict the computer revolution before it was upon us, and we have only some idea of where that may be taking us.

The next hundred years are likely to bring about changes comparable to those of the industrial revolution. A combination of space explorations and computer developments (not to mention biogenetics) may bring about situations and contexts akin to those of the Age of Discovery and the industrial revolution combined, as a forcing ground for the human sciences. Who knows what changes in human consciousness will occur, or what new machines and instruments will appear? And insofar as the human sciences have arisen out of attempts to achieve knowledge about recent emergent social phenomena, who knows what shape they will then take?

These imaginings may seem fantastic. Let us look at their underpinnings in reality, taking our first example from geology. Its emergence was shaped forcefully both by the needs of the industrial revolution and by the religious arguments going on at the time in British society. Once established in its own right, geology contributed not only to the further development of industry but also to the undermining of religious knowledge and the substitution for it of scientific knowledge about the earth and humans, culminating in the great work of Darwin and his evolutionary theory. Thus, our awareness of evolution and emergence is an unintended consequence — one that has further such consequences.

Martin Rudwick puts the economic side of the matter concretely:

The growth of new forms of industrial production, though hardly yet perceived as the "industrial revolution" it was to become in retrospect, was transforming parts of Britain and spreading to parts of the Continent too. Heavy industry was increasingly dependent on expanding supplies of coal and metal ores; mines were meeting the demand by driving ever deeper below the surface. . . . In doing so, the owners and managers of mines found their traditional empirical rules of operation increasingly inadequate. The extension of old workings and the discovery of new ones demanded new methods of prediction and exploration; geology was looked to as a potential source of methods that would be soundly based on scientific principles. Geology was therefore esteemed as much for its prospective contribution to the economy as for the enlarged view of the natural world.[12]

Our next example, map making, moves closer to the human sciences but is still on the cusp. Here, the argument is that modern maps emerged as a result of the same development of the state that was requisite for the coming of capitalism. "In growing societies, the continuing need for increasing hierarchic integration produces first a simple enlargement of the mapping function, but then its ceaseless branching. Thus the state, in its premodern and modern forms, evolves together *with* the map as an instrument of policy, to assess taxes, wage war, facilitate communications and exploit strategic resources." We can even make a connection with our earlier example, geology, for, as Denis Wood continues, the modern state requires mapping of "subsurface geology likely to contain oil."[13]

The science of geology, the making of maps — these can be placed in the context of unintended consequences and emergence. Political science, to choose a social science example, can be viewed in the same manner. Here let me make reference to Keith Baker, who describes how the workings of the absolutist state gave rise to one aspect of modern political science. Writing about Condorcet's definition of the social field, Baker declares that it was "formed in the context of a crisis of government and society that developed as the centralizing tendencies and universalizing implications of bureaucratic absolutism came into open conflict with the particularistic and corporatist conceptions that were the traditional foundations of the French monarchy."[14] Neither the monarchy nor the parlements intended what happened — the events of 1789. Nevertheless, mirrored and focused in the theories of such philosophes as Montesquieu, Turgot, and Condorcet, the triumph of the bourgeoisie came about as an unintended consequence, emerging from the combined political and ideological struggles of the French Revolution — and with it emerged a new version of political science concerning democracy.

Once new forms of scientific knowledge have emerged, whether in the shape of geology, economics, historicized political science, or whatever, they illuminate phenomena that had previously been viewed without such insights. This is nicely stated by Karl Marx when he remarks that "bourgeois society is the most developed and many-faceted historical organisation of production. The categories which express its relations, an understanding of its structure, therefore, provide, at the same time, an insight into the structure and the relations of production of all previous forms of society the ruins and components of which were used in the creation of bourgeois society. . . . The anatomy of man is a key to the anatomy of the ape. On the other hand, indications of higher forms in the lower species of animals can only be understood when the higher forms themselves are already known. Bourgeois economy thus provides a key to that of antiquity, etc."[15]

I cannot refrain from adding one more example to what is already a lengthy list of quotations: Foucault's explanation of how the human sciences have arisen from bourgeois society. In his case, the emphasis is on the prison system. "The carceral texture of society assures both the real capture of the body and its perpetual observation; it is, by its very nature, the apparatus of punishment that conforms most completely to the new economy of power and the instrument for the formation of knowledge that this very economy needs." Continuing, Foucault says, "I am not saying that the human sciences emerged from the prison," but then undercuts his reservation by adding, "But, if they have been able to be formed and to produce so many profound changes in the episteme, it is because they have been conveyed by a specific and new modality of power: a certain policy of the body, a certain way of rendering the group of men docile and useful."[16]

It is important, in this flurry of quotations, not to lose sight of the fundamental point to be made. Emergent phenomena, partly explicable in terms of unintended consequences, must necessarily be the starting point of the human sciences. The human sciences, in turn, seek to understand, in a scientific manner, the phenomena that have produced them. Humans cannot jump over their own shadow. We cannot, I am suggesting, truly evaluate the prospects of the human sciences because we cannot foresee with any certainty the future of the human species.

What we can do with some success is to turn the spotlight backward. Whereas philosophers of antiquity, for example, had no resources with which to understand the emergent form of a capitalist market economy, today's social scientists grappling with that now emergent phenomenon are in a position to utilize their newfound knowledge to understand the overall system. This is no small achievement. Yet Hegel intuited correctly that we can have this sort of knowledge only when the owl of Minerva has taken flight, that is, when the past has occurred.

Human as Machine

Having recorded this sobering thought, I now propose that we go on something of a drunken spree. I suggest that we do so in a manner that combines science fiction and a thought experiment by once again thinking of human beings as machines.

The idea of human as machine is not fantasy but science. Ever since Descartes and the seventeenth-century debate over the animal-machine, the view that humans could be thought of mechanistically has been seriously advanced. Needless to say, such a view arouses bitter enmity in those who believe in special creation of human beings and their unique soul (Descartes himself held such religious beliefs, although they did not stop him from going to the edge of the concept of human as mechanism). Yet the application of mechanistic principles of explanation to humans is solidly grounded in the natural sciences and is legitimately extended even to human consciousness. Consciousness, in this view, is an emergent property, arising from the interaction of innumerable cells and nerves, which are themselves composed of tiny mechanisms.[17]

The thinking about humans as machines that I have in view, however, is at a much grosser level. Until the late twentieth century, those who wished to turn humans into machines were thinking basically about an eighteenth-century mechanical being, devoid of autonomy. Typical was the demand of Josiah Wedgwood concerning his workers that in his pottery factories they must "make such *machines* of the *Men* as cannot err." Expressed as a fear, we find it in Thomas Carlyle, who declares that his age was making men "mechanical in head and heart, as well as in hand."[18] The view persisted into the twentieth century and was espoused, for example, by the Russian scientist Ivan Pavlov. For him, humans (and all other animals) were simply mechanisms moved by conditioned and unconditioned reflexes. Many behavioral scientists seem to flirt with, if not embrace, the same idea.

The importance of this idea for our consideration of the human sciences is that it does pose the possibility of a positive science. If humans are merely an old-fashioned type of machine, readily conditioned and deterministically manipulatable, then, in principle, a certain and predictive human science could be constructed. (In this definition of the machine, machines are assumed to be without motives, desires, fantasies, and so forth.) In such a world, a human being becomes a particle like any other in the universe, and in the nineteenth-century view of particles, before Werner Heisenberg and the uncertainty principle, such mechanistic particles were Laplacean in their movements, that is, deterministic.

Such was the case at least in principle. With the coming of artificial intelligence in the late twentieth century, another version of this idea has become floatable: the human being as programmable — and therefore readily reprogrammable. In this conception, as in the mechanical, humans possess no autonomy, and another, more computerlike form of positive human science is possible. But the important consideration for our purposes is that, from this perspective, humans can be looked at as machines and, in general, as machines of an early, primitive nature (only later being equipped with a low level of artificial intelligence).

This perspective can be articulated in organic terms as well as mechanistic. Aldous Huxley, in *Brave New World*, envisions a completely controlled and predictable world whose denizens are mass-produced in test tubes, as alphas, betas, and so on. Nongenetic stratification is unnecessary; each person is genetically fashioned for exactly the role he or she is to play. Feelies and soma pills are additional reinforcements. Time has been made to stop (in theory), and nothing in the way of emergent phenomena or unintended consequences is meant to exist.

Even a moment's consideration tells us that this is a world of fantasy. Such a world (in either its mechanical or organic version) is also a simplistic and reductionist notion of what is involved in the mechanistic approach to phenomena. The subject is, in fact, a tricky one to handle for these reasons, and the imaginer must proceed carefully. Those who resolve the problem of the human sciences by postulating a mechanical man, whose actions can be controlled and predicted with certainty, ignore with their *reductio ad automaton* all that has been said about hermeneutics and meaning, not to mention common sense.

Nevertheless, even if human beings are not simple machines, they are becoming more mechanical. Freud called the human a prosthetic god, and the extension of human senses, via telescopes, microscopes, and ever more sophisticated instruments, is increasingly being matched by an intension, to coin a usage, whereby mechanical hip replacements, pacemakers, and other devices are turning humans into what elsewhere I have called *Homo comboticus*. Such beings are, however, still human beings, not freed from the human condition and still wrestling with the problems of the human sciences as I have been outlining them.[19]

A true escape from the dilemma of the human sciences can nonetheless be envisioned, in principle, by means of a machinelike development: the emergence of a new species, a computer-robot, that can learn, can move about and have new experiences, and can reproduce with variations. Because it is an emergent being, we cannot predict with any certainty what its development path will be or what sort of society might result from computer-robots' interactions. Indeed, we can say little about it, except as a present possibility with an unknown future.[20]

Even if this new species is possible — and the reader may be gasping with incredulity at this point (wrongly, I believe) — the problems faced by the human sciences would persist, for the human species would remain human. I have mentioned positivistic fantasies simply to take one line of thought to its conclusion. With humans continuing to be humans, though made even more prosthetic than at present and possibly coexisting with a new computer-robot species, we are left with the same problems that we have been grappling with up to now. A human science dealing with humans in a world that includes computer-robots must still be human in its intentions. A science dealing with humans but undertaken by computer-robots would presumably not be.

Prescription, Renewed

Let us return to the here and now. In Chapter 1 we touched not only on the machine issue but on the notion of prescription as well. Again, we need to re-examine the subject in the light of all that we have said up to this point.

A prime example of prescription, that proffered by modern neoclassical economics, is advanced formally, as a postulate: economists assume as the actor a rational, self-interested, profit-maximizing human in a condition of perfect information. Economic analysis consequently works if enough real people behave something like the postulated "economic man." We then have a market economy, subject to the laws of economics.

As we know, such a market and such an economic man have not always existed (and only exist imperfectly even now). The history of capitalism shows that the emergence of market values and profit-maximizing actors was a long, slow process. The postulate had first to become a prescription, preached and practiced by enough individuals to become an approximation of the modeled reality. A glance at nineteenth-century manufacturers shows how hard it was for them to harden themselves as required and to ignore promptings of the heart and earlier traditions. Dickens's Gradgrind in *Hard Times* is unable to hold to the new prescription.[21]

One of the accomplishments of Max Weber in *The Protestant Ethic and the Spirit of Capitalism* was to give historical specificity to the prescription. He showed how religion changed to preach a self-monitoring, calculating, deliberate mode of rational reckoning that transferred readily to the requirements of emerging capitalism. The Protestant character became an economic one. Changed people changed the dominant culture, and the culture continued to produce changed people. We can see now, in the light of anthropology's discovery of the concept of culture, that a given culture can itself be viewed as a prescriptive system. It tells the members of a culture how to think, feel, and behave.

In a primitive or small society, the cultural voices are more or less united, and the prescription self-enforcing. The prescriptive system is part of that whole that anthropologists claim to be analyzing. In more developed societies, with more varied institutions and prescriptions, a struggle appears constantly to be going on among rival physicians to the body-and-mind political and cultural. To still the discordant voices and to achieve a single prescription, some have tried to establish a totalitarian society, with one prescriptive form of thought and behavior that is, in principle, enforced totally. The Soviet Union and Nazi Germany are the obvious modern instances. The former, lasting for more than seventy years, prescribed a "New Man" in place of capitalist economic man and sought to mold him (and her) by any and all means.

The Soviet system failed, as we know. As an instance of prescription, it stands as a kind of reductio ad absurdum, as hideous in its consequences and as simplistic and dogmatic in its ambitions as Huxley's brave new world. But such a reduction ought not to obscure the importance of the notion of prescription to the human sciences. It reminds us, in a peremptory manner, that desires, values, and actions are part of any attempt at social science. It revives the hoary "what is/what should be" debate in modern form. Humans not only adapt to the existing world but remake it to suit their values and visions. Prescription is another way of stating this intention. We tell others how to live as we do, or would like to, and, where possible, we require them, by persuasion or coercion, to do so.

Prescription often poses as a claim to knowledge; and the claim to knowledge often presents itself as scientific. Isaiah Berlin points out that a social science would have heightened empirical reach and predictive power in a society that unanimously agreed on its own goal. "In a society dominated by a single goal there could in principle only be arguments about the best means to attain this end — and arguments about means are technical, that is, scientific and empirical in character: they can be settled by experience and observation or whatever other methods are used to discover causes and correlations; they can, at least in principle, be reduced to positive sciences. In such a society no serious questions about political ends or values could arise, only empirical ones about the most effective paths to the goal."[22]

Michel Foucault focuses on the same truth, though from an entirely different angle and with a drastically different intention. As he argues, expressing an old idea in postmodern terms, knowledge is power. The human sciences are not disinterested efforts at knowledge but constitutive elements of a ruling hegemony. In his philosophy, the aim of science, to control phenomena, is realized in actual life through power over people. This power expresses itself prescriptively, using various semiotic forms.

Foucault aside, how does or might prescription further relate to positivistic aspirations and hermeneutic intentions? One view, already noted, is that humans are passive particles, to be analyzed and prescribed for mechanically and positivistically. Another view abandons analysis and prescription and retreats to a purely hermeneutic interpretation. But there is an intermediate position (our own, in fact): accepting the role of prescription, accepting that human intentions and motives (analyzed hermeneutically) can enter into the effort at scientific understanding of human actions while giving due recognition to both existing circumstances and the human desire to affect them.

Consider sociology as a case in point. Its object need not be a Durkheimian fact but an inquiry into society. The inquiry may lead toward a specification of new institutions more appropriate to the conditions, say, of industrial society, as well as toward a clear indication of the means by which new arrangements might be brought into existence. This is what Marx was groping toward: a recognition that human aspirations enter into the composition of any human science.

One result is that close prediction and certainty fly out the window (a fact that Marx, the revolutionary, was unable to accept). The implications are that any law in the social sciences is part of a process including prescriptions that fosters change, which may then create new conditions in which that law no longer effectively applies. The process itself is neither all or nothing, on or off; the change is gradual, involving many interacting causal relations. Unlike in some of the natural sciences, where a precise passage from one state to another, from water to ice at thirty-two degress Fahrenheit, is identifiable, the human sciences are generally forced to deal with vague, historical transformations.

A simple Lindblom-Cohen analysis, with an emphasis on political activism, is helpful but not sufficient, even if we add unintended consequences. Prescription can and must take its rightful place in the context of changing processes. It figures as part of the active political process and as part of the scientific knowledge that we try to achieve about that process, as well as about additional, larger human processes.

The Conditions of Human Science

Having considered characteristic features, or problems, of human science — unintended consequences, emergent phenomena, the desire for mechanical perfection, prescription — we need to examine some of the conditions under which scientific knowledge of humanity is possible. In inquiring into this set of conditions, we will be resuming the discussion of Chapter 2 about witnesses and witnessing, though in a more complex context, then extending it to the more general issues of scientific communities, their relations to civil society and to democracy, and the future possibilities of such communities for the human sciences.

Witnesses

We start with the natural sciences. The idea of scientific witnesses, as we saw, figured largely in the work of Robert Boyle and his friends in the Royal Society. A scientific experiment or observation was confirmed by either direct or virtual witnessing. Both experiments and observations were for public viewing, with the latter made additionally available in reports of the original experiments, allowing for their reenactment in the mind or by a repetition. The result of witnessing was an early version of positive science, where the knowledge obtained was verifiable and predictable.

At the time, the question arose as to how to evaluate the witnesses. Might they report falsely? One safeguard was to repeat their experiments and look for the same results. But testing was often inconvenient and not feasible for most people. Science may be cumulative, but one cannot test all the building blocks for oneself. Trust is necessary. But on what basis is one to trust? The answer in the seventeenth century, it has been argued, was in terms of class structure: one trusted "gentlemen," for they would not lie. In this formulation, science was an elite form of knowledge. As Steven Shapin points out, "The identification of trustworthy agents is necessary to the constitution of any body of knowledge." [23] In this case, gentlemen were perceived to be disinterested, that is, without any incentive to fudge or misrepresent the results.

The idea of gentlemanly witnesses persisted into the nineteenth century. Martin Rudwick shows much the same pattern in early nineteenth-century geology. There, "the treatment of Knowledge claims depended on their point of origin." Indeed, one of his chapters is entitled "Arenas of Gentlemanly Debate." [24] In such arenas, gentlemen were to engage in restrained and civilized debate, and the science that emerged was to be presented to the public as a consensus that could be trusted because of its class origins.

The reality was quite different. No less a figure than Charles Darwin spoke of "contemptible quarrels" and remarked that "I am out of patience with the Zoologists, not because they are overworked, but for their mean quarrelsome spirit. I went the other evening to the Zoological Soc. where the speakers were snarling at each other, in a manner anything but like that of gentlemen.[25] Zoologists or geologists, the debates were contentious and sometimes ill-mannered. Still, as Rudwick notes, "that the animosities of those rivals were so well concealed from the general audience was not only a tribute to their gentlemanly restraint and [William] Buckland's astute chairmanship, but also a small triumph for the image of science that the leaders of the British Association were striving so hard to disseminate."[26] In fact, by the mid-nineteenth century the genteel image of science had begun to fade. Science was becoming professionalized, divided into disciplines, and pursued by all manner of persons (including women), not just virtuosi and gentlemen. It needed to have its results justified by a more universal kind of witness.

Fortunately, there was another option, this one rooted in modernity. It corresponded satisfactorily with a science that was formed by the rejection of both ancient authority and the newer tradition of gentility. This other option involved individualism, and took its rise first in the humanism of the sixteenth century and then in the natural science of the same seventeenth century that fostered the gentlemanly tradition of witnessing. The origin of the tradition of individualism was most dramatically proclaimed in humanism by Montaigne and in natural science by Descartes. "Montaigne's great originality in his day was to see himself as an individual personality worthy of an attempt at definition, just as Descartes's complementary originality in the following century was to see himself as an individual mind with the right to think for itself from scratch."[27]

In principle, individualism was, and perhaps is, essential to science; the ideal of personally witnessing evidence and making inferences runs counter to the groupthink embodied in criteria of birth, blood, or faith. Instead, a different form of groupthink, based on other criteria, arises, for scientific thinking postulates a scientific community to which any person can belong out of individual conviction, publicly arrived at by a set of scientific procedures.

Does the reality match the ideal? Can it do so perfectly, or must the ideal be qualified? John Locke, attacking the crabbed world of scholastic learning, held that "the floating of other Mens Opinions in our brains makes us not one jot the more knowing. . . . what in them was Science, is in us but Opiniatrety."[28] Locke's view is in accord with Descartes's rejection of all authority except one's own reasoning. Such a view, however, does not accord with the original Baconian rejection of authority: Bacon attacked antiquity in order to make way for a newer form of authority, that achieved by the collective efforts of trustworthy scientists. It is the Baconian position, I would argue, not the Lockean, that leads to the modern theory of communicative action in regard to science, whether that action is in Habermasian or related form.

Such a Baconian-Habermasian position (which we shall shortly discuss) has had added to it in more recent times large doses of linguistic philosophy and advanced notions about the community of witnesses. The philosopher Michael Welbourne suggests what is involved when he writes, "The evidence-theorist maintains that we (implicitly) reason our way from the testimonial evidence to a belief, which may amount to knowledge if the evidence is good and the reasoning sound. I hold that the mechanism which gets us from testimony to knowledge is quite different from this. What is required of the listener is not an ability to weigh evidence justly and to reason well; what is required is an understanding of the language and the will to believe." [29]

In our terms, the witness need not weigh each piece of scientific assertion individually—there is neither the need nor the possibility of inventing all the wheels of science again. Instead, witnesses must understand the language of those addressing the issues of what is involved in reaching modern scientific knowledge, what is acceptable scientific method, and who are acceptable practitioners of that method, and they must be willing to obey the requirements of the scientific community. In addition, and here I do take a slightly different tack from Welbourne's, a witness must be able to weigh evidence and reason well, even if not about specific technical details, as a preliminary basis for being willing to be persuaded by evidence and theory.

Even though an individual has not gone through the evidence and proofs concerning atomic phenomena, for example, he or she must be prepared to believe the assertions of nuclear physicists. Acceptance of expert testimony carries with it an awareness that the physicists may debate among themselves and come to revise their data and theories. In fact, it is exactly such awareness that persuades me, as an individual, that I can, at second hand, be a witness to their witnessing. I am, I believe, still exercising individual judgment, even though it is not of the Lockean kind.

We have become more aware today of the social construction of the individual; we accept the truth of the saying "No man is an island unto himself." Thus, the emphasis on an individual's using reason to the extent possible and prescribed by scientific method must be situated in the general socialization process that is itself part of modern scientific culture. As we have noted, one of the cardinal rules of this culture is that the individual may challenge and criticize any finding or pronouncement within science; but he or she must engage in this activity within the framework of the sort of scientific community that we have been discussing. Such, at least, is the ideal, and as an ideal, it frequently and perhaps generally produces real-life consequences that match its proclamations. The new witness is potentially everyman. Instead of the gentleman, any rational individual accepting scientific method should be able to serve as a legitimating authority.

This view has often been presented in distorted form, and it has frequently been thought that a disembodied, purely rational witness is best. Free of class or group bias of any sort, such an individual would serve "objectively," reporting only the truth. He or she would be a Martian of sorts, and "it is the conception of an observer from a distant planet, which has always been a favorite with positivists from Condorcet to Mach."[30] Such an ideal witness is assumed to be totally objective.

For our own version of the ideal witness we assume instead that he or she is as objective as possible. Our witness is not from a distant planet, but from here on earth. He or she is also a member of a scientific community. Individuals make up individualism, but they are real individuals who live in an actual society, not on Mars or a Crusoe-like island. In this real setting, society itself will either be a scientific community or have within it prototypes of such a community. In scientific communities, objectivity can be neither certain nor mechanically achieved. It can only be a human aspiration, fitfully approximated and always subject to criticism and renewed witnessing.

A word more about objectivity is perhaps necessary. There may still be hard and fast adherents of the notion that complete Martian-like objectivity is achievable; if so, with such friends, one hardly needs enemies. But there are many declared enemies to the idea of objectivity, however cast or qualified. Calling on the insights of the sociology of knowledge and cultural studies, they lay siege to the notion of objectivity. Yet the defense of objectivity lies exactly in the study of culture and sociology of knowledge itself. As we have seen, science is an emergent phenomenon in modern culture. Would-be scientists are shaped by their position in society and their participation in culture. One part of that culture being science, the aspiration to objectivity follows. Of course, objectivity cannot be total, but it can be asymptotically approximated. Witnessing is one effort at approximation; every branch of science then seeks to work out its own methods of ascertaining a realizable objectivity.

A doctor working to achieve a cure for AIDS may not like AIDS victims; yet he or she will follow objective procedures in seeking an inoculation and will be required to subject the results to objective verification methods. A historian studying the decision to drop the atom bomb on Hiroshima may have to evaluate the competing claims of army, navy, and air force proponents (and may favor one or the other in general) but must seek to do so objectively and impartially and will be held to these standards by other historians. Just because two historians may come out with different interpretations, as may well happen, does not mean that they have not each operated in as objective a fashion as possible. It simply means that they are subject to the testing grounds of perspective and hermeneutics, with their uncertain results.[31]

Even given our particular defense of objectivity, stemming from the remarks about witnesses, it must be noted again that the positivist version of the purely objective witness is deeply flawed. Before the positivist view of witnesses is discarded, however, the kernel of truth must be extracted. Simply put, it is that the individual — shaped by culture and society but seeking along with others to be rational and to strive for as much objectivity as is humanly possible — must bear witness to scientific truth.

How do we extend the notion of witnesses to the human sciences? What does it mean to be a witness in its various fields? Because a number of the human sciences — anthropology, sociology, scientific history — emerged in the second half of the nineteenth century, when specialization and professionalization were in the ascendant, those human sciences from their inception have seemed to operate outside the tradition of the gentleman witness. (If we include literature in the human sciences, as I tend to — vide our discussion of narrative — class bias of the study of the classics, which requires Latin and Greek, figures as an exception.) The problem of witnessing and legitimating results in the human sciences was fairly democratic socially from the start (which is not to say that their practitioners were not drawn disproportionately from the middle and upper classes).

Nevertheless, the general problem of witnesses remains. Although Boyle-like experiments are hardly to be expected in the human sciences, creditability is at the heart of reported observations, whether ethnographic or historical. Linked as observations are to interpretation, they must be verified by witnesses. Writers of travel accounts go to great pains to convey a sense of verisimilitude to their readers, because, as the author of *Mandeville's Travels* professed, "many men trow not but that as they see with their eyes, or that they may conceive with their own kindly wits." [32]

Wit — that is, the use of reason — and witnessing are closely connected, and not just by the accident of words. In fact, so-called eyewitnesses can rarely be trusted in their accounts. Many learned witnesses pronounced the appearances in Galileo's telescope delusive. [33] As we know, our eyes conjure many strange monsters; our wits, many wild tales. In the end, however, it is Mandeville's kindly wits, or reasoning persons, who must sit in judgment on travel and other accounts and declare them true or false. Whatever the claim to seeing microbes or primitive men firsthand, it is reason (exercised by an individual who shares the tenets of the scientific community) that must decide on the validity of what was seen. And reason must do so, whether in the natural or the human sciences, in accordance with an appropriate scientific method, one of whose aims is objectivity.

Let me sum up. In the previous chapter, we secured some idea of how representative human sciences carried out the injunction to use the scientific method. Looming over all such positivist aspirations is hermeneutics. The result is that in the human sciences the witness is generally an interpreter, though going far beyond the kind of interpretation that applies in the natural sciences. Thus, the subject of the witness in the human sciences is of a different (though related) dimension from that we have encountered in the natural sciences and is best considered further in terms of the concept of the scientific community.

Scientific Community, Renewed

There is some reason to believe that the concept of scientific community was first coined by Michael Polanyi.[34] Before the concept, however, the thing itself existed in the scientific academies of Renaissance Italy and in the royal societies of England, France, and other countries. What connected the members of such societies, or scientific communities, was an agreement to think and discourse according to the rules of scientific method.

As we know, this method emerged in its major features around the seventeenth century. Historians of science have sometimes lost sight of this central fact in their admiration for the particular scientific achievements of the time: those, for example, of Descartes, Newton, and the lesser virtuosi. By mistakenly identifying particular theories as being at the core of the scientific revolution, they have lost sight of the fundamental breakthrough: the establishment of a general method for thinking and of a sociocultural framework in which such thinking can occur.

The theories have subsequently changed, been called into question, discarded, revised, or shown to be vulnerable to criticism and needful of qualification. That is exactly what one would expect from the continuing use of scientific method. But this state of affairs has led some writers on the subject to question the cumulative nature of scientific knowledge and to speak of paradigm shifts, ruptures, and relativistic knowledge. In this light, the scientific community is a form of political power, reflecting the current social arrangements and offering ephemeral, parochial knowledge. The question of a constant, consistent method is ignored.

We have already dealt with this matter, especially in a short discussion of Thomas Kuhn's paradigm in Chapter 2. We can therefore continue with the argument that places method, not particular theories, as the constituting force of scientific communities. The method, it must be reiterated, though characterized by certain constants, is flexible in its application. Indeed, it must adapt itself to the materials being studied and not be seen as a mold to be imposed on all phenomena. Physics envy is a pernicious disease. It has created many an Othello in the human sciences, leading them to perpetrate unhappy and often suffocating acts of misapplied imitation.

Scientific method has to adapt and make itself fit for the new wholes that must be studied. Facts and theories are, by the very injunctions of scientific method, demanding of criticism and revision. That they change does not cast doubt on the validity of scientific method. As Steven Jay Gould puts it, discussing Lyell and early geology, "Scientists then and now have recognized that their profession is defined by its distinctive modes of inquiry, not by its changing perceptions of empirical truth."[35]

Another author makes the same point, but less felicitously, in writing of Claude Bernard. The French historian of science Georges Canguilhem confirms our approach here when he writes, "What Bernard wanted was a way of doing research in physiology based on assumptions and principles stemming from physiology itself, from the living organism, rather than on principles, views and mental habits imported from sciences as prestigious, and as indispensable to the working physiologist, as even physics and chemistry."[36]

Scientists must be careful not to be slavishly bound by a particular kind of scientific procedure, a local variant of scientific method, that has worked well in regard to some phenomena; rather, they must view it as a tradition possibly obscuring the facts and correlative theories of a new discipline. As Gaston Bachelard, another French historian of science, wrote, "Concepts and methods alike depend on empirical results. A new experiment may lead to a fundamental change in scientific thinking. In science, any 'discourse on method' can only be provisional; it can never hope to describe the definitive complexion of the scientific spirit."[37]

What is true for the natural sciences is true for the human sciences. They are bound by the same allegiance to scientific method that we are treating as the essence of science—the combination of sustained observation, accumulation of data, experimentation, testing of results, establishing of theories, logical inferences, use of mathematics if suitable, return to data gathering, weighing and interpretation of materials and findings, and so on—although the forms and details of that method must be adapted to the phenomena under consideration. Alfred Schutz made the point before me when he rejected the claim that the "methods of the social sciences are *toto coelo* different from those of the natural sciences," and affirmed that "a set of rules for scientific procedure is equally valid for all empirical sciences whether they deal with objects of nature or with human affairs."[38]

In practice, we find the natural and the human sciences on a continuous spectrum, where borrowings, exchanges, and inspirations characterize their relations as much as the imitation that can be so dangerous. A reading of Hans Blumenberg's *Genesis of the Copernican World* confirms this statement in exhaustive depth, but even a moment's consideration of nineteenth-century biology and social science illustrates the traffic in ideas between them; consider Malthusian spectres from economics and Social Darwinism from biology.[39]

Biologists and economists may be said to be members of separate scientific communities; and in a limited sense this is true. If we focus, however, on their adherence to something called scientific method, however flawed in practice, we can view them and any who join them in this adherence as part of a single, broader scientific community. Whatever our final judgment in this matter, what is clear is that scientific method is the binding force holding together the members of a scientific community, which, in turn, is what makes possible the reception and acceptance of any real science. Over time, our earlier witnesses, whether in the natural or the human sciences, gather together in a community bound by a credo of method.

Scientific Method, Encore

The scientific community, I shall now argue, is a special form of a Habermasian truth community. As Anthony Giddens describes it: "When we say something is true, we mean we can back up what we say with factual evidence and logical argument. . . . Truth refers to agreement or consensus reached by such warrants. A statement is 'true' if any disputant faced by those warrants would concede its validity. Truth is the promise of a rational consensus."[40] Truth communities are often conceived in regard to political life. Thus, we are told that the political theorist Carl Schmitt outlined a model for public discussion as "an exchange of opinion that is governed by the purpose of persuading one's opponent of the truth or justice of something, or allowing oneself to be persuaded of something as true and just. . . . To discussion belong shared convictions as premises, the willingness to be persuaded, independence from party ties, freedom from selfish interests."[41] An ideal parliament might be construed in this fashion.

The problem with such a community is that the reality often falls desperately short of the ideal. This is one reason why political science offers so little reasoned discourse. Jefferson might argue that "truth is great and will prevail if left to herself," but truth never is left to itself.[42] The authors of *The Federalist Papers,* more realistic than Jefferson, knew that truth had to be provided with crutches — the balance of powers, and checks and balances — in order to hobble forward.

Interests, values, the heterogeneity of humans, make it difficult to achieve consensus, even with the best will in the world. Politics, as a result, is the constant stumbling toward a result that might be agreed upon (see Chapter 5). Anyone who has engaged in a discussion, say, of the Vietnam War will realize how difficult it is to secure even provisional agreement on the facts — on what happened, on whether it was a civil war, an extension of Soviet power, of Chinese power — let alone on an interpretation of the facts. Even if, for argument's sake, a consensus could be achieved, how might one build on it to make a decision about the next case — the Bosnian crisis, say? Rarely is accumulation possible.

In the next chapter, I shall be stressing the limitations of the Habermasian approach: its neglect of desire, its ignoring of passion, and so forth. Nothing, however, that I have said so far about the political should discourage us from the attempt to approximate the Habermasian ideal. Like democracy, of which it is a part, the Habermasian ideal may be a poor system, but it may be the best we have.

I highlight Habermas's communicative ideal here because I want to contrast it with a scientific truth community. In such a community the procedures of the search for truth are what provide it with a unique character: the achievement of cumulative, scientific knowledge. Although scientific method is not one unchanging procedure, it is based on a number of constant features. In the pursuit of particular scientific knowledge, these features will be found in different degrees and with different emphases and often in different balances. What are these features? Let me summarize again the most important ones here. One is the continued questioning of authority and tradition — the Baconian challenge — and the bringing of their edicts to the bar of individual judgment or, more precisely, the scientific community. The scientific community is made up of those individuals who agree to abide by the conventions and rules — the features — of scientific method. Findings must always be open to public scrutiny and future revisions; there can be no dogma shielded from constant, critical scrutiny.

The leading norm of scientific method is "universalism," to use Robert Merton's terms. By this he means that science is required to be impersonal: "truth claims, whatever their source, are to be subjected to preestablished impersonal criteria: consonant with observation and with previously confirmed knowledge. The acceptance or rejection of claims entering the list of science is not to depend on the personal or social attributes [so much for the gentlemanly thesis] of their protagonist; his race, nationality, religion, class and personal qualities are as such irrelevant."[43] We cannot reject relativity theory on the grounds that Einstein was a Jew.

Science is an inquiry, a search for facts, which are then related to theories, and vice versa. Darwin's particular definition was that "science consists in grouping facts so that general laws or conclusions may be drawn from them."[44] Scientists go about their tasks in many ways, but especially by employing observations and experiments, imposing classifications, and drawing inferences from the evidence to help in the creation of further explanatory theories.

Facts, with all their attendant philosophical difficulties, are basic. Much of natural science, as noted, has been the extension of human sensory apparatus to expand the range of relevant facts. It was the proud boast of Richard Hooke, in his *Micrographia, or Some Physiological Descriptions of Minute Bodies Made by Magnifying Glasses with Observations and Inquiries Thereupon,* that *"The first thing to be undertaken in this weighty work,* is a watchfulness over the failings *and an* enlargement of the dominion, *of the Senses."*[45] The telescope, as well as the microscope, were pioneering efforts in this widening and deepening perception of facts, and the subsequent development of X-rays, resonance devices, and so forth, were steps in the same direction.

The sensory extensions of natural science are obvious. One must reflect for a moment before coming up with comparable extensions in the human sciences. Here what we find are explorations and discoveries of new worlds, which are more cultural than physical (though necessarily dependent on physical evidence). Archaeology unearths previously unknown bones and shards. Hermeneutics — textual interpretation — discredits some old and supplies some new facts about humanity's past. Statistics create and correlate data previously unobserved — for example, about population and the spread of the human species. Psychoanalysis, though more controversial than archaeology, likewise unearths new materials — in this case, about the human psyche — and it deepens our factual knowledge about unconscious as well as conscious human thought processes, using, as noted earlier, free association and other instruments to probe the human mind.

Where possible, especially in regard to the natural world, the scientist resorts to experiments, both to test old facts, along with theories based on them, and to secure new facts. To quote Hooke again, "So will all those Notions be found to be false and deceitful, that will not undergo all the Trials and Tests made of them by experiments."[46]

Unfortunately, experiments are rarely available in the human sciences (or, if available, rarely are they ethically acceptable). When Frederick Barbarossa, in the twelfth century, reputedly tried to determine the original language spoken by humankind by isolating two children and waiting to hear what they would say, his subjects died on him. Generally, human science must seek, not experiments, but equivalents to experiments. Yet, as an ideal, experiment is present as part of a generalized scientific method.

Interpretation is another essential part of scientific method. Its role, though sometimes obscure in the natural sciences, is all-encompassing in the human sciences. What the ideal of scientific method adds to the usual views about interpretation are certain strictures as to how its results are to be judged: not by private, intuitive, unverifiable, dogmatic procedures but by a subjection to public, logical, proof-laden and arguable means.

In this sense, objectivity achieves partial realization, even in regard to hermeneutics. Objectivity, however unreachable as a totality in practice, sits well with the public nature of scientific method. In contrast to private experiences, such as trances and revelations, which are basically noncommunicable and must be taken on faith, human scientific experiences are public, and their results demonstrable, with relevant facts and explanations available for proof and rational testing.

Method and Community

The characteristics of scientific method are constantly being overlooked, certainly by the public at large and even by many scholars who should know better. With this said, we must complete our circle of thought and add what should also be obvious: a scientific community based on the acceptance of the scientific method is essential for science. There must exist a community of believers, believers in the scientific method, who test their beliefs according to it. If someone insists that the moon is made of green cheese, and refuses to believe the evidence refuting this belief, as well as the logical objections against it, nothing much can be done. Similarly, if someone believes that a certain ethnic group plots world domination or controls world capital, in spite of overwhelming evidence to the contrary, nothing much can be done (at least in terms of argument).

It is in this regard that the natural and human sciences find themselves in very different situations. In a natural science community, the number of adherents can be small. Often they share linguistic conventions, scientific jargon, not available to those outside. The technical language keeps out those deemed unqualified to understand the matters under consideration. Of course, the community is democratic in the sense that anyone who undergoes the required training and is capable of understanding the technical language can become a member. Within the community, all members are rationally equal (though generally differing in prestige). To the outside world, the community can be seen as elite.

Thus, a natural science community can follow scientific method freely and convincingly in the problems that it addresses. Is cold fusion possible? Consult the experts. Can humans be cloned? You and I have no scientific grounds on which to weigh that question (unless we immerse ourselves in the detail). Let the molecular biologists speak. They may differ among themselves, but they do so on the basis of accepted rules. Their immediate "public," as a result, is fellow scientists, not the mass of their fellow human beings.

It can be argued that the more artificial, that is, reified, the type of knowledge, the faster its rate of change. I would add that the more numerous the adherents of a type of knowledge, the slower its rate of change. Thus, a religious belief embraced by large numbers of people is likely to change more slowly than scientific convictions held by a few. The smallness of particular scientific communities, such as physicists, fosters their built-in tendency to weigh new facts and theories, to adapt accordingly, and to change rapidly.

When we think of natural scientific communities — associated with the disciplines of physics, chemistry, biology, geology, and all the rest — as sharing a common allegiance to scientific method, suitably nuanced for their own researches, we may then speak correctly of a broader natural scientific community. William Whewell, the mid-nineteenth-century British philosopher of science, caught hold of a piece of this truth when he introduced the term *science*, in place of *natural philosophy*, to embrace all of the proliferating specializations and professionalizing societies of his time.

If adherence to scientific method depends, as it must, on a scientific community, how stand the human sciences? Their condition, we must acknowledge, is precarious. Economists, sociologists, anthropologists and so forth, have their professional societies, their truth communities. Unprotected by effective technical language (with the possible exception of economics, though the operative term is "effective"), they strive to live by the scientific method adjudged proper to their fields. In almost all cases, however, their findings are vulnerable to attack by the actors whom they are classifying, describing, prescribing for, and seeking to understand in theoretical terms.

These actors *should* be a scientific community. The human sciences need humanity as their community.[47] But they don't have it. Natural scientists, all things being equal, can go about their business, gathering facts and postulating theories about, say, big bang theory, cold fusion, evolution by natural selection without having to offer their findings for review by untrained, uninformed, unscientific others. (Creationists can refuse to accept Darwinian theory, but this hardly affects the course of biological science, aside from possible funding difficulties.)

The same is not true in regard to the human scientists. They do have scientific knowledge, arrived at by methodical means, but they have no real scientific community, the mass of human beings (not just other social scientists) thinking and acting scientifically, *of the sort needed* for the useful reception of their work. What good does it do to identify the sources of ethnic conflict in the mistaken ideas and unconscious passions of human beings if the actors involved refuse to abide by the methods of factual accumulation and critical rationalism that social scientists bring to bear in the inquiry?

The individual who insists that the moon is made of green cheese can be dismissed by the natural scientist. The person who insists that the earth is flat because it is so stated in a sacred text can also be dismissed by the natural scientist; but when this believer claims that anyone who thinks otherwise is an atheist and deserves punishment, then the human scientist must take him or her seriously as an object of study.[48]

The natural sciences have a great advantage in imposing the findings of their scientific community on the rest of the populace. Natural science works; it has proven utility: it has put a man on the moon and produced a genetically altered tomato. (Questions of normative judgment, obviously, are a different matter.) In contrast, the utility of the human sciences is not obvious, except in the Lindblom-Cohen sense. Their uncertain nature is endemic to them as scientific enterprises. Even while carrying out inquiries in accord with scientific method and achieving the substantial results discussed in the previous chapter, they suffer from difficulties in comparison with the natural sciences.

One set of difficulties involves internal factors: verification, experimentation, replication, prediction, and so forth. The difficulties with these, as I have argued, can be partially compensated for by the development of acceptable methods. Another difficulty is external: acceptance of the findings of the human sciences by the greater public. Broad acceptance involves us in the problems of prescription, intention, and all the rest that we have wrestled with earlier. Without acceptance, the phenomena subvert the very effort at science and thereby vitiate the use of scientific method as such. Moreover, as I have argued, whereas both forms of science, natural and human, deal with emergent phenomena, the human sciences impose additional burdens on human inquiries.

Given these difficulties, the absence of a scientific community adequate to the needs of the human sciences renders the task of those sciences almost impossible. Is this condition surmountable? Can we even envision a scientific community that includes most of humanity, that can agree on Habermasian-like rules, that will abide by the requirements of adherence to evidence and critical rationality? The prospects look gloomy. But let us continue to look at the subject, first in terms of two possibly widening communities — civil society and the global society — and then at a community whose claims are often ecumenical but are not scientific: the religious community.

Civil Society

The velvet revolution of 1989 in eastern Europe and the subsequent fall of the Soviet Union has made the topic of civil society a popular one among political scientists. Although it would be easy to succumb to the temptation to add to the literature on the subject, our concern is limited. It has two parts. One requires us to raise the question whether science, natural or human, can flourish only in a democratic setting, which incorporates civil society. The other pushes us to ask whether civil society is a model for the scientific community required by the human sciences. In the historical development of the notion of civil society, these two aspects have been combined. Jean Cohen and Andrew Arato hint at the connection when they argue the necessity of a democratic political culture: "For it is through political experience that one develops a conception of civic virtue, learns to tolerate diversity, to temper fundamentalism and egoism, and to become able and willing to compromise."[49]

What exactly is civil society? Cohen and Arato offer a definition: "We understand 'civil society' as a sphere of social interaction between economy and state, composed above all of the intimate sphere (especially the family), the sphere of associations (especially voluntary associations), social movements, and forms of public communication." But the definition is formal. Words mainly become meaningful as they carry their history with them.

As historical accounts of civil society indicate, the term was originally synonymous with the state or political society. Its classical origins were in Aristotle's *koinonia politike* and Cicero's *societas civilis*. It was closely tied to the notion and practice of the polis. It meant citizens taking an active role in public life. It also stood in contrast, as Krishan Kumar points out, to "the 'uncivilized' condition of humanity — whether in a hypothesized state of nature or, more particularly, under an 'unnatural' system of government that rules by despotic decree rather than by laws."[50] Formulated as a contrast, the concept of civil society emphasized the human achievement of civilization, distinguishing human from beast by the use of reason. It could also be used by the Greeks, for example, to separate themselves from other people — barbarians — who had not yet reached the rational state. As a concept, civil society was a specific part of political science (one of its achievements) then and now.

It takes on a new and vastly more extended meaning around the seventeenth and eighteenth centuries in western Europe. The concept of society emerges about that time, embodying an awareness that societies can change over time and are and can be constructed anew by human beings. Society, as an artificial construct, stands opposite the individual, whose individualism is also newly constructed as an entity.

Thus, in modern times society in "civil society" has a meaning different from that which it had in antiquity. The equation of civil society and the state is broken. Nevertheless, historically, modern civil society arose under the shelter of but in opposition to the absolutist state. And it arose with two major emphases.

The first was economic. Only with the triumph of the market economy could modern civil society come into its own. When Adam Ferguson writes his *Essay on the History of Civil Society* (1767), he has in mind the material conditions of civilization, not its political organization. Private property, originally protected by the absolutist state (which civil society would eventually shake), underpinned the existence of civil society. The bourgeoisie, with its capital possessions, was the prime mover and shaker of civil society, and now it comes to occupy a space between the individual and the state.

The other emphasis was free discourse, whose realm was the public. As Anthony J. La Vopa notes, "It was in eighteenth-century Europe, and particularly in England, France, and the German states, that the 'public' first assumed a recognizably modern shape and became a powerful ideological construct. That construct was a characteristic product of the Enlightenment, and it marked one of the critical zones of intersection between Enlightenment discourse and a broad range of socioeconomic and institutional changes." [51] The "public" and its "opinion" henceforth became a tribunal before which even the state was tried.

Discourse, the rational exchange of opinion — conceived no longer as fickle expressions of partial views but as statements about unchanging universal truths — was mostly carried out in the medium of print. The philosophes spoke of the "Republic of Letters," in which all voices could be heard freely and equally. An invisible republic, like the invisible college before it that became the Royal Society, was intended to serve as the model for political as well as philosophical life. Opinion was tested and made scientific by classical probability theory, as well as in the agora of letters. [52]

Civil discourse in western Europe in the eighteenth century institutionalized itself in the salons, the cafés, the Masonic lodges, and the academies that were devoted to either science or belles lettres (or sometimes both). In such settings, truth and equality were to reign, with neither particular interest nor prestige and status to play a role.[53] Here, if anywhere, Kant's adage "Dare to Know" might be realized, and the critical, rational spirit roam without restrictions other than civility.

In the realm of philosophy, the great summarizing theorist of civil society was Hegel. As Cohen and Arato remark, "While Hegel's *conception* of civil society may not be the first modern one . . . his is the first modern *theory* of civil society.[54] In *Philosophy of Right* (1821), Hegel proclaims "the creation of civil society" to be "the achievement of the modern world."[55] Aware of Adam Smith and the emergence of the market economy, placing civil society midway between the family and the state, Hegel can be said to have set the terms for all future discourse on civil society.

Our intent is not to get too involved in the discourse about discourse.[56] For our purposes, the questions about civil society as a context and civil society as a model for the natural and the human sciences are the central ones.

The first question can be restated thus: Is democracy — the political expression of civil society — essential for the pursuit of science? A number of scholars have argued in the affirmative. Indeed, a case can be made that, historically, modern science has flourished in conjunction with democracy, that scientific community can exist only in a larger democratic community.

In my view, that conclusion is too simple. A glance at the former Soviet Union should be enough to refute it. Although the free, rational examination of scientific ideas was distorted there — as in the Lysenko affair — by and large specialized scientific communities were allowed and encouraged to produce real science. Soviet achievements in space science, fusion, and other areas support such a conclusion. In the natural sciences at least, a scientific community following rigorous scientific method existed under totalitarian rule and got on with its research.

The same was not true in the human sciences. Utility did not protect them; they were fully under the sway of ideological commands. Nothing in the way of open, rational discourse was permitted, and rationality meant only Marxist Reason. Unlike in the case of the natural sciences, the conclusion seems inescapable: the human sciences require the space afforded by civil society and the protection accorded by a democratic political system.

Effectively, this means a daunting mental jump: whereas natural science can function in terms of a bounded scientific community (thus repeating the boundedness of many of its own experiments), human science requires an approximation of the whole of humanity as its community. To put natural scientific knowledge into effect, to activate it, is comparatively easy, although social acquiescence is required; utility usually carries the day. By contrast, to consciously activate human scientific knowledge is extremely difficult, apparently requiring the assent of the great majority of humans, and they are hardly disposed to heed the calls to evidential and rational discourse emanating from a professional community of human scientists.

Civil society holds the potential of becoming a larger professional scientific community, a Habermasian truth community. In fact, civil society is impure even in conception. When identified with the free market economy, it is tainted by its pursuit of self-interest and its commitment to ethical blindness. When identified with democracy and parliamentary institutions, it all too often succumbs to the lure of rhetoric rather than truth: politicians have to get elected. In short, when the concept of civil society is detached from economic and political reality, it runs the risk of becoming simply another utopian dream.

In spite of all the criticisms that can be made of civil society, it remains a crucial part of efforts to achieve scientific community. It fosters vibrant independent associations and an active civic life, in whose context scientific communities can flourish. Civil society also serves as a model for how discourse can proceed in Habermasian terms: pursuing truth by rational argument, calling upon sifted evidence, and inviting thoughtful revision. Actually, after serving as a proper social setting, civil society appears to be modeled on a scientific community, rather than serving first as such a model.

Globalization

Throughout this book, I have been emphasizing the importance of emergent phenomena. Their role is especially important in the human sciences, making any long-term prediction in the precincts of the latter highly precarious. We have seen, however, how a free market economy, capitalism, and now civil society all benefit from being viewed retrospectively as emergent entities. With this established, our task now is to seek a prospective perspective and ask: Can we glimpse a new form of scientific community, favorable to the human sciences, emerging in the centuries ahead?

The provisional answer appears to be yes; the form of community, global. Optimism must immediately be dampened by our recognition that the process of globalization is at present vague, and its future development uncertain.[57] As an emergent phenomenon, it necessarily bears unintended consequences.

We must start with what we know. Globalism is coming into actuality as a result of a number of correlative factors: our thrust into space, which imposes on us an increasing sense of being in one world — "Spaceship Earth" — as seen from outside the earth's atmosphere; satellites that link the peoples of the earth in an unprecedented fashion; nuclear threats from weapons and utility plants, showing how the territorial state can no longer adequately protect its citizens from either military or ecologically related invasions; environmental problems that refuse to conform to lines drawn on a map; and multinational corporations that increasingly dominate our economic lives.[58] Much could be added to this list: music, both popular and classical, that has found a global audience; computer networks, linking individuals and firms; television and film, which display the same visual images across the globe; and the human rights movement that has become everywhere the conscience of all peoples and regimes (even when spurned).

What is especially important is the synchronicity and synergy of all these factors, making for a new whole. Both extension and depth could be added to the list, and detail provided, but just the bare list invites an obvious conclusion: we are entering a new phase of human history, which can usefully be labeled "the global epoch."

What are the implications of a global epoch for the human sciences? Earlier thinkers, too, such as the philosophes, boasted of being citizens of the world. The Republic of Letters, it was thought, knew no boundaries. But whatever Voltaire, for example, might say, he was a typical eighteenth-century Frenchman, limited in his knowledge of human nature and inhabiting a Republic of Letters more or less restricted to a literate elite in western Europe.

What is striking about the present globalization, in contrast, is that the factors outlined above are having a more or less immediate impact on all peoples of the globe. Willy-nilly, for better or worse — and it may well be worse in many ways — humans are, or are on their way to becoming, denizens of a global community. Multinationals increasingly order their work life and dominate their consumption patterns. SCUD missiles descending on Jerusalem are seen on the evening news in Jerusalem, Damascus, New York, and Sydney. Our existence has become synchronous. Our space has become almost equally collapsed: leaving Tokyo on one day, we arrive in New York the same day.

The implications of what is happening are enormous, even if shadowy. Although globalization is not inevitable — at least its effects are not — there do seem to be strong tendencies at work. If these tendencies persist, more and more of human life is likely to slip out of local control (local can also mean the nation-state). One result already, perhaps predictably, has been an anxious return to older verities (or so perceived) and a resurgence of local affiliations, for increased globalism is necessarily in tension with increased localism. As the possibility of global community increases, the return to local community takes on greater allure.

The history of the next few decades will likely be, in large part, the story of how local communities deal with the unavoidable impact of the factors of globalization. Each local history will be different, though dealing with the same global forces. Ideally, multiple identities will be forged whereby individuals will retain ties to families, even tribes, regions, states, and strengthen their ties to the globe.

Will the global community be civil society writ large — and more equitable and broad-minded? There is some reason to hope so. It is certainly true that natural science, say, physics, is the same for a Japanese, a Russian, an Indian, an American, or a European practitioner (although style of practice may vary). The particles in particle physics do not respect national or ethnic boundaries. Is the same true for human rights? As an intrinsic aspect of civil society in its political garb, civil rights are now being claimed for all humans. That claim is being contested, either by competing "globalisms," such as Islam, or by local customs. Clearly, the sort of globalism to which we are referring is neither a certain development nor one predictable in its details.

Nevertheless, collapsing a huge argument, we can speculate that if human sciences can truly flourish solely in the setting of a civil society, a global community would mark a potentially enormous step forward. Globalism itself, as an emergent phenomenon, might then give rise to new or drastically revised human science. The new science, if it were to arise, would allow us better to understand the new happenings.

For a hint about the possible shape of progress, let us turn to John Locke, that incomparable pioneer in staking out the grounds of true knowledge, who presents the experience of the king of Siam. "To a man whose experience has been always quite contrary, and who has never heard of anything like it, the most untainted credit of a witness will scarcely be able to find belief. As it happened to a Dutch ambassador, who entertaining the King of Siam with the particularities of Holland . . . told him, that the water in his country would sometimes, in cold weather, be so hard, that men walked upon it, and that it would bear an elephant, if he were there. To which the king replied, 'Hitherto I have believed the strange things you have told me, because I look upon you as a sober fair man; but now I am sure you lie.' " [59] The king of Siam was thinking as a good scientist: empiricism spoke against the ambassador's story. Only by an expansion of shared experience could the king make a more informed judgment. With globalization, experiences are becoming sharable by all humans.

The scientific community is, in principle, potentially becoming a humanity that is the subject of and the believer in both the natural and the human sciences.

The Other at the heart of the hermeneutic approach represents the hand that must grasp and be grasped to form part of the hermeneutic circle. Now we can see that the concept of the Other has been largely the result of incipient globalism (for globalism has a history of its own, in which the New World discoveries play a major role). The king of Siam and the Dutch ambassador were others to each other; nowadays, they are increasingly members of the same global community. Here, then, is an emergent phenomenon that gives some grounds for optimism concerning the sort of community required by the human sciences.

Alas, we must take note of the negatives. The situation is more favorable for the natural sciences; even Islamic fundamentalists join hands with American fundamentalists in accepting the principles of computer science and chemistry. But whatever empirical practice and theoretical rationality is brought to bear in these areas can be rigorously curtained off from the domains of the human sciences. Christian fundamentalists using the Internet may deny that the earth has existed for more than six thousand years. Similarly, Islamic and Jewish fundamentalists may remove the female sex from the category of humans who have human rights (this is how Westerners might see it). No heed need be paid by them to either the assertions of fact or the rational arguments about human prehistory offered as knowledge by human science.

The human sciences can truly succeed only if their subject humans are prepared to live by the admonitions of scientific method and a Habermasian truth community. Even a global truth community would have its share of problems (not least of which is the inequity of North-South economic relations). All we can say at this point is that a global scientific community appears essential for the human sciences (as well as the natural sciences), it exhibits some glimmer of rising, and it looks shadowy in detail.

Religion

A community other than the scientific rivals its claims and, indeed, goes well beyond them: the religious community or communities. In fact, the religious community existed long before the scientific community and has been far more extensive in its domination.[60]

In talking of religion, I will necessarily be generalizing and constructing a kind of ideal type. Religion comes in an incredible array of shapes and expressions. So, too, the list of needs that may be satisfied by religion, in its various definitions, is extraordinarily long, covering all aspects of human existence. Religion offers solace or solution for the human desire for transcendence, especially in regard to death; explains the existence of suffering and injustice; binds people together; offers rituals and ceremonies that embody its various claims; gratifies, at least in imagination, all sorts of other wishes, for milk and honey, sloe-eyed beauties, wings. But we need not go into detail. Our specific, limited question is whether religion also offers acceptable truth statements. Does it, in fact, create a rival "scientific" community whose claims are to be given credence? Or does religion, while satisfying so many needs, exist in spite of being "untruthful"? (More than one saint has said, "Believe even though—or because—it is absurd.")

Most major religions claim to have witnesses, to test miraculous events, to place events, both mundane and miraculous, in a structure of theory, to offer long-range prediction—in short, to appear as a religious version of a scientific community. As such, they claim validation of and adherence to their truths in rivalry with the communities of the natural and human sciences. How shall we judge such claims? We could pussyfoot about the subject, but in matters of fact, religious claims are insupportable. On any rational grounds of persuasion, they are literally unbelievable. As David Hume pointed out a few centuries ago, much of religious belief requires us to dismiss the findings of the physical sciences and to suspend their lawlike explanations. Much religion is based on revelation, yet the witnesses to sacred discourses would hardly stand up to the most perfunctory judicial testing and probing. Miracles and similar phenomena are regularly reported but, like the tricks of magicians, do not pass the test of strict scrutiny.

All of what has just been said is the stuff of philosophy and science—and is denied by true believers in religion. The needs that religion satisfies are too strong to be doused by reason and empiricism.[61] Survey after survey show that large numbers of people are either ignorant or unaccepting of scientific knowledge and are completely convinced by the religious alternatives. In a typical opinion poll of Americans, presumably members of a scientifically literate society, "a majority of American adults do not know that humans evolved from animal species or that the Sun and Earth are in the Milky Way galaxy. And one third think that humans and dinosaurs existed at the same time." In the opposite direction, a recent poll finds that "64% say religion can answer all or most of today's problems," 59% say religion is very important in their lives, another 29% say that it is "fairly" important, and 70% belong to a church of some sort.[62]

Clearly, a scientific community suitable to the development of the human sciences does not exist and, further, is hampered from coming into being by religious belief. The natural sciences, for reasons already spelled out, can exist and progress irrespective of general acceptance or knowledge. The human sciences either cannot or have great difficulties in doing so. Yet the nature of their phenomena seems to demand a sustaining scientific community not limited to professional associations.

This summary is stark. It needs to be added that religion, like science, is usefully viewed as a human effort to deal with what Hans Blumenberg called the absolutism of reality, that is, the situation faced when prehistoric ancestors of *Homo sapiens* came down from the trees and confronted the struggle for existence with instincts that were no longer adapted to the environment.[63] The origins of culture lay in this situation. Seen in this context, religion, and myth are great human constructs, tremendous inventions of the human mind, enormous forces of creative imagination.

Religion gave, and gives, humans a means of coping with both the natural and the human surround. It answered the human need to predict, and thus control, the cosmos and human society by assigning powers to the gods. In the end, in its monotheistic versions, it seems to say, in effect, that all is in the hands of God, whose purpose is inscrutable (but whose messages can be interpreted by properly authorized representatives). The transcendental yearning is ultimately answered by a response that transcends human understanding.

Secular science, vulnerable to the charge of hubris and subject to the requirements of ordinary reason, springing up from its roots in religion, has sought to substitute scientific method for religious conviction. Accumulating slowly over the centuries, it, or, more precisely, natural science, has had spectacular successes since the seventeenth century. In area after area, it has emancipated itself from the religious community and affirmed its own truths. It has offered viable alternatives to the question of origins in the form of cosmological and evolutionary theories. Comte was right: in physics, chemistry, biology (and, I would add, geology), natural science has freed itself from religious modes of thought and triumphantly (at least in the realm of theory) advanced its own.

The human sciences have attempted to do likewise. Paradoxically, their hyperbolic claims, as in the case of Marxism or even Comtism, have obscured their legitimate accomplishments. Still, even with their accomplishments acknowledged, the human sciences have failed to have their alternate mode of thinking widely accepted. Although religion's validity as a truth claim has been undermined, this finding has little affected the phenomenon itself: people still espouse and act in accordance with religious beliefs, rather than the laws and methods of the human sciences.

There is little likelihood that the scientific community necessary for the human sciences will manifest itself in the near future. Such a community would accept fully, but continue to test, the findings of human science in regard to the cultural evolution of the species; it would acknowledge the observations of the human sciences, especially the social ones, about ethnic, racial, and similar matters and act accordingly. The truths, however spotty, exist; the community does not.

For the human sciences to develop, the average person would have to become knowledgeable about the human sciences. Such a development would be an emergent phenomenon, a necessary accompaniment to the emergent human sciences. It would be a result of the further evolution of the human species, adapting in new and even unforeseen ways to the environment that it itself is creating. Neither certain nor predictable, such a development exists merely as a remote possibility.

7 / "Da Capo," or
Back to the Beginning

O
ur conclusion is necessarily inconclusive. This befits the uncertain sciences. Humanity is still in the process of cultural evolution. Many emergent phenomena are still in the womb of time. In pursuing our inquiry into the nature and meaning of the human sciences, we have had to proceed as if in a fox hunt, not chasing our quarry in a straight line but over hedges and ditches and through the trees.

To pursue the metaphor, however, we must not lose sight of the quarry for the trees. The one clear call sounded by the hunting horn is that of scientific method. That method is not a single, fixed procedure but a varying scientific form of the Habermasian truth discourse; it requires a correspondent scientific community as well as a specific suitability to the particular human science or sciences to which it is applied. Here, it would seem, we have reached a resting place. Yet there are problems with the Habermasian truth discourse as a model, which must be faced squarely.

The first is that the German philosopher has in mind primarily a political community, rather than a scientific one. In this regard, his faith in reasoned discourse seems naive in its underestimation of desire, passion, and fantasy. His response would be that his recommended community tries to overcome these human traits by setting up boundary conditions of reason to keep them out. Applied to politics, this seems unrealistic. In any effort to extend Habermas's speech community to all of humanity, the same strictures would apply forcefully.

Another problem arises out of critiques in literary theory. We can sum up this line of criticism by quoting Stephen Greenblatt, who writes, "When, in *Knowledge and Human Interests*, Jürgen Habermas called for clear and unimpeded communication, I thought the goal was utopian, and I think so still. The literary theory of the past decades has insisted with extraordinary trenchancy and power on the complex, ineradicable mediations between sign and referent, and textual pleasure is bound up with a capacity to tolerate ambiguity, irony, and indirection."[1] In short, communication is necessarily unclear and is impeded or enriched by problems inherent in language. As a result, all of the problems of hermeneutics — and the challenges of translation — need to be faced directly rather than set aside in a Habermasian model.

Habermas could defend the utility of his "utopia" by claiming its function as an ideal to be asymptotically approached. Indeed, his discourse model does apply with relative success to scientific communities studying natural phenomena. Well might he invoke, for example, Newton as a witness, who declared during his dispute with Hooke, "Your Designes and myne I suppose aim both at the same thing wch is the Discovery of truth and I suppose we can both endure to hear objections, so as they come not in a manner of open hostility, and have minds *equally inclined to yield to the plainest deductions of reason from experiment*" (my italics).[2]

Utopias leave little room for change. Indeed, if everyone were as rational as Newton's image of himself, a certain stultification might result (as in the Royal Society under his dictatorial presidency).[3] Further, as Mario Biagioli subtly insists, "the situation that results from everyone being willing to learn the 'other's' worldview would not be characterized by a perfectly ecumenical and consequently totally rational science, but by the absence of *different* groups, disciplines, paradigms, and — consequently — by the absence of science itself."[4]

Yet, in spite of the limitations and oversights, there is insight to be derived from Habermas's inspiration. I have labored in that spirit, seeking to understand truth discourse in both natural and human science while focusing on scientific method as the form in which truth is created, constructed, and tested. Only after achieving some understanding of the nature of truth discourse have I sought to extend this model to the political realm, to the human sciences at large, and to humanity itself.[5]

The notion of a truth community is a useful guide as we inquire into the future possibilities of the human sciences, with the suitability of such a community to be found on a spectrum running from the aspirations of positivism to the inspirations of hermeneutics. A comment made about Robert Merton captures much of the matter: "Merton's version of the belief in the ultimate unity of sciences is far from the extreme of positivism, as it rejects the validity of copying natural-science methods and procedures. But it is also far from the opposite extreme of radical humanism, as it accepts the validity of general scientific standards for the domain of man, society, history."[6]

Seeking to avoid the extremes, we have been hanging on to a central thread: scientific method in a scientific community. That method embodies one of humanity's best efforts to adapt rationally to its world. That world is constantly changing, both independently of humans and partly because of their activities. Humans are animals that have lost many of their instinctual guidelines and must increasingly survive by the use of their wits. The symbolic activity of humankind both permits and fosters the use of those wits, but it also causes the species to go seriously astray. The clear fact is that humanity is at least slightly mad, much given to fantasy, subject to erratic bouts of anxiety, and mainly irrational in behavior and thought. Reason is a straw, or at best a thin plank, stretched across the abyss of human existence.[7]

In the shape of science, with all the misuse to which that approach can be subjected, reason represents a high aspiration to know reality. We seek scientific knowledge about both natural and human phenomena, with the human necessarily being part of the natural, and the natural conceivable only in human terms. Scientific method is the shared form taken by reason in what has all too often been an artificially divided domain.

In spite of the overwhelming irrationality of humanity, the achievements of science have been extraordinary. Those in the natural sciences are evident, but in actual life they have created perhaps as many problems as they have solved. The advances in knowledge in the human sciences have also been of great significance, but they have had limited effect in ameliorating the human condition (including the problems created by the natural sciences).

The reason often given for the inadequacy of the human sciences is that they are latecomers to the realm of science. There is some truth to this explanation. In earlier days, appeals were heard for a Galileo or a Newton of the social sciences. But it is now clear that the human sciences differ from the natural in that the acquisition of knowledge takes place in more piecemeal, less giant-step-like fashion. It is also dogged by two more serious and, I believe, fundamental problems: one, the absence of the requisite scientific community, humanity itself, and, two, the limitations to its positivist-like accumulation of knowledge.

Is there an alternative to humanity as a scientific community? I believe that a precarious affirmative can be given to this question. Before going on to this final challenge, however, at least one other vessel of knowledge needs further examination: the arts.

Music, Visual Art, and Literature

There are many other ways than the scientific by which the human spirit expresses itself. Music, visual art, and literature are a few, and they pose particular problems to the inquirer after truth. Still, a few words about them will help place in context our findings about the human sciences and what we will be saying about consciousness.

It is generally asserted that the arts are not cumulative, in contrast to the sciences. The arts, instead, are believed to have traditions, with all traditions nowadays being seen as essentially of equal worth. In this view, the tradition of Greek drama is equivalent in value to Elizabethan drama, each being merely different expressions of the human spirit at different times and in different settings. Shakespeare's knowledge of human nature is no greater than Sophocles'. By implication, then, human nature remains essentially the same, which is why we can still appreciate Sophocles and Shakespeare.

There is a good deal of truth to this view. Artistic products are, in fact, judged in terms of a tradition — for example, a Piet Mondrian painting takes its meaning from the fact that its object is to cleanse the canvas of the clutter of its predecessors — but the artistic tradition cannot be said, generally, to be cumulative in the way the search for scientific knowledge claims to be.

Nevertheless, the view that all artistic products have equivalent value needs to be qualified. If one looks at Western classical music, what does one see? There is, first of all, the invention of musical notation. This accomplishes in music what writing does in literature. It permits the transmission of musical ideas in other than oral form. It allows for a new kind of accumulation within the tradition. After it was invented in the Middle Ages, harmony, counterpoint, and other devices developed that allow for an extraordinary complication of composition.

There is also technical development in instruments. Specialization in woodwinds, percussion, brasses, and strings takes place, allowing for a previously unknown range of effects. The symphony orchestra comes into existence, specialized and stratified according to instrumental choirs and eventually led by a conductor. With such a unified "instrument" (for so we can regard the new organization), composers like Haydn and Mozart could create music of a level and complexity unknown earlier.

Can it truly be said that no accumulation has taken place — that Mozart's music is not a greater accomplishment than the piping of a shepherd? True, tastes differ, and some may prefer the panpipe. In this matter of reception, a difference from accumulated scientific knowledge suggests itself. But a deeper potential difference between accumulation in music and in science springs to our attention. The symphonic form as we know it, may be exhausted. The sciences, though subject to great shifts, do not seem to face exhaustion.[8]

What I am saying is simply that the arts do not accumulate in the manner of the sciences—the arts seek a different result, a different kind of truth—but having acknowledged this fact, I must enter the necessary qualifications and admit that the arts are constantly subject to interpretation in a way and to a degree that the sciences are not. Constant interpretation, the hermeneutics of the arts, moves in the opposite direction from the positivist accumulation of the sciences.

In the end, the human sciences, too, may be subject to hermeneutic ephemerality, and we must recognize as well that the search for truth by means of scientific method neither exhausts nor satisfies fully the human attitude toward experience. The arts, along with myth and religion, speak to enduring human needs. Fantasy must take its place alongside reason. Ecstatic feelings coexist with sober ratiocination. All of these aspects of humanity's encounter with reality constitute legitimate parts of historical consciousness. The continuing dilemma of the human sciences is that they must take such aspects of humanity as facts—and yet seek to transcend these facts by placing them in a scientific context.

Consciousness

Some thinkers persistently pooh-pooh changes in mentality, stressing only the advances in material life and in technology. This deprecation appears to accord well with the downplaying of intellectual and cultural history currently fashionable in some of the groves of academe. Typical is the comment of Fernand Braudel, so brilliant in his own field: "If we called on Voltaire at Ferney, we should find him to all intents and purposes our contemporary *in his ideas*, his intelligence, his passions; but all the details of material existence, even the care which he took of his person, would astonish us" (my italics).[9] His ideas the same as ours! Did Braudel really believe that Voltaire's mind was filled with the ideas of Darwin's theory of evolution by natural selection or Freud's theory of unconscious mental processes? That the follower of Newton was also acquainted with quantum mechanics, indeterminacy theory, and Einstein's theory of relativity? Is Braudel simply ignorant of science while wonderfully informed about the minutiae of everyday life?

If so, the same excuse cannot be made for a valued colleague of mine at MIT,

Michael Dertouzos, who seems to suffer from a partial amnesia when it comes to mental contents. He is reported to have said that "technology has shifted dramatically . . . but we have not changed. You are the same human being. You may watch TV and eat with a spoon and fork, but you have the same brain, same musculature, *the same aspirations and beliefs you had 1,000 years ago*" (my italics).[10] On the contrary. Aspirations change along with ideas. For example, in the year 1000 the idea of progress and the desire for it were nonexistent. In 1776, when Adam Smith wrote the *Wealth of Nations*, the notion that the poor might aspire to widespread consumerism was a faint emanation from the pages of his book. The very concept of happiness was an eighteenth-century invention — "a new Idea," as Saint-Just claimed in 1793.[11] The same would be true for beliefs, although the concept is vague and perhaps overlaps with ideas.

Such bright people utter such fatuities partly because of the lure of universals. A reminder of what is involved comes from evolutionary psychology and sociobiology. "These scholars hold that certain universal mental principles guide our behavior — that we are, in effect, prewired to act in certain ways because that is how we have been molded by millions of years of evolution. Thought processes, they maintain, and the complex patterns of behavior those thoughts engender, are just as programmed into our species as are bipedalism, hairlessness, or the ability to communicate by language."[12] The existence of universals can be no excuse, however, for ignoring changes in ideas, which arise cumulatively from our ability to use language. It is our symbolic activity, along with our technical abilities, that underlies cultural evolution. At the heart of that evolution sits our consciousness, itself an emergent property, now taking the form of historical consciousness.

What possibilities for the future of the human sciences are contained in the development of consciousness? Can the concept of historical consciousness, that is, accumulated consciousness, offer a possible escape from the impossible requirement for a scientific community composed of most of humankind? It is a conundrum that we must gradually unroll.

The general idea is obviously not entirely new. Georges Gusdorf, in his magisterial eight-volume work on the human sciences, announces at the beginning that "the most urgent problem in the domain of knowledge would be that posed by the absence, still today, of a logic and epistemology of the human sciences." His conclusion is that such a logic and epistemology must be reached historically, that is, as part of "la conscience humaine."[13] Norbert Elias builds an entire book, *The Symbol Theory*, around the thesis that with humans there has been a shift in the balance of learned and unlearned (that is, genetically fixed) forms of conduct and, consequently, an increase in consciousness in the course of evolution.[14] And before them there was Hegel.

I do not, therefore, claim a unique emphasis on consciousness. Rather, it is the way the importance of consciousness emerges from the context of this inquiry into the human sciences that must be considered significant. With Gusdorf, consciousness seems to end up being equated with the history of *mentalités*, a modern version of historicism and cultural history. With Elias, the social sciences and the problem of scientific method are given little attention. With Hegel, consciousness is viewed as having a foreordained end.

The view of consciousness being advanced here seeks to build on the work of many others.[15] I am emphasizing the way consciousness can be envisioned as supplying the sort of scientific community whose lack keeps the human sciences from being solidly scientific. Consciousness is, ideally, always being constituted and judged in terms of scientific method, although that hardly composes the whole of its content, as our observations on music, art, and literature have shown. Such consciousness is not doomed, or promised, to a certain goal, for it is rooted in evolutionary needs and tied to emergent phenomena.

I have been driven to this conclusion because our inquiry into the existing social sciences, as a subset of the human sciences, suggests their limited ability to develop further in the direction of so-called positive science. Economics, sociology, and the others will certainly change, and they will produce work of value, especially along the lines already indicated. Such change will deepen our consciousness of what is involved in human social behavior and consequently our behavior as well. As one philosopher remarked, "Since consciousness, unlike earthquakes and diseases, is constituted by the beliefs you have about it, changing what you think changes the phenomenon itself."[16] But such change will not provide accumulated laws and generalizations or much direct additional control over social phenomena. We must turn in another direction.

One type of inquiry provides the context for all investigations into the human sciences, including the social. It is itself a form of social science; though with methodological limits, it transcends itself. I am referring to history. History is the counterpart in the human sciences to evolutionary theory in the natural sciences. Unlike the evolutionary theory, however, its subject is cultural evolution. Unfortunately, though sharing the belief in chanciness at the core of Darwin's theory, history has no precise understanding of selection as a driving mechanism. As a result, history leans more toward the hermeneutic than toward the positive approach. Thus, the emphasis must be on consciousness.

What, in fact, is this consciousness of which we speak? From what does it arise? It is, from the point of view of science, an evolutionary development. It is not implanted in humans, fully blown and unchanging, by some god. It is not a unique possession of humanity, a Cartesian soul, but rather the result of a fitful process of development in the natural world. Although we know much about it, human consciousness is the "last surviving mystery."[17]

Our starting point must be material life (unless we abandon science and find our answer in divine creation at a single moment in time). Consciousness emerges out of cellular connections over aeons of time (which does not preclude mutations or jumps within that emergence). It can be seen in other animals besides humans. What is unique to the human species is its degree, which depends on the development of a special kind of language and symbolic activity. Other species, like ants, may have highly developed chemical systems of language, but they do not have the cumulative possibilities present and realized in human culture.

I have spoken of consciousness as restricted to animals. An early theorist, the philosophe Denis Diderot, in *D'Alembert's Dream*, went further, rooting consciousness in materialism and pursuing his logic relentlessly by allowing it even in stones. Consciousness is, of course, a matter of definition and the ability to perceive what is interior to another entity. Being aware of something within oneself; being conscious of an external object, state, or fact; being in a state characterized by sensation, emotion, volition, and thought—these are some of the traits that the dictionary ascribes to consciousness.

It is fitting to instance the progenitor of evolutionary theory, Charles Darwin, in this attempt to explain consciousness. As he cryptically jotted down in his notebooks (around 1837), consciousness "bears . . . relation to time & memory." In another note he deepened his reflection, observing that "a Planaria must be looked at as animal, with consciousness, it choosing food — crawling from light. — Yet we can split Planaria into three animals, & this consciousness becomes multiplied with the organic structure, it looks as if consciousness as effects of sufficient perfection of organization & if consciousness, individuality."[18] In the notebooks, privately, and in *The Descent of Man*, publicly, Darwin continued his ruminations on consciousness. Indeed, he must be seen as a pioneering figure in psychology, for he even approached an awareness of the unconscious and its importance. But psychology was not Darwin's main interest, and he turned his notebooks over to his protégé George Romanes. Romanes was very clear that, as Howard Gruber puts it, "consciousness evolved gradually, making its first rudimentary appearance in the coelenterates (e.g., jellyfish, coral)."[19]

In its earliest stages, consciousness appears to be characterized by sensation, volition, and something perhaps resembling emotion and thought. It is an emergent property, arising unexpectedly from the movement of molecules within the animal and having no precise physical location (cutting up the animal will not disclose it).[20] It is still a very low-level phenomenon. What is clear, however, is that consciousness is on a continuum; it is shared by a number of animals and is based on the idea that physical evolution moves toward mental evolution.

With humans, consciousness seems to take a quantum leap forward. In adapting to the environment, the human species learns to use symbols and to acquire a substantial past and future sense and, possibly, hindsight and foresight. It begins to acquire culture, a symbolic accumulation of experience as regards both nature and fellow species members. We must pay attention now, not just to the brain, but to the mind, which is filled not only with firing neurons but also with meaning.

Advanced consciousness is not a reified Hegelian entity; it emerges from the nervous system of individuals who inhabit cultures. It was in 1843, a decade or so after Hegel, that Emil Du Bois-Reymond demonstrated that electricity, not a vague soul-force, traveled through the nervous system and, in some then-mysterious biological fashion, produced consciousness. Conscious individuals must engage in a supersophisticated learning process, which can go haywire at any moment.

Human learning is embedded in institutions that serve to transmit and to transform culture, for human consciousness and memory die with the individual. Consciousness is stored in institutions—the family, the tribe, the religion, the centers of learning—where it can survive and become accumulated. Learning is essential. Unlike the memory boards of computers, individual consciousness cannot be lifted out of the brain and placed in another body, at least not yet (if ever). It can only be "off-loaded," not onto disks, with perfect replication, but into the brains of other individuals by oral transmission, books, and other means—including computer files.

When did conscious mental activity of the sort we are describing arise? Unfortunately, although fossil remains may inform us about brain structure, they cannot reconstruct brain content based on that structure. We must guess. As far back as we can go in prehistory, early humans are found with tools. Do primitive flints indicate an advance in consciousness? I suspect so but cannot tell with surety. The best guess seems to be that a form of consciousness that we would recognize and acknowledge is first made evident in the cave paintings of thirty thousand or more years ago. Here, in art, human consciousness akin to self-reflexivity seems to proclaim itself.

The next or accompanying site of consciousness appears to be in religion, manifested in burial remains, which suggests a sense of future life. The variations in the treatment of the bodies and in the materials buried with them also suggest the existence of stratified society even in the earliest times, along with the consciousness required for such a nongenetic social arrangement.

An intriguing thesis has been advanced by the psychologist Julian Jaynes. Based on laboratory studies and archaeological findings, Jaynes hypothesizes that ancient humanity was unable to introspect as we do today; instead, its members experienced auditory hallucinations, which, coming from the brain's right hemisphere, told them what to do in times of stress and novelty. He calls this ancient mentality the bicameral mind. Internal voices were mistaken for external instructions from the gods. Inner and outer reality were confused, and there could be no true self-awareness, no human consciousness. In what might be regarded as Jaynes's version of a "Just So" story, he goes on to argue that the breakthrough to consciousness occurred gradually, starting about three thousand years ago. It was a form of evolutionary adaptation. Whether right or wrong in its its details, Jaynes's thesis is bold and suggestive and probably points in the right direction.[21] It is a very imaginative attempt to bridge what he calls the "awesome chasm" between inert matter and conscious being.

Ought we also to look at Karl Jasper's notion of axial periods, including the appearance in the fifth century B.C. of a number of the great world religions? Does this mark a great leap forward in human consciousness? Can we look at the discovery of new worlds in the fifteenth century and at the strides forward in the awareness of the Other as major steps onward in consciousness? What about the development in the nineteenth century of the scientific approach embodied in the field and fieldwork of anthropology? Can we also so regard the earlier Copernican revolution, which made us conscious of our peripheral place in the planetary system? And the subsequent changes in perspective embodied in the so-called scientific revolution from the seventeenth century onward? These all represent changes in human consciousness that are accumulated in our institutions, ideas, and sense of personal identity. They are the fruit of our cultural evolution and are known to us by our historical sense, which is itself incorporated in our culture. With the globalization taking place today more sharing of our consciousness is possible, and such sharing would be the foundation for a true human-scientific community.

Historical Consciousness

The great philosopher of consciousness as a historical accumulation was Hegel. He was lyrical on the subject and held steadfast to his vision. He was also arcane, erecting his Easter Island–like figures of Reason, Idea, Spirit, Freedom, and the Absolute. Declaring that Freedom is present when one's existence depends solely upon oneself, Hegel writes that "this self-contained existence of Spirit is none other than self-consciousness — consciousness of one's own being. Two things must be distinguished in consciousness; first, the fact *that I know;* secondly, *what I know.* In *self* consciousness these are merged in one; for Spirit *knows itself.* It involves an appreciation of its own nature, as also an energy enabling it to realize itself; to make itself *actually* that which it is *potentially."* [22]

Hegel readily admits that this definition is abstract. It is also idealistic, with Hegel attaching his progress of the Idea to an informed voyage through the great world religions. His image of the Egyptian sphinx as symbolizing Reason emerging from the brute body is one of the great poetic expressions of his Idea. In spite of his often crabbed prose, his teleology, and his overabstraction, Hegel is a towering figure in the land of human consciousness. He was aware of the centrality of the idea.[23]

If Hegel was the great philosopher of consciousness, Karl Marx was its great sociologist. Marx was inspired by Hegel but stood him on his head and rolled him off in a materialist direction. From Marx, we can pick up one piece in the ongoing development of knowledge about consciousness. It involves Marx's focus on the question What are the determinants of consciousness? His general answer is, labor and economic conditions (in fact, Hegel earlier had also glimpsed their importance). Making the subject concrete, where Hegel had left it abstract, Marx broke up consciousness, so to speak, into class consciousnesses: we are aware of what our class permits us to know (except for those few rarefied souls who can rise above their origins).

Marx was the pioneer of the sociology of knowledge. He himself spoke of false consciousness, implying a true one, supposedly immune from the distortions of class — that is, a scientific consciousness. Few believe any longer that Marxism is science rather than ideology. Nonetheless, almost all the work concerning the social conditions of consciousness derives from Marx, whether for or against him. Max Weber, Karl Mannheim, the Frankfurt School, Antonio Gramsci and his notion of hegemony, Michel Foucault and his archaeology of knowledge, the current theorists of the social construction of science — all bear the imprint of the first sociologist of consciousness.

Sigmund Freud makes the third member of the triumvirate. He is the great psychologist of consciousness, who added the unconscious to it.[24] Consciousness is the surface of our mental processes, shaped by what goes on in our depths. Only by analyzing the unconscious and the processes by which it works can we truly understand the workings of the conscious mind, or what Freud came to call the ego.

We need not go into details about Freud's work and the surrounding controversies. After him, however, human consciousness could not be looked at in the same old way. Now we know that the rational part of the mind sits on top of a great deal of the irrational (to use a particular spatial metaphor) and that most of our consciousness is intimately related to our unconscious. Thus, myth is not merely the opposite of reason (although Freud betrays his own insight by often treating the pair as battlers to the death) and consequently fated to disappear; rather, myth is rooted in the workings of our unconscious and finds its way recurrently to our conscious being.

Here our special concern is with historical consciousness—the culturally accumulated self-awareness about our self and society—especially as derived from the results of the use of scientific method. Because in the human sciences the hermeneutical is a necessary part of that method, in seeking to understand ourselves and the knowledge that constitutes us, the reflective lens ground by Hegel, Marx, and Freud, and others, has become a requisite instrument of perception.

Self-awareness, too, has a history. It captures a consciousness about consciousness that has not always existed. We have already noticed the way the self, around the seventeenth century, required society in order for it, conceptually, to distinguish itself from what it was not. Even earlier, human beings had begun to accept themselves as an object of study, first in anatomy, then in ethnography, thus distancing themselves from their selves. Yet in this distancing, humans, while potentially also alienating themselves, also have come closer to themselves.

As in so much else, Hans Blumenberg adds insight. For him, it is the Copernican revolution that frees the self to truly see itself. I quote him at some length: "The naturalism that gathers man's position in the world from the fact that the Earth, as his home, occupies the center of the world, is pushed aside by the reflection that the 'center' of the world, as the reference point of its rationality, is wherever man finds himself. This is [a] suspension of the indicative role of cosmology for man's self consciousness. . . . [God, in this Renaissance awareness] has given Adam no fixed abode, no fixed form, no definite task, so that he himself can choose and adopt an abode, an appearance, and a task. . . .

Man is supposed to determine his nature himself on the basis of his freedom; which now means to see the world as serving his self-definition." [25]

As part of that new self-definition, humans have become separated in a special sense from nature. When they construct a linear time perspective for themselves and their knowledges — the beginnings of an idea of progress — they depart from a view of themselves as subject to the cyclicity evident in nature. Copernicus provides a turning point in the new self-definition, but not so much in his book, *De revolutionibus,* as in the myth of his achievement. "For the modern age Copernicus is less a figure in the history of science than a figure in the history of consciousness." [26] What we become conscious of is the supplanting of one system of thought, the Ptolemaic, by another; henceforth, all findings are made potentially problematic, and so is their creator, man.

Man, in short, has become perspectival man. He sees everything, himself included, from a particular, self-conscious perspective. The accumulation of such perspectives is the historical consciousness. We see the world and one another and our selves in terms of the knowledges that we have inherited or, rather, learned from the consciousness embedded in our culture. A change in consciousness creates a different person, a changed self, who acts differently in regard to the environment, natural and human. [27]

Consciousness, therefore, has significant effects. It is not the same as the contemplative consciousness, the Aristotelean sciential, conceptual grasp; it is active. It is, in fact, part of doing: an active knowledge. [28]

The deep-seated changes in consciousness of the kind described are fundamental to human science. As one observer glimpses, the main stock in trade of human scientists "is not prescriptions or laws or definitive assessments of proposed actions; we supply concepts and these alter perceptions. Fresh perceptions suggest new paths for action and alter the criteria for assessment." [29] Let me amend this statement. Prescription, as I have insisted, is more or less built into the human sciences. They do not come unfreighted with our intentions, no matter how much we may strive for objectivity. Prescriptions emerge from our consciousness. Change the latter, and the former are changed. Nor are altered perceptions simply the result of changed concepts; indeed, the case can be made that the flow is the other way.

Qualifications aside, there is much truth in the statement quoted. An example will help: the movement to establish and expand human rights. Let me single out those of women. What laws, concepts, positivistic inquiries, figure here? The answer is, if they exist (for example, in the shape of statistics), they function only in a very minor way. What counts most significantly, as women themselves have realized, is consciousness. Changes in it come about only in small part by deliberately "raising" it (though that is important). Changes are mostly slow, often imperceptible, sometimes almost unidentifiable and are mainly the result of unintended consequences.[30]

A major task of the human sciences is, in fact, to identify the elements in changed consciousness and to participate, by their findings, in making changes. The very word *human* is defined by the inquiries into the long evolutionary past of the species. In this context, we come to realize, for example, that gender roles are rooted in a sociobiological matrix, which itself then becomes subject to evolutionary change—but now in the form of cultural evolution. Women's roles change fundamentally in a society characterized by such developments as industrialization and birth control by artificial means. Only in such a climate of consciousness can prescription and action begin to figure.

I have only touched on some of the elements that have entered into our increased consciousness, in this case, about women. I have made much of the Other in this book and the way it started in a European study of the new worlds and peoples encountered in the course of imperialistic exploration. It is only with the incorporation of the Other into ourselves that a true concept of the human comes about.[31] Males, as well as females, come to see that females are as human as themselves and are deserving of the same human rights.

I have also alluded to stratification. The awareness that it is nongenetic, that it is a matter of social arrangement, allows us to rethink the presumed sociobiological basis of gender roles. As this awareness permeates our consciousness almost unconsciously, we are freer to shape our lives accordingly.

The social sciences may tell us something of the unintended processes involved in human behavior. They do not tell us how we should act upon that knowledge. The intention to shape what is otherwise unintended emanates from our consciousness, that is, from our historically formed, action-oriented understanding. Humans sharing this consciousness may be viewed as a scientific community that theoretically validates the knowledge derived from the human sciences by behaving in terms of such knowledge. We have witnessed one example: how acceptance of the Other as having rights equal to one's own emerged from the human sciences and from the gradual incorporation of their inquiries into consciousness.

Consciousness as a Scientific Community

We are able now to see a possible way out of our dilemma concerning the need in the human sciences for a scientific community. It requires to be made completely explicit. In the spirit of Habermas, we explored earlier the need for humanity as a whole to become a kind of truth community adhering to the scientific method. In the end, we had to acknowledge the utopian character of such a community. For the foreseeable future it appears unrealistic to expect the emergence of a humanity prepared to act in accord with the dictates of positivist aspiration and hermeneutic inspiration and to exercise the requisite attention to empirical evidence, understanding of the requirements of theory, willingness to employ logic, and the rest, in regard to both the natural and the human world.[32]

We must think, then, along somewhat different lines. I have been hinting all along at the alternative. Historical consciousness, vague and unscientific as it is, can and does embody the general results of the natural and human sciences reached by their methodical procedures. Large sections of humanity have become a different species; other voices, those of the natural and human scientists, speak within their minds. And this in spite of their unawareness of scientific method.[33] We have all become Marxists and Freudians, even when we are anti-Marx and anti-Freud, for we live within their mental categories. After Copernicus, we all became Copernicans, even if unthinkingly so.

Let us couple this idea of a historical consciousness — which does not itself operate according to a scientific method but which does unconsciously incorporate the results of such methods as are deployed in more limited scientific communities — with other observations that we have made about the human sciences. For example, I have emphasized the way action, including action by prescription, is a vital element in human science. I have also insisted on the hermeneutic component in our efforts to construct the uncertain sciences.

It follows that the human sciences, both in their aspect as prescription and by their analysis of human desires, thoughts, and values, enter into the actions that we take, thus providing the further phenomena that we attempt to study scientifically. Desires are heterogeneous, though heavily shaped by society. Thoughts are not just consciously held but unconsciously as well, in terms of historical consciousness. We have become aware that values shape our thoughts and actions and are reflections of our desires (or the repression of our desires, vide Nietzsche and Freud).

The value issue is a complicated one. In the view, for example, of Max Weber, science can have nothing to say about values. We must admit that this sweeping statement is true in the sense that Weber had in mind: values are not given as part of reality, waiting merely for the scientist to discover and verify them. Does this throw us into the arms of pure relativists? Are values totally remote from a scientific scrutiny?

It would appear not. Even though values are constructed under different circumstances and change with changing conditions, they are experienced as real within given societies. As the human species becomes more conscious, however, and especially more self-conscious, it increasingly can take into its own hands the construction of a value system created in terms of its own historical development. Such values shape our "project" for the future, the creation of a more humane and enlightened species. Of course, the value system is at best a fought-over intention whose realization can be only partial, uncertain, and subject to countervailing forces. The creation of a better humanity is not derived from thin air but takes shape in response to our evolving historical consciousness.

With Bacon, humankind claimed to be inventing a philosophy of invention.[34] Invention included new self-invention as well as technological and scientific innovations. As numerous authors have pointed out, the modern period has been a time of self-fashioning and self-awareness. Salman Rushdie, for one, peers deeply into the heart of our modernity. He says, "Given the gift of self-consciousness, we can dream versions of ourselves, new selves for old."[35] We can also dream of new societies for such selves to inhabit.

This is what I mean by the human sciences becoming part of the project that humanity has set for itself. It does not proceed by creating a formal scientific community. Rather, humanity shares a historical consciousness from which, with increased self-consciousness, it draws its values. On this basis, humankind seeks to understand itself. Understanding interacts with unintended consequences and forms the phenomena that the human sciences attempt to comprehend. In this strange, uncertain, and unexpected manner, humanity begins to resemble a scientific community adequate to the demands placed upon it by the nature of the human sciences.

The historical consciousness of which I am speaking is not a mystical conception, a Hegelian-like spirit moving through time to an inevitable end. It draws upon Habermas's notion of a truth community, elevated to include all humankind, but recognizes that scientific method cannot be expected of all humans. Therefore, we are driven to posit a historical consciousness many of whose elements result from the natural and human sciences, which do employ a suitable scientific method. While there are other elements, as we have noted, than the scientific in our historical consciousness, they are not our prime concern here. Our concern is especially with human science and its search for truth about human phenomena, notably how that search becomes incorporated in historical consciousness.[36]

Diseased Consciousness

My emphasis so far has been on the affirmative features of increased consciousness. For many modern observers, these are overridden and obscured by the deleterious effects, among others, of a self-awareness that has become sick. In their eyes, the awareness of self has become a fixation, and excessive self-consciousness a destructive force. As the Devil puts it in *Paradise Lost*, "Which way I fly is Hell, myself am Hell."[37] This, too, is one result of the Copernican revolution. The self-absorbed soul can find no support in anything outside itself. Forced to create its own meaning and its own values, it may feel itself on the edge of an abyss.

Hell seems to threaten many contemporary thinkers, from Nietzsche to the postmodernists. Nietzsche, whom I quoted in support of the notion of historical consciousness, says that the self-critical tendency leads to "the most extreme form of nihilism," to the view that there is no true world because all that exists is "a *perspectival appearance* whose origin lies in us."[38] Anthony Cascardi speaks of discourse as having become the search for, and affirmation of, self, rather than of "truth."[39] The aim, it too often appears, is a plunge into Narcissus's pool rather than immersion in the human sciences. We are drawn into the narcissistic pool by words, by our fixation on our own discourse. Our symbol-making ability turns upon itself, losing all contact with the external environment for which it has been adaptively devised.

In this case, the human sciences become our salvation. Unlike the discourse described (though not necessarily favored) by Cascardi, the discourse of the human sciences is concerned with the self mainly because of its intimate ties to the development of consciousness. In the human science discourse, although there is awareness that words can make us self-aware to the extent of being overaware and thus filled with angst, the more general concern is with words as repositories of historical meaning. In this perspective, consciousness grows and develops out of the interchange between expanding language and the self and between the self and historical experience. One marker of this increasing consciousness is the now ubiquitous dictionary, an invention showing the increased vocabulary available to people living in modern times. The dictionary is a repository of historical consciousness.

A Deepened Consciousness

What we can hope for in regard to the human sciences, along with an increase in positive knowledge, is, where obtainable, an increase in consciousness as I have defined it. Another way of putting this, already broached in the chapter on hermeneutics, is to speak not of increase but of deepening.

Deepening has two levels. On one is the desire of eighteenth- and nineteenth-century Western thinkers to see behind appearances and below the surface. In the field of literature, the romantics "mined" this lode of consciousness, finding many of their metaphors in the advancing technology concerned with the underground.[40]

We encounter the imagery of depth everywhere. Almost at random, I quote the English poet Matthew Arnold, writing in the second half of the nineteenth century. In his poem "The Buried Life," he tells us:

> There rises an unspeakable desire
> After the knowledge of our buried life.
>
> .　　　.　　　.
>
> A longing to inquire
> Into the mystery of this heart which beats
> So wild, so deep in us.
>
> .　　　.　　　.
>
> Our hidden self.

In Western culture consciousness of this drive to the interior depths appears to be a relatively late occurrence. As Christopher Herbert points out, the first recorded usage of *deep* in the English language, in the figurative sense of "lying below the surface; not superficial; profound," appears at the "surprisingly late date of 1856."[41]

I cannot help noting that this is also about when the first bones of Neanderthal people are discovered. But it is not just archaeology, or prehistory, anticipating, accompanying, or following on literature, that flourishes in regard to depths. The human sciences in the shape of Marxist inquiry, for example, see themselves as not only piercing the veil of illusion but also going down into the infernal pit of capitalism. With Nietzsche and Freud, the direction downward is "deepened" by their descent into the underground caverns of the human mind. To go deep for them means to get closer to reality.

In fact, the effort to go deep was under way long before the nineteenth century. Adam Smith preceded Marx in going behind and beneath the appearances of the market and discovering its hidden meaning and mode of operation. Even earlier, Montaigne and Diderot plumbed the depths of the human spirit and touched upon its irrational stratum. What is new in the nineteenth century is the consciousness of this deepening.

That is one level on which depth is to be found. Another, more pertinent to our thesis, is the deepening of historical consciousness through the acquisition of knowledge. Hegel, for one, was aware that his knowledge extended only to the time, his present, when the owl of Minerva took flight. Think of the advances in the human sciences since his time. We now know much more about the two to three million years of pre-hominoid origins. We know not only about human prehistory in much detail but also about evolutionary and historical development in agriculture and industrialization. What is more, we have the social sciences of economics, sociology, anthropology, political science, and history (to name only the ones considered here in any detail), with their data and conceptualizations, all only in incipient stages in Hegel's day. Our knowledge — our consciousness — in the shape of human science is deeper than that of the great German philosopher. It is this deepened consciousness, no matter how differentially diffused, that constitutes the individual being of each of us, planted as it is in our societies, and that shapes our intentions, conscious and unconscious.

Seen in this light, the task of the human sciences is not so much increased control and prediction of what is outside a person's mind but deepened knowledge of what is inside it. That very knowledge helps constitute what the person is. It should not be overlooked that we have acquired, and are acquiring, more and more positive knowledge about the physical, chemical, and biological basis of consciousness. But this is not the same as knowledge about the content of consciousness, which remains the province of the human sciences.

We can speculate that the deepening of historical consciousness will continue. Is there any evidence to support that prospect? I have already referred to Jaynes's thesis. Even before Jaynes, there was the work of Vico, the Higher Critics, the Marxists, the Durkheimians, and many others (not to mention Freud). All of these have contributed to an understanding, a knowledge, of the way humans form conceptions of the gods, erect the symbolic structures we call cultures, and relate to one another in societies. We hardly realize what a distance we have come from earlier levels of consciousness, where both consciousness and the hermeneutics to interpret it remained in a nonexistent or primitive state.

Now we are faced with a new task. Difficult as it is to look back thousands of years and correctly estimate the development of consciousness and its hermeneutical interpretation, it is undoubtedly even more difficult to look ahead a couple of hundred, a thousand, or two thousand years and speculate on a future consciousness and the human sciences that will accompany it. Earlier, a change in time sense was needed for advances in the sciences. In geology, Lyell needed hundreds of thousands and then millions of years beyond the biblical allotment of six thousand in order that the uniformities of change in the earth might have sufficient time in which to display their lawlike regularity. Similarly, in biology Darwin believed that he needed his aeons for natural selection to bring about the evolution of new and perhaps more complex species.

The human sciences are somewhat different. The recent pace of cultural change has been so fast that we need, mentally, to reverse our time perspective and compress the extended vistas of time unrolled by the natural sciences. In the human sciences, we must note that human consciousness has emerged relatively recently and that historical consciousness has emerged only in the past few thousand years. Where Lyell and Darwin stretched the time frame on which human existence has been drawn, the human sciences may suddenly have to contract it.[42]

Maybe in a few hundred years or a few thousand we will be able to see the further dimensions of a deepened and developed human science. I have already stated that the likelihood of an increased positivistic human science — whatever one's views on its desirability — appears strong. So, too, does an expanded hermeneutics in regard to this science. How will the two trends relate? We can only guess that humans will survive, however ambiguous and uncertain their condition. And we can guess that their condition will be measured and calculated more scientifically, or naturally, at the same time as it will be comprehended more hermeneutically, or culturally. Both developments, concurrently, I suspect, will be part of human science.

In any event, the accumulation of consciousness is unpredictable and relatively slow, even given the time scale that we are presupposing. We cannot tell for sure whether a scientific community of the kind most favorable for the human sciences might arise even over extended time.

What we do know at the moment is that the average person does not need to know the findings of natural science or understand the workings of technology in order to enjoy their benefits (and suffer their deficits).[43] The same, however, cannot be said about the human sciences. As we asked earlier, are we therefore faced with an impossible situation in which humanity as a scientific community is essential but unrealizable? Can this situation be rescued by our concept of a shared historical consciousness that embodies scientific results? The answer is inconclusive. It depends in part on whether our overall conclusion about the centrality of consciousness is right: that human beings cannot share in such a consciousness without entering into it, even if unconsciously or half-consciously. In the end, this is the true dilemma of the human sciences.[44]

A Summary

Where does this leave us? Charles Babbage, speaking of his early computer, referred to it as "the engine eating its own tail."[45] He meant that the results obtained by computing one table of numbers could affect all the other tables computed, and so on. In effect, this book has been eating its own tail. What has been said in one part affects all the rest. This has meant recurring to topics taken up in one chapter in the context of another. It has meant a constant cycling of messages back and forth. We have been engaged in a hermeneutic circle concerning the subject of the human sciences and efforts in those sciences to use the scientific method to gain knowledge.

We started by looking at the problem confronting the human sciences. In stating the problem or, rather, some of its aspects, I touched upon definitional issues, possible aims, the nature of the subject/object—that is, the human species—possible methods of studying it scientifically, and historical preconditions for the emergence and development of the sciences devoted to the study of humanity.

It then became important to examine one claimant to correct method— positivism—in depth. I attempted this task in historical fashion, going back to Francis Bacon and following the career of positivism as it moved from him to natural scientists, such as Boyle and Hooke, with their notion of witnesses, to thinkers in later centuries, with positivism unfolding both as sociology and as a prominent aspiration of the human sciences in general. In the course of our review, we evaluated philosophically the pros and cons of the positivist method, especially as it applies to our subject.

Before turning to the other major method applicable to the human sciences —hermeneutics—we returned to the subject studied by the human sciences: the human animal. I emphasized two major aspects of the development of that animal into what we recognize as humankind: evolution and emergence. The evolutionary perspective is the single most important one for the human sciences, for this way of seeing must be part of all our efforts at understanding our strange species. The concept of emergence—of phenomena emerging unexpectedly, unpredictably, and unintentionally as wholes from simpler and earlier forms—is similarly crucial. An outstanding emergent factor in human evolution is symbolic activity, which allows, in turn, for the emergence of culture. All of this and more was illustrated and connected to the concept of classification in science by using a specific example in the human sciences: social stratification.

With these notions in mind, we turned to hermeneutics. Here we sought, by a historical and philosophical examination of the notion of Geisteswissenschaften, that is, the moral sciences, to explore what we would later recognize as the hermeneutic circle — the way any interpretation proceeds (thus imitating Babbage's engine swallowing its own tail). I opposed a sharp division between the natural and the human sciences, insisting that the former also had to employ interpretation and that both used a common scientific method. That scientific method is on a spectrum ranging from positivist to hermeneutic and differs in degree according to the nature of the phenomena being studied.

In dealing with hermeneutics, perspective is important. In highlighting perspective, we were able to return to our consideration of the Other, first taken up in the opening chapter as a precondition for knowledge in the human sciences. The coupling of myth and narrative (and madness) with hermeneutics then occupied us. Last came the critical subject of criteria — how do we judge the validity of an interpretation? — addressed with uncertain results.

The next question was the problems of content or method in the human sciences, the argument being that their accomplishments have been great. Both as general human science, whose focus is on the human species and its evolution, and as individual human sciences — here, economics, sociology, anthropology, political science, and history — the achievements have been highly significant.

Having reviewed the achievements of human sciences, we started an inquiry into their future prospects. Our path took us through the thickets of unintended consequences (and their relation to the concept of emergent phenomena), then, slightly drunk, staggering past the possibility of humans becoming machines and thus positive creatures, and then back to the straight and narrow to look at the notion of prescription, this time in connection with culture and consciousness.

Thus sobered, we returned to a reconsideration of the conditions of science, first reexamining the subject of witnesses, but this time in regard to ideas about gentlemen and individualism. This examination led us back to the question of scientific community and how it is defined by the strictures of Habermasian truth claims. The Habermasian community, however, is not itself a scientific community, we decided, for a scientific community must adhere to specific scientific methods.

After looking at scientific methods, the realization came home that the scientific community needed by the human sciences was, in fact, humanity at large. Civil society, globalization — were these leading us closer to such a community? On the other hand, was religion, a rival claimant to the truth, forever blocking the emergence of a true scientific community, as required by the human sciences?

Alas, the notion of humanity as a possible scientific community would not hold. Reexamining the Habermasian truth community, we were forced to conclude that it was probably vulnerable to serious criticisms and certainly utopian. We sought an alternative in the concept of a historical consciousness that, though not scientific in method, could encompass the findings of the natural and human sciences. If this notion is tenable, a major task of future inquiry must be to explore the workings of such a historical consciousness without losing sight of the foundations of positivism and hermeneutics on which particular human sciences stand.[46]

Looking back over this book, we can say that it had a beginning, a middle of sorts, and now an end. Our end is, however, another return to the beginning, a return required in order that this end, or lack of one, be better understood. Our end is an inconclusive conclusion, an interminable deepening of our knowledge and understanding.

Such a "conclusion" may feel anticlimactic in comparison, for example, with Darwin's final evocation in *The Origin of Species*, where he speaks of "grandeur" in the view of life that sees the emergence of the higher animals. In contrast, I am calling, in this view of the human sciences, for Sisyphean labors along with moments of Promethean rebellion, for humility as well as hubris. A more appropriate word than *grandeur* for us may be *ardor*, with its etymological roots in the Latin for "burning" and its image of unrequited passion. It at least partly conveys the way humans seek an understanding of themselves and a universe that can be known only through those selves — and do so especially through the self-abnegation and self-emancipation to be found in science.

Appendix: Statistics

n the social sciences, statistics is ideally employed to promote or ensure objectivity. In fact, statistics is not just a social science technique but a whole branch of study. The word *statistics* is from the same Latin root as *state* (the Latin *status* meaning "standing or position"), and statistics as a putative science first manifested itself in the service of the rising absolutist states. Data relating to population, taxes, trade, and so forth, were collected by the cameralists and mercantilists of sixteenth-century western Europe, especially in the German states, to inform state policy, state building, and trade.

By the seventeenth century, statistics, often in the form of political arithmetic, was considered above politics; that is, it was thought to offer knowledge that was objective because it was free of party bias and untainted by court strife. These statistics can be seen as an early form of positive knowledge. Work in statistics took strength from the religious wars, seeking to rise above them and to offer a scientific ground for authority in place of the contending religious claims. In England, for example, the same effort at such objectivity can be found in John Graunt and William Petty, based on their statistical collections, as in Thomas Hobbes, who, however, sought it in deductive certainty.

The eighteenth century was the age of calculators, as Burke called statisticians (along with economists and sophisters). Statistics were often collected by private individuals—such as Burke's archenemy, the Dissenting minister Richard Price—frequently in the service of democratic intentions. Not subjects but citizens are studied statistically. Scientists like Condorcet introduced notions of probability, now in the service of a possible welfare rather than absolutist state. Statistics was well on its way to becoming a major form or adjunct of the social sciences, with significant innovations in techniques, such as regression analysis, to come later.

A basic postulate is that aggregates exhibit regularities that are unobservable by the mere study of individuals. In analyzing aggregated data, statisticians often lay claim to what, in the event, turn out to be false precision and false objectivity. Controversy swirls about the subject, and arguments are launched asserting that the neglect of causal analysis vitiates the approach. Still, it is clear that statistics is invoked by all sides in the political and social world and that agencies and departments to collect figures about almost any imaginable subject sprout like mushrooms in the rain. It is hard to imagine modern society without censuses and gross national products (a statistic developed by Simon Kuznets in the 1930s), batting averages and weather reports.

Without evaluating the scientific status of statistics, whether in mechanics, biology (two areas on which we have not even touched), or the social sciences, let me simply underline their increasing importance and highlight two aspects. One is the claim to objectivity. Statistics is a form of mathematics, and as the physicist William Thompson, Lord Kelvin, announced, "When you cannot measure it, when you cannot express it in numbers, your knowledge is of a meagre and unsatisfactory kind." (This kind of positivism has, as noted in the text, been thoroughly challenged.)

The other aspect is the use of statistics in support of policy recommendations. Obviously, this is partly because of the persistent claim that numbers have some sort of objectivity. Critics of statistics, ranging from Jonathan Swift to Charles Dickens, have made fun of the enterprise — one might recall the Mudfog Association, Dickens's mockery of the Statistical Society of London, and Mr. Gradgrind's lecture to "girl number twenty" on the importance of figures — while pointing out, ad hominem, that "figures don't lie; liars figure." Yet, such telling comments aside, one cannot imagine modern society without its statistics — both the study and the data collected and analyzed provide a vital form of knowledge with claims to being scientific.

For some useful reading on statistics, see the works of Theodore M. Porter, especially *The Rise of Statistical Thinking, 1820–1900* (Princeton, N.J.: Princeton University Press, 1986), and Stephen M. Stigler, *The History of Statistics: The Measure of Uncertainty Before 1900* (Cambridge: Harvard University Press, 1986). I found two articles by Peter Buck most helpful: "Seventeenth-Century Political Arithmetic: Civil Strife and Vital Statistics," *Isis* 68, no. 241 (1977), 67–84, and "People Who Counted: Political Arithmetic in the Eighteenth Century," *Isis* 73, no. 266 (1982), 28–45.

Edward R. Tufte, in *The Visual Display of Quantitative Information* (Cheshire, Conn.: Graphics Press, 1983), links statistics and graphics in an intriguing manner. Speaking of data maps, he tells us. "It was not until the seventeenth century that the combination of cartographic and statistical skills required to construct the data map came together, fully 5,000 years after the first geographic maps were drawn on clay tablets" (20). One of the great inventors of modern graphic designs was, as Tufte reminds us, William Playfair (1759–1823), an English political economist, a fact that dramatically exhibits the links between mapping (and its discoveries), statistics (and their service to the state and society), and economics (as an emerging science).

Susan Buck-Morss helps us understand what is involved in the links that Playfair embodies when she writes:

The discovery of the economy was also its invention. As Foucault told us (and neo-Kantians before him), every new science creates its objects [and vice versa]. The great marvel is that once a scientific object is "discovered" (invented), it takes on agency. The economy is now seen to act in the world; it causes events, creates effects. Because the economy is not found as an empirical object among other worldly things, in order for it to be "seen" by the human perceptual apparatus it has to undergo a process, crucial for science, of representational mapping . . . the map shifts the point of view so that viewers can see the whole as if from the outside, in a way that allows them, from a specific position inside, to find their bearings. Navigational maps were prototypical; mapping the economy was an outgrowth of this technique.

For Buck-Morss, incidentally, the first such map was that of the French Physiocrat François Quesnay, with his "Tableau économique" (Buck-Morss, "Envisioning Capital: Political Economy on Display," *Critical Inquiry* 21 [Winter 1995], 439–40). In the end, indeed, one cannot envision mapping the economy without joining the effort to statistics. The same can be said of sociology, with quantitative history adding its claims.

Notes

Introduction

1. Joseph Schumpeter, *History of Economic Analysis* (New York: Oxford University Press, 1954), 184.

2. For some indication of how I believe this particular study might be undertaken and in what conceptual framework, see my introduction to *Conceptualizing Global History*, ed. Bruce Mazlish and Ralph Buultjens (Boulder, Colo.: Westview Press, 1993). Incidentally, in alluding here and elsewhere to my own previous books or articles, I do so not out of hubristic celebration but because it is in them that further elaboration of some of my arguments can be found, arguments that cannot be repeated here for reasons of space or composition.

3. For one account of how this subject has been argued about in terms of philosophical hermeneutics by Hans-Georg Gadamer and Jürgen Habermas, see Thomas McCarthy, "Rationality and Rationalism: Habermas's 'Overcoming' of Hermeneutics," in *Habermas: Critical Debates*, ed. John B. Thompson and David Held (Cambridge, Mass.: MIT Press, 1982), 57–78.

4. I recognize that other kinds of knowledge are also possible—see Chapters 6 and 7.

5. In violation of my own injunction, let me say a few words about Foucault. The topic of the human sciences and their history resonates throughout much of his work, but it is most famously addressed in his *Order of Things* (*Les Mots et les choses*), published in 1966, whose subtitle in English is "An Archaeology of the Human Sciences." In *The Order of Things* (I will be citing the translation [New York: Vintage Books, 1973]), the last chapter, 10, is entitled "The Human Sciences." In this brilliant and enigmatic book, which one scholar, James Miller, has called Foucault's "most fiendishly intricate text" (*The Passion of Michel Foucault* [New York: Simon and Schuster, 1993], 18)—its intricacy, however, did not prevent its becoming a bestseller when it was published—Foucault refuses to believe that "anything like man" (344) appeared as early as the eighteenth century. While this may be a matter of

terminology, Foucault seems to be arguing that what he calls the human sciences is strictly a product of postindustrial and French Revolution society.

If I understand Foucault, and what follows is hardly a full account (many other books do take on this task; see the Critical Bibliography), he is principally attacking the positivist version of human science(s), with its view of man as an object of presumed "objective" inquiry, when in fact man is a subject. In Foucault's own words, "It is useless, then, to say that the 'human sciences' are false sciences; they are not sciences at all; the configuration that defines their positivity and gives them their roots in the modern *episteme* at the same time makes it impossible for them to be sciences." He continues, "Western culture has constituted, under the name of man, a being who, by one and the same interplay of reasons, must be a positive domain of *knowledge* and cannot be the object of *science*" (366–67). A bit later, however, he uses the term *human science* loosely when he declares that "psychoanalysis and ethnology are not so much two human sciences among others . . . they span the entire domain of those sciences." (379); in fact, for Foucault they are "counter-sciences." At the end, he asks, "Ought we not rather to give up thinking of man, or, to be more strict, to think of the disappearance of man — and the ground of possibility of all the sciences of man — as closely as possible in correlation with our concern with language?" His answer appears to be: "As the archaeology of our thought easily shows, man is an invention of recent date. And one perhaps nearing its end" (386–87).

With this Spenglerian (Nietzschean?) shudder, Foucault appears to bring the inquiry into the human sciences crashing down. Man is just a linguistic innovation, not a real subject/object (which, admittedly, can only be approached through language). The ambiguity of what Foucault intended is made more complicated by his own statement that the book "is a pure and simple 'fiction.' It's a novel, but it's not me who invented it" (quoted in Miller, *Passion of Foucault*, 162). At one point, he apparently thought seriously of withdrawing the book and asked his publisher to stop printing it (ibid., 158).

6. *Die Fröhliche Wissenschaft* is generally translated as "The Gay Science." But the word *joyous* captures Nietzsche's intent as well or better.

7. Quentin Skinner, *The Return of Grand Theory in the Human Sciences* (Cambridge: Cambridge University Press, 1985). See further comments in the Critical Bibliography.

8. The director of the Institute for Human Sciences, Krzysztof Michalski, a Nietzsche and Heidegger specialist, kindly acknowledged my inquiry about the nature and title of his institute.

9. "Aims and Scope," on the back cover of *History of the Human Sciences* 9 (August 1996). One of the editors of *HHS*, Irving Velody, kindly responded to my request for a definition. "Effectively," he said, *human sciences* "is taken as an all inclusive term for all non-natural science disciplines and inter-disciplinary researches and investigations. To put it more simply we do not offer or attempt to offer any tightly

formulated definition of the theme" (E-mail message, January 15, 1997). *The Oxford English Dictionary* has no entry under either "human science" or "human sciences."

10. *Inventing Human Science,* ed. Christopher Fox, Roy Porter, and Robert Wokler (Berkeley: University of California Press, 1995). See my review of this book in the *American Historical Journal* (April 1997).

11. Carlo Ginzburg, "Morelli, Freud, and Sherlock Holmes: Clues and Scientific Method," in *The Sign of Three: Dupin, Holmes, Peirce,* ed. Umberto Eco and Thomas Sebeok (Bloomington: Indiana University Press, 1983), 103; see further comments in the Critical Bibliography. For Ginzburg the reference point for all the human sciences was medicine, especially anatomical studies. (In addition to what is said in the Critical Bibliography about abduction in *The Sign of Three,* see also the entries under William Whewell, who is perhaps the first to advance the theory.)

12. For a present effort in this direction, see John H. Weiss, "Interpreting Cultural Crisis: Social History Confronts Humanities Education," *Journal of Interdisciplinary History* 26 (Winter 1996), 459–74. Reviewing books by Fritz K. Ringer and Peter N. Stearns, Weiss summarizes the latter's argument by saying: "Stearns holds that this [humanistic] scholarship has produced general theories that guide its inquiries, that it constantly produces new knowledge that represents usable progress, and that it provides unique possibilities for understanding social and cultural phenomena. This resurgent humanistic learning in such fields as history and literature possesses the educative attributes of a highly developed social science, able to impart crucial analytical skills, attuned to further discovery, and linked to contemporary social and cultural issues" (461).

13. The merely suggestive nature of the notes and bibliography must be underlined. Given the range of topics dealt with in the book, it would require another book of equal size to list the required citations. My few pages on the Age of Discovery, for example, tap into an extraordinary field of scholarship, which is itself constantly expanding. Fortunately, much of such discipline-specific bibliography is readily available elsewhere, and many of the books on various topics that I do cite carry extensive further references.

Chapter 1. The Problem of the Human Sciences

1. In a foreword to Arnold Gehlen, *Man in the Age of Technology,* trans. Patricia Lipscomb (New York: Columbia University Press, 1980), Peter Berger describes philosophical anthropology as "philosophical reflection about the nature of man." He adds that "German philosophical anthropology has a long history, going back to Herder, Hegel, and the nineteenth-century modifications of Hegelianism (notably those of Feuerbach, Marx, and Nietzsche). Its immediate ancestors are in phenomenology. . . . The first systematic utilization of . . . biological work for philosophical

anthropology was made by Helmuth Plessner, who, like Gehlen, was trained both in philosophy and biology" (viii). More will be said about some of these figures in Chapter 4.

2. In addition to what was said in the Introduction, we can note that, for example, if anthropology is accepted as a social science, its rapprochement with literature pulls the latter toward the same camp. In this vein, a typical comment is James Clifford's: "The development of ethnographic science cannot ultimately be understood in isolation from more general political-epistemological debates about writing and the representation of otherness" (*The Predicament of Culture: Twentieth-Century Ethnography, Literature, and Art* (Cambridge: Harvard University Press, 1983), 24. See the whole of his chapter 1, "On Ethnographic Authority." The literary scholar Stephen Greenblatt notes the same tendency, when he declares that "ever since anthropologists began to speak of the cultures they study as textual, and literary critics began to speak of the texts they study as cultural, one of the dreams of cultural poetics has been to link the text of anthropology and the text of literary criticism." ("The Eating of the Soul," in *Representations* 48 [Fall 1994], 97. Greenblatt takes a critical, though sympathetic, stance toward the linking possibility. My own consideration of literature as a possible part of the human sciences will circle around an analysis of narrative.

3. Quoted in Stephen Jay Gould, "A Special Fondness for Beetles," *Natural History* (February 1993), 6.

4. Whewell first proposed the word *scientist* at the 1833 meeting of the British Association for the Advancement of Science. For a full account of Whewell's usage, see Richard Yeo, *Defining Science: William Whewell, Natural Knowledge, and Public Debate in Early Victorian Britain* (Cambridge: Cambridge University Press, 1993).

5. Keith Baker, in *Condorcet* (Chicago: University of Chicago Press, 1975), tells us that the earliest use of the term *science sociale* "appears in a pamphlet addressed to Condorcet by his younger friend, Garat, at the end of 1791," but Baker adds, "Whether Garat actually coined the term 'science sociale' is still uncertain" (391). Baker's appendix B, "A Note on Early Uses of the Term 'Social Science,'" is valuable. Subsequent researchers have argued, however, that the first recorded usage was in the first edition of Abbé Sièyes's "Qu'est-ce que le tiers-état?" (What is the Third Estate?). See Robert Wokler, "Saint-Simon and the passage from political to social science," in *The Languages of Political Theory in Early Modern Europe*, ed. Anthony Pagden (Cambridge: Cambridge University Press, 1987).

6. A. R. Radcliffe-Brown, "On Social Structure," *Journal of the Royal Anthropological Institute* 70 (1940), 2, quoted in Christopher Herbert, *Culture and Anomie: Ethnographic Imagination in the Nineteenth Century* (Chicago: University of Chicago Press, 1991), 10.

7. Louis A. Sass, *Madness and Modernity: Insanity in the Light of Modern Art, Literature, and Thought* (New York: Basic Books, 1992), 329.

8. Bronislaw Malinowski, *The Sexual Life of Savages* (New York: Harcourt, Brace, 1929), 506.

9. Bronislaw Malinowski, *Argonauts of the Western Pacific* (Prospect Heights, Ill.: Waveland Press, 1984; orig. pub., 1922), 83, 84. Malinowski's early studies were in mathematics and physics, and he wrote his doctoral dissertation on the physicist and positivist Ernst Mach.

10. Ibid., 2. For exactly how subjective that self is, see Bronislaw Malinowski, *A Diary in the Strict Sense of the Term* (Stanford, Calif.: Stanford University Press, 1989), with its frequent venting of the anthropologist's "rage" and expressions of "boredom."

11. For a magisterial treatment of its subject, see Lorraine Daston, *Classical Probability in the Enlightenment* (Princeton, N.J.: Princeton University Press, 1988).

12. Isaac Asimov, *Foundations* (New York: Ballantine Books, 1951), 3. The next quotation is from p. 92.

13. Quoted in Mark Seltzer, "The Aesthetics of Consumption," MS prepared for a talk at the Center for Literary and Cultural Studies, Harvard University, n.d., p. 17.

14. Charles E. Lindblom and David K. Cohen, *Usable Knowledge: Social Science and Social Problem Solving* (New Haven: Yale University Press, 1979), 18.

15. Thomas L. Haskell, "Persons as Uncaused Causes: John Stuart Mill, the Spirit of Capitalism, and the 'Invention' of Formalism," working paper, August 15, 1991, p. 32.

16. By a piece of good fortune and research, empirical evidence is at hand to support this conclusion. One study of students demonstrated that "economics students 'performed significantly more in accord with the self-interest model' than non-economists." In a related study researchers asked "1,245 randomly selected college professors how much they gave to charity each year. About 9% of the economics professors gave nothing; the proportion of professors in other disciplines ranged between 1.1% and 4.2% (despite generally lower incomes than the economists)." It appears that, at least in economics, teachers practice what they preach; and, for better or worse, the students actually pay attention (*Economist*, May 29, 1993, p. 71).

17. Quoted in Bruce Mazlish, *The Fourth Discontinuity: The Co-Evolution of Humans and Machines* (New Haven: Yale University Press, 1993), 67. In large part, this book was undertaken to understand better the human aspiration to be machinelike and thus predictable.

18. Quoted in John M. Staudenmaier, "Denying the Holy Dark: The Enlightenment Ideal and the European Mystical Tradition," in *Progress: Fact or Illusion?* ed. Leo Marx and Bruce Mazlish (Ann Arbor: University of Michigan Press, 1996), 193. The Chicago Exposition motto is from Chicago Century of Progress International Exposition, *Official Book of the Fair* (Chicago: Century of Progress, 1932), 11.

19. Quoted in Charles Norris Cochrane, *Christianity and Classical Culture* (New York: Oxford University Press, 1957), 468.

20. Lindblom and Cohen, *Usable Knowledge,* 23.

21. Malinowski, *Argonauts,* 9–10.

22. Thomas McCarthy, *The Critical Theory of Jürgen Habermas* (Cambridge, Mass.: MIT Press, 1982), 273.

23. Richard J. Bernstein, *The Restructuring of Social and Political Theory* (Philadelphia: University of Pennsylvania Press, 1990), 5. Incidentally, this situation makes grant making in the human sciences a more precarious and often more personalized enterprise than in the natural sciences.

24. Alan Ryan, "Foucault's Life and Hard Times," *New York Review of Books,* April 8, 1993, p. 16.

25. See Anthony Pagden, *The Fall of Natural Man: The American Indian and the Origins of Comparative Ethnology* (Cambridge: Cambridge University Press, 1982).

26. Bernstein, *Restructuring,* xiii, 53.

27. Edward O. Wilson, *On Human Nature* (Cambridge: Harvard University Press, 1978), 1–2.

28. William James, *The Principles of Psychology* (Cambridge: Harvard University Press, 1983), 332.

29. Two exceptional works that explicate the historical relations of geology and religion are Nicholas A. Rupke, *The Great Chain of History: William Buckland and the English School of Geology (1814–1849)* (Oxford: Clarendon Press, 1983); and Martin J. S. Rudwick, *The Great Devonian Controversy: The Shaping of Scientific Knowledge Among Gentlemanly Specialists* (Chicago: University of Chicago Press, 1985).

30. Herbert, *Culture and Anomie,* 136. At the risk of possible inconsistency, when I speak of a capitalized Other I have in mind the conceptualization of other peoples, especially in postmodern studies. The lowercase other refers to groups seen by the viewing group as importantly different and alien in terms of lived experience.

31. On attitudes toward "barbarians," see Anthony Pagden, *European Encounters with the New World* (New Haven: Yale University Press, 1993).

32. Stephen Greenblatt, *Marvelous Possessions: The Wonder of the New World* (Chicago: University of Chicago, 1991), 122. The other is met with constantly in the course of human evolution. The Greeks were neither the first nor the only ones to encounter others; the Chinese knew of subcontinent Indians. Following Greenblatt, however, we are on the track of the concept of the "Other."

33. Carl N. Degler, *In Search of Human Nature* (New York: Oxford University Press, 1991), 5.

34. Timothy Hampton, "Turkish Dogs: Rabelais, Erasmus, and the Rhetoric of Alterity," *Representations* 41 (Winter 1993), 62. In fact, as a result of the Crusades many Europeans brought back knowledge of the Arab world; trade networks were constructed or expanded, and so forth. Such activities, it must be assumed, also contributed to the sense of the strangely familiar, although Hampton's emphasis is on the psychological. Anthony Pagden, in *European Encounters,* suggests a context broader than that of the New World for contact with the Other. The Portuguese had

encountered "advanced" civilizations in Africa in the sixteenth century, and other Europeans had had direct contact with Indonesia. Even with these facts noted, however, the case for the special importance of the New World discoveries in regard to awareness of the Other seems strong.

35. Tzvetan Todorov, *The Conquest of America,* trans. Richard Howard (New York: Harper Perennial, 1984), 17, 165.

36. Cf. Margaret T. Hogden, *Early Anthropology in the Sixteenth and Seventeenth Centuries* (Philadelphia: University of Pennsylvania Press, 1971); Ronald L. Meek, *Social Science and the Ignoble Savage* (Cambridge: Cambridge University Press, 1976); and William Brandon, *New Worlds for Old: Reports from the New World and Their Effect on the Development of Social Thought in Europe, 1500–1800* (Athens: Ohio University Press, 1986).

37. Phillips Verner Bradford and Harvey Blume, *OTA: The Pygmy in the Zoo* (New York: St. Martin's Press, 1992), 3.

38. Gyan Prakash, "Science 'Gone Native' in Colonial India," *Representations* 40 (Fall 1992), 156. The quotations that follow are from pp. 165 and 172.

39. An essential book is *Inventing Human Science,* ed. Christopher J. Fox, Roy Porter, and Robert Wokler (Berkeley: University of California Press, 1995). Along with many others, David Hume is treated in this book as a key figure in the eighteenth-century attempt to think through a science of man.

40. It is useful to compare Bentham's Utilitarian as the chosen legislator for mankind with the poet Shelley's "poets and philosophers" as "the unacknowledged legislators of the world," first stated in his treatise "A Philosophical View of Reform" (written in 1820, though not published for one hundred years after its composition).

41. See J. W. Burrow, *Evolution and Society: A Study in Victorian Social Theory* (Cambridge: Cambridge University Press, 1966), for a useful treatment of evolutionary anthropology.

42. For elaboration of this point, see my "Darwin, the Bedrock of Psychoanalysis," in *The Leader, the Led, and the Psyche* (Hanover, N.H.: Wesleyan University Press, 1990), especially the notes. Frank J. Sulloway, *Freud, Biologist of the Mind* (New York: Basic Books, 1979), is an outstanding work on the connection between Darwin and Freud.

43. Quoted in Pagden, *European Encounters,* 154.

Chapter 2. Positivism

1. Paolo Rossi, *Francis Bacon: From Magic to Science,* trans. Sacha Rubinovitch (London: Routledge and Kegan Paul, 1968; orig. pub., 1957), xiv. Bacon's position as the fount of positivism is challenged by Peter Manicas, who argues that "in seeing positivist/industrialized science as science tout court, we look for its first voice, find Bacon, make him into a positivist, elevate him to an overblown prominence,

and explain science as *inherently* aimed at domination" (*A History and Philosophy of the Social Sciences* [Oxford: Basil Blackwell, 1987], 206–7). Ronald Levao, in "Francis Bacon and the Mobility of Science," *Representations* 40 (Fall 1992), on the subject of domination, argues that Bacon, assaulting other's "magistrality," himself aimed at "a totalizing knowledge of and dominance over intellectual disciplines and over nature" (3). Levao also professes (echoing Timothy J. Reiss, in *The Discourse of Modernism* [Ithaca, N.Y.: Cornell University Press, 1982]), to see "intriguing analogies between Bacon's science and his royalist politics, even suggesting the former as a species of the latter" (2). A serious and impressive disquisition supporting this latter view is in Julian Martin, *Francis Bacon, the State, and the Reform of Natural Philosophy* (Cambridge: Cambridge University Press, 1992). B. H. G. Wormald, in *Francis Bacon: History, Politics and Science, 1561–1626* (Cambridge: Cambridge University Press, 1993), on this same topic, seems pedestrian. In *Dialectic of Enlightenment* (1944), Max Horkheimer and Theodor W. Adorno begin their rethinking of the Enlightenment with Bacon, whom they see as the fount of the patriarchal attitude to nature, disenchantment of the world, and the structure of scientific unity. They more or less end up equating the Enlightenment with Nazism, or as at least leading to Nazism, which certainly seems to place a good deal of blame on the early English thinker and statesman.

2. Rossi, in *Francis Bacon*, writes that Bacon's "modernity does not lie in his championing induction against deduction, but in his courageous rejection of pre-established limitations to scientific enquiry" (223).

3. Charles Webster, *The Great Instauration: Science, Medicine and Reform, 1626–1660* (London: Duckworth, 1975), xv.

4. Francis Bacon, "Advancement of Learning," in his *Advancement of Learning and Novum Organum*, rev. ed. (New York: Colonial Press, 1900), 139.

5. Bacon, "Novum Organum," 335.

6. In fact, Descartes requested assistants to help him in his work, but the task was still basically his alone. For Descartes's views on science, see his *Discourse on Method* (1637) and *Meditations* (1641).

7. See W. von Leyden, "Antiquity and Authority," *Journal of the History of Ideas*, 19 (October 1958).

8. As Bacon remarked, "Our method, though difficult in its operation, is easily explained. It consists in determining the degrees of certainty" ("Novum Organum," 311); and, "But our object is not uncertainty but fitting certainty" (364).

9. The first idols are the idols of the tribe, those incident to humanity at large; the second are the idols of the cave, those incident to the particular constitution of the individual; the third are the idols of the marketplace, those resulting from the influence of words on the mind; and the fourth are the idols of the theatre, those arising from received and misleading systems of philosophy.

10. Bacon, "Novum Organum," 319.

11. Ibid., 4.

12. Ibid., 156

13. Quoted in Rossi, *Francis Bacon,* 162.

14. Bacon, "Novum Organum," 340.

15. Ibid., 388.

16. Bacon, *Advancement of Learning and Novum Organum,* v.

17. Bacon, "Novum Organum," 364.

18. Rossi, *Francis Bacon,* 61. See also Antonio Pérez-Ramos, *Francis Bacon's Idea of Science and Maker's Knowledge Tradition* (Oxford: Clarendon Press, 1988), which, in spite of its unwieldy title, contains many interesting observations on this subject as well as on Bacon in general. For example, Pérez-Ramos treats Robert Boyle as a direct follower of Bacon, with Boyle writing that "there is no certain diagnostic agreed on whereby to discriminate natural and fictitious bodies" (quoted on p. 176). See also my *Fourth Discontinuity,* 8–9, for a long quotation from Bacon and the context in which he treats artificial nature, that is, the nature created by Man's arts, as ontologically equal to so-called natural nature.

19. Bacon, "Advancement of Knowledge," 105.

20. See ibid., 157.

21. Ibid., 227–28.

22. Ibid., 275.

23. Ibid., 298.

24. Ibid., 299.

25. See ibid., 412.

26. Pérez-Ramos, *Francis Bacon's Idea of Science,* as its full title makes clear, emphasizes the maker's knowledge, so prominent in Bacon as well as Vico, more than the beholder's or user's knowledge. Implicit in Bacon's conception of science as maker's knowledge is the view that science is a matter not of fixed formula but of scientific method adapted to the phenomena with which it deals. Pérez-Ramos quotes Gilbert Ryle to the effect that "there is no such animal as science" (36) and declares that Thomas Kuhn "seems to ignore the logical skeleton of the cognitive project . . . [and] fails to connect the evolution of each approach with the forms of reasoning (deductive, inductive, analogical, or whatever) which was favored at each stage" (37). Pérez-Ramos also reminds us that the debate over the "historicist" or "logicist" theories of science was carried on vigorously in the nineteenth century. William Whewell "regarded the history of science as the source and warrant of evaluative criteria that should be extracted from the historical record, whilst John Stuart Mill held that some kind of atemporal logic was the true fountain-head of such evaluative criteria, which can be justified by considerations independent of the actual history of science" (22). In spite of my great admiration for John Stuart Mill (see my *James and John Stuart Mill* [New York: Basic Books, 1975]), I believe that in this debate Whewell has the right of it.

27. Steven Shapin and Simon Schaffer, *Leviathan and the Air-Pump: Hobbes, Boyle, and the Experimental Life* (Princeton, N.J.: Princeton University Press, 1985).

28. Ibid., 60.

29. Ibid., 40.

30. Thomas Sprat, *History of the Royal Society of London* (London: S. Chapman, 1722), 111–13. Even Hobbes switched from the luxuriant prose of his work on Thucydides to the more spare style of *Leviathan*, where he sought to rescue the civil world from Descartes's doubt and to make it a fit subject of demonstration. See further my *Riddle of History* (New York: Harper and Row, 1966), 24–26.

31. Shapin and Schaffer, *Leviathan*, 80.

32. Cotton Mather, "A Brand Pluck'd Out of the Burning," in *Narratives of the Witchcraft Cases*, ed. G. L. Burr (New York: C. Scribner's Sons, 1914), 104.

33. Quoted in Shapin and Schaffer, *Leviathan*, 315.

34. Svetlana Alpers instances a 1628 print by Pieter Saenredam that "represents . . . several cross sections cut through an old apple tree. . . . [Saenredam's] image was made to repudiate an earlier one that had represented the widespread belief that the dark core of the apple tree represents the miraculous appearance of Roman Catholic priests. . . . Saenredam's print . . . shows us the shape and thickness of the cuts through the trees and then isolates the dark cores for our view [i.e., it shows how the supposed picture of the priests grew naturally]. . . . It argues, in effect, that the miracle is bound to a mistake in image-reading or to a mistake in interpretation" (*The Art of Describing: Dutch Art in the Seventeenth Century* [Chicago: University of Chicago, 1983], 80–81).

35. On the human being as increasingly a "prosthetic God," see my *Fourth Discontinuity*.

36. For example, Julian Martin, in *Francis Bacon*, argues that the great lord chancellor's scientific community was to be in the service of an imperial sovereign.

37. See Chapter 5. Cf. J. Bronowski and Bruce Mazlish, *The Western Intellectual Tradition: From Leonardo to Hegel* (New York: Harper and Row, 1960), chap. 3, "Machiavelli." The notes, though now somewhat outdated, suggest other readings on Machiavelli.

38. Descartes used his method of doubt to achieve certainty of knowledge. As he discovered, in human affairs, however, "we are much more greatly influenced by custom and example than by any certain knowledge" (*Discourse on Method*, trans. Laurence J. Lafleur [New York: Liberal Arts Press, 1950], 11). Unguided by apodictic knowledge, humans as a subject was itself outside the scope of Descartes's method for achieving scientific knowledge. Although Descartes reached the admirable conclusion that "those who held opinions contrary to ours were neither barbarians nor savages" (11), it followed that their uncertain examples and customs offered us no guidance and that we were better off obeying the "laws and customs" (15) of our own country, including its religion. Since shifting laws and customs were unworthy objects of study, Descartes resolved "to seek no other knowledge than that which I might find within myself" (6); and the *Discourse on Method* is written, in fact, as "autobiography" (3). On the bedrock of self, untouched by culture, Descartes be-

lieved that he could establish certainty, but only "in the great book of nature" (6). Thus, dualism in Descartes is not just of body and soul but also of the natural and human sciences (with the latter being for him a contradiction in terms).

39. Quoted in Ernst Nagel, "The Origins of Modern Science," in *Introduction to Contemporary Civilization in the West*, prepared by the CC Staff of Columbia College, 2d ed. (New York: Columbia University Press, 1960), 515–16.

40. Stuart Hampshire, "The Last Charmer," *New York Review of Books*, March 4, 1993, p. 15.

41. Ibid., 18.

42. Ibid., 16. A fuller account of the intellectual development under regard here would have more on Buffon as the leading expositor of natural history. See the excellent article by Claude Blanckaert, "Buffon and the Natural History of Man: Writing History and the 'Foundational Myth' of Anthropology," *History of the Human Sciences* 6 (February 1993), 13–40. An interesting selection from Buffon is to be found in *From Natural History to the History of Nature: Readings from Buffon and His Critics*, ed. and trans. John Lyon and Phillip R. Sloan (Notre Dame, Ind.: University of Notre Dame Press, 1981).

43. David Hume, *A Treatise on Human Nature* (Garden City, N.Y.: Doubleday, 1961), xiii, 247.

44. A good beginning can be made with a book mentioned earlier in a different context, *Inventing Human Science: Eighteenth-Century Domains*, ed. Christopher Fox, Roy Porter, and Robert Wokler (Berkeley: University of California Press, 1995).

45. Jürgen Habermas discusses these points in his *Knowledge and Human Interests*, trans. Jeremy J. Shapiro (Boston: Beacon Press, 1968), 71–81. His entire book, as he announces in the preface, is "a historically oriented attempt to reconstruct the prehistory of modern positivism with the systematic intention of analyzing the connections between knowledge and human interests" (vii).

46. For details on this level of criticism, see my *Riddle of History*, 211–18.

47. Quoted in Diana Postlewaite, *Making It Whole: A Victorian Circle and the Shape of Their World* (Columbus: Ohio State University Press), 1984), 39.

48. In addition to Postlewaite's book *Making It Whole*, which deals brilliantly with positivism in terms of such figures as Samuel Taylor Coleridge, John Stuart Mill, George Combe and Charles Bray (both phrenologists), Robert Chambers (an early evolutionist of sorts), Harriet Martineau, George Henry Lewes, Herbert Spencer, and George Eliot, along with Auguste Comte, see Peter Allan Dale, *In Pursuit of a Scientific Culture: Science, Art, and Society in the Victorian Age* (Madison: University of Wisconsin Press, 1989). Dale also deals with some of Postlewaite's figures, though in a nonrepetitive fashion, and adds others, such as Leslie Stephens, Thomas Hardy, and Sigmund Freud. For a broad treatment of the crisis in which sociology, as a specific part of positivism, found itself in the nineteenth century, see Wolf Lepenies, *Between Literature and Science: The Rise of Sociology* (Cambridge: Cambridge University Press, 1988); and my review of this book in *Social Forces* 69 (June 1991), 1249–56.

49. George Eliot, *Daniel Deronda* (Penguin Books, 1983), 572.

50. Dale, *In Pursuit of a Scientific Culture*, 157.

51. For one account, see H. Stuart Hughes, *Consciousness and Society* (New York: Alfred A. Knopf, 1958). Frank Miller Turner, in his book *Between Science and Religion: The Reaction to Scientific Naturalism in Late Victorian England* (New Haven: Yale University Press, 1974), deals with six figures, some well known, some less known, and states that their lives and works "disclosed the failure of scientific naturalism to fulfill the much-vaunted promise of its adherents to provide a complete guide to life" (7).

52. John Locke, *An Essay Concerning Human Understanding* (any edition), bk. IV, chap. 1 (2). The quotations that follow are from bk. IV, chap. 10 (1), and bk. 1, chap. 1 (5). For more on Locke and his views on sense perception and consciousness, see Chapter 7, n. 23.

53. Quoted in Gerald Holton, "Ernst Mach and the Fortunes of Positivism in America," *Isis* (1992), 83, 28.

54. Leszek Kolakowski, *The Alienation of Reason: A History of Positivist Thought*, trans. Norbert Guterman (Garden City, N.Y.: Doubleday, 1968; orig. pub. in Polish, 1966), 119.

55. Quoted in Christopher G. A. Bryant, *Positivism in Social Theory and Research* (London: Macmillan, 1985), 110. Bryant's book is a fundamental work on its subject.

56. I have used here the paraphrase of Neurath's views by Bryant in ibid., 113.

57. See Holton, "Ernst Mach," 39.

58. Quoted in ibid., 45.

59. Thomas S. Kuhn, *The Structure of Scientific Revolutions* (Chicago: University of Chicago Press, 1962), x. According to Steve Fuller, Kuhn "openly solicited that doyen of positivists, Rudolf Carnap, to have *Structure* published as part of the International Encyclopedia of Unified Science" ("Being There with Thomas Kuhn: A Parable for Modern Times," *History and Theory* 31, no. 3 (1992), 244. His article contains an interesting discussion of the reception of Kuhn's theory, with much bibliographic material.

60. Kuhn himself was shocked by reactions to his work and argued that he was not calling into question the "progressive" accomplishments of modern science but only the rationalist accounts of scientific growth. Cf. Barry Barnes, "Thomas Kuhn," in *The Return of Grand Theory in the Human Sciences*, ed. Quentin Skinner (Cambridge: Cambridge University Press, 1985), 85–100.

61. I. P. Pavlov, "Reply of a Physiologist to Psychologists," in his *Selected Works* (Moscow: Foreign Language Publishing House, 1955), 446–47.

62. Quoted in Carl N. Degler, *In Search of Human Nature* (New York: Oxford University Press, 1991), 153.

63. Quoted in Holton, "Ernst Mach," 34.

64. William James, *The Varieties of Religious Experience* (New York: Collier Books, 1961), 29.

65. Quoted in Bryant, *Positivism*, 139. I am deliberately ignoring the American tradition of pragmatism embodied in such figures as George Herbert Mead. This admirable thinker, as well as a number of others, would have to be given sustained attention in any full account of American sociology.

66. William J. Thomas and Florian Znaniecki, *The Polish Peasant in Europe and America* (New York: Dover, 1958), 1.

67. Quoted in Bryant, *Positivism*, 137.

68. *The Positivist Dispute in German Sociology*, trans. Glyn Adey and David Frisby (London: Heinemann, 1976; orig. pub. in German, 1969), 11, 104. Cf. Agnes Heller, "The Positivism Dispute as a Turning Point in German Post-War Theory," *New German Critique*, no. 15 (Fall 1978).

69. *Positivist Dispute*, xxix.

70. Ernst Mach, *Popular Scientific Lectures* (La Salle, Ill.: Open Court, 1986; orig. pub. in German, 1894), 339, 348.

71. Bryant, *Positivism*, 122.

72. For further arguments, see my "Progress: An Historical and Critical View," in *Progress: Fact or Illusion?* ed. Leo Marx and Bruce Mazlish (Ann Arbor: University of Michigan Press, 1996), where also some attention is paid to postmodern critiques. The entire collection of essays should be consulted on its overall topic.

73. Kolakowski, *The Alienation of Reason*, 7.

74. It has been suggested that the sociology of knowledge might be viewed as a form of response by positivism to complaints as to its lack of insight about values. Perhaps one might look at the work of Karl Mannheim from this perspective.

Chapter 3. The Human Species as an Object of Study

1. Some humans are born or become blind; here, however, I am talking about the entire species. Those born blind come to "know" through the accumulated use of sight by their fellow humans.

2. Marvin Harris, *Our Kind* (New York: Harper Perennial, 1990), 45–46.

3. Alan Macfarlane, "Ernest Gellner and the Escape to Modernity," in *Transition to Modernity*, ed. John A. Hall and I. C. Jarvie (Cambridge: Cambridge University Press, 1992), 129.

4. Quoted in Robert Nisbet, *Social Change and History* (New York: Oxford University Press, 1969), 104. The trope of a single mind is persistent in modern Western thought, connecting with notions of culture and education. Compare Hegel's take on the subject:"The particular individual, so far as content is concerned, has also to go through the stages through which the general mind has passed, but as shapes once assumed by mind and now laid aside, as stages of a road which has been worked over and levelled out. Hence, it is that, in the case of various kinds of knowledge, we find that what in former days occupied the energies of men of mature mental ability sinks

to the level of information, exercise, and even pastimes, for children; and in this educational progress we can see the history of the world's culture delineated in faint outline" (*The Phenomenology of Mind*, trans. J. B. Baillie [New York: Harper Torchbooks, 1967], 89–90). In this formulation, the particular German notion of *Bildung*, the development of mind, is given a central role.

5. My colleague Michael Fischer reminds me that the Nuer are not simply a bounded Other but part of colonial and neocolonial society and thus intimately related to "us." My general point, however, is clear.

6. Eric R. Wolf, *Europe and the People Without History* (Berkeley: University of California, 1982), 98.

7. Ibid., 91.

8. Jared Diamond, "What Are Men Good For," *Natural History* 5 (1993), 24.

9. Norbert Elias, *The Symbol Theory*, ed. Richard Kilminster (London: Sage, 1991), 95. While many other works have followed theirs, I have found both Elias's book and Susanne K. Langer's *Philosophy in a New Key* (New York: New American Library, 1949) especially helpful in thinking of humans as symbolic animals.

10. Elias, *Symbol Theory*, 96, 133.

11. Jean-Jacques Rousseau, *Emile*, trans. Allan Bloom (New York: Basic Books, 1979), 186.

12. Ernst Mayr, *The Growth of Biological Thought: Diversity, Evolution, and Inheritance* (Cambridge: Belknap Press of Harvard University Press, 982), 63. The concept of emergence implicitly calls into question the validity and usefulness of reductionism.

13. See Stephen Jay Gould, *Wonderful Life* (New York: W. W. Norton, 1989), 321–23.

14. Another version of emergence, "Emergent AI [Artificial Intelligence]," comes from work on computers. Early computers had, and have, rules rigidly programmed into them and can process information in ways that go beyond human computational skills. However, they cannot go beyond what is programmed into them. In emergent AI, instead of rules as such, a network of independent elements is set up inside the computer, and the interaction of these elements produces intelligence; in short, learning takes place. Gerald Edelman, in his version of the theory, called Neural Darwinism, takes the view "that the brain is a selective system more akin in its workings to evolution itself than to computation" ("Through a Computer Darkly: Group Selection and Higher Brain Function," *Bulletin of the American Academy of Arts and Sciences* 36 [October 1982], 21). See also my *Fourth Discontinuity: The Co-Evolution of Humans and Machines* (New Haven: Yale University Press, 1993), 193.

15. Bert Holldubler and Edward O. Wilson, *Journey to the Ants* (Cambridge: Harvard University Press, 1994), is an enthusiastic treatment of its subject.

16. On fire in human life, see Johan Goudsblom, *Fire and Civilization* (London: Allen Lane, 1992).

17. The connection of this assertion with the theories concerning the social con-

struction of science should be obvious. For interesting observations on Linnaeus's missionaries, see Lisbet Koerner, "Peregrinum in Patria: Cross-Cultural Encounters in Early Modern Natural History," MS presented to the Department of the History of Science, Harvard University, February 6, 1995.

18. Clifford Geertz, *Works and Lives: The Anthropologist as Author* (Stanford, Calif.: Stanford University Press, 1988), 115. Geertz himself is very alive to the conditions of the modern world. Some anthropologists claim that modern anthropology does not ignore change. As in all such debates, we are faced with a question of degree.

19. See Donald E. Brown, *Human Universals* (Philadelphia: Temple University Press, 1991). In addition to psychological phenomena, Brown includes among his universals "the use of fire and tools, a division of labor by sex, and much more" (5).

20. For a balanced treatment of sociobiology, see especially Mary Midgley, *Beast and Man: The Roots of Human Nature* (Ithaca, N.Y.: Cornell University Press, 1978); and Melvin Konner, *The Tangled Wing: Biological Constraints on the Human Spirit* (New York: Holt, Rinehart, and Winston, 1982); and almost all the writings of Edward O. Wilson. The literature on the subject is voluminous. The trick with sociobiology is to retain its valid findings while avoiding its extreme and overreaching statements.

21. Hans Blumenberg, *The Genesis of the Copernican World*, trans. Robert M. Wallace (Cambridge, Mass.: MIT Press, 1987), 6. The next quotation is from p. 3.

22. In fact, in the course of human development all forms of knowledge proceeded, and proceed, conjointly, if erratically; that is, the Comtean progression is a false rigidification.

23. One ought at least to note Stephen Jay Gould's objections and qualifications to the view that evolution is going toward increased complexity.

24. Johan Goudsblom, chapter 1, in *Human History and Social Process*, by Johan Goudsblom, E. L. Jones, and Stephen Mennell, Exeter Studies in History no. 26 (Exeter, England: University of Exeter Press, 1989), 23.

25. Paolo Rossi, in *Philosophy, Technology, and the Arts in the Early Modern Era*, trans. Salvator Attanasio (New York: Harper and Row, 1970), instancing Leibniz and Pascal, describes their view that the history of the world is comparable to the history of an individual and comments that "the cognizing subject is the whole of mankind that operates in history" (133). Such a statement seems also in accordance with Hegel's conception of Reason. In a different vein, Freud has written of the "omnipotence of thought"; should we, therefore, view the philosophes and Hegel as examples of the narcissism of intellectuals, who identify their own reified thought with the history of the human species?

26. One might speculate whether the use of the computer to create cyberspace might be a new instrumental way of creating a new phenomenal world.

27. Edward O. Wilson, *The Diversity of Life* (New York: W. W. Norton, 1992), 38. The next quotation is from p. 39.

28. We know this evolutionary story only because we have developed culturally,

with one part of the culture being science—in this case, geological and biological science. Though building on perhaps twelve thousand years of human mentation, this knowledge is a very late possession, coming within the past few hundred years. Not until the 1850s can archaeology be said to have emerged as a discipline, with the publication of Daniel Wilson's *Archaeology and Prehistoric Annals of Scotland* (1851). Slightly earlier, the emergence of geology and, slightly later, the discovery of Neanderthal remains (1857) and the publication of Darwin's *Origins of Species* (1859)contributed to the great expansion of knowledge about the human species and especially its existence over extended time. As for Neanderthal man (a sexist expression), there is a certain looseness in assertions about him. Sometimes the name is spelled Neandertal (Marvin Harris, *Our Kind* [New York: Harper and Row, 1989]), and sometimes the date of discovery of the first skull in Europe is given as 1856 (J. W. Burrow, *Evolution and Society* [Cambridge: Cambridge University Press, 1966], 116). As for Cro-Magnons, their dates stretch from thirty thousand to forty-five thousand years ago. None of this impreciseness, which is inherent in the subject, affects the points being made here.

Chapter 4. Hermeneutics

1. A different history of hermeneutics could be written, as my colleague Michael Fischer suggests, emphasizing the tradition of intersubjective understanding and invoking more centrally the Dilthey-Weber-Schutz tradition, along with the linguistic one associated with such names as Ferdinand de Saussure and Benjamin Lee Whorf. So, too, greater overt stress could be placed on the efforts of some thinkers to be both positivist and interpretive. Although I shall pay some attention to the combinatory effort, especially in Dilthey, my main intention is to treat hermeneutics as a theoretical counterpoint to positivism and to inquire into its validity in the human sciences.

2. Cf. Erich Rothacker, *Logic und Systematik der Geisteswissenshaften* (Bonn: H. Bouvier, 1947), 6. Mill's collected works were translated into German under the direction of Theodore Gomperz, a classical philologist and philosopher who, upon reading *System of Logic,* immediately enrolled himself as a disciple of the great English Utilitarian. For the twelfth and final volume of the collected works, a twenty-four-year-old Austrian physician was assigned to translate four of Mill's essays: Sigmund Freud, whose contributions to interpretation—which we are calling hermeneutics—are not insignificant.

3. John Stuart Mill, "On the Definition of Political Economy," in *Essays on Economics and Society,* ed. J. M. Robson (Toronto: University of Toronto Press,1967), 316–317, for this and the following quotations.

4. Quoted in Keith Michael Baker, *Condorcet; From Natural Philosophy to Social Mathematics* (Chicago: University of Chicago Press, 1975), 197, 86.

5. Hegel constantly used the notion "Spirit of the Times" in his lectures on the

history of philosophy, delivered nine times between 1805-6 and 1829-30. The passage cited is quoted in Theodore Ziolokowski, *German Romanticism and Its Institutions* (Princeton, N.J.: Princeton University Press, 1990), 13-14.

6. Quoted in ibid., 14. I am reminded here of Oswald Spengler and his use of this "spiritual" idea in his *Decline of the West*. Partly influenced by Spengler, Ruth Benedict later brought the idea back to the smaller circle of anthropology with her *Patterns of Culture* and the notion of a whole.

7. John Stuart Mill, "The Spirit of the Age," in *Essays on Politics and Culture*, ed. Gertrude Himmelfarb (Garden City, N.Y.: Doubleday, 1963), 1. William Hazlitt published his book *The Spirit of the Age* in 1825; in it he referred to William Wordsworth as "a pure emanation of the Spirit of the Age" (quoted in M. H. Abrams, "English Romanticism," in *Romanticism Reconsidered: Selected Papers from the English Institute*, ed. Northrop Frye (New York: Columbia University Press, 1963).

8. Thomas Carlyle, "Signs of the Times," in *Thomas Carlyle: Selected Writings*, ed. Alan Shelston (Penguin Books, 1971), 68, 64; Mill, "Spirit of The Age," 43.

9. See further my *James and John Stuart Mill: Father and Son in the Nineteenth Century* (New York: Basic Books, 1975), chap. 16.

10. In Germany, as in other countries, there are numerous schools of thought — a strong materialist tradition in the nineteenth century, for example, and positivism, represented by Mach, in the twentieth. But the hermeneutic tradition does seem to have flourished especially in German thought.

11. Quoted in Richard E. Palmer, *Hermeneutics: Interpretation Theory in Schleiermacher, Dilthey, Heidegger, and Gadamer* (Evanston, Ill.: Northwestern University Press, 1969), 84. This is an indispensable work. In all matters related to hermeneutics, the numerous writings of Paul Ricoeur are classic. For interesting short characterizations of Schleiermacher and Dilthey, see, for example, Ricoeur's *Hermeneutics and the Human Sciences: Essays on Language, Action and Interpretation*, trans. John B. Thompson (Cambridge: Cambridge University Press, 1981).

12. As Alvin Gouldner remarks, "There is little doubt that hermeneutics' roots in the modern era are traceable to Romanticism." He is quoted in Dmitri Shalin, "Romanticism and the Rise of Sociological Hermeneutics," *Social Research* 53 (Spring 1986), 77. This is an excellent article, filled with many suggestive thoughts.

13. Hans-Georg Gadamer, *Truth and Method*, 2d rev. ed. (New York: Crossroad, 1992; orig. pub. in German, 1960), 179. Although I disagree with much of his argument, I am in his debt: he is a wonderful source for the history of hermeneutics. Beyond that, although it may not be obvious, Gadamer lays down fundamental ideas about the public sphere. He is concerned with the same arena as Habermas, with whom he entered into dialogue. In the present book I quote Habermas frequently. I admire his work, although our approaches are often quite different. His notion of evolution is heavily tinged by Marxism (even though he is critical of it), mine more by Darwin and biological science. His empiricism is developed in terms of linguistics and cognitive development, mine in regard to historical studies and especially the

history of science. He comes out of the Frankfurt School, with its emphasis on critical theory; I do not. He seems, rather surprisingly, to embrace a kind of teleology— "mankind's evolution toward autonomy and responsibility" (*Knowledge and Human Interests* [Boston: Beacon Press, 1971], 315); I am prepared to agree that this is a worthwhile goal, but I cannot accept the deterministic implications of his statement. And so on. In the final analysis, he thinks primarily like a philosopher—a great and brilliant one—and I primarily like a historian, though of a "philosophical" sort.

14. A comparison with Erik Erikson's treatment of evidence and inference in psychoanalysis is in order here. See his *Insight and Responsibility* (New York: W. W. Norton, 1964), chap. 2.

15. Quoted in Shalin, "Romanticism," 104.

16. Wilhelm Dilthey, *Pattern and Meaning in History,* ed. H. P. Rickman (New York: Harper and Brothers, 1962), 125. A six-volume translation of Dilthey's key writings is being published by Princeton University Press. See especially *Introduction to the Human Sciences,* vol. 1 of *Wilhelm Dilthey: Selected Works,* ed. Rudolf A. Makkreel and Frithjof Rodi (Princeton, N.J.: Princeton University Press, 1989).

17. Dilthey, *Pattern,* 77.

18. Dilthey apparently embraced the model of empathy as a way of entering into the thoughts of others. Jürgen Habermas, critical of this model, sought to rescue Dilthey by recasting his final position: "Different regions of facts cannot be conceived of ontologically but only epistemologically. They do not 'exist'; rather, they are constituted. The difference between the natural and cultural sciences must therefore be reduced to the orientation of the knowing subject, its attitude with regard to objects" (*Knowledge and Human Interests,* trans. Jeremy J. Shapiro [Boston: Beacon Press, 1971], 141). See, however, Michael Ermarth, *Wilhelm Dilthey: The Critique of Historical Reason* (Chicago: University of Chicago Press, 1978), which offers a different reading of Dilthey's use of empathy. There is also a heated controversy among Dilthey scholars over the meaning of *Verstehen* (understanding) and *Erlebnis* (experience).

19. Quoted in Palmer, *Hermeneutics,* 28.

20. See Chapter 4, n. 13.

21. Gadamer, *Truth and Method,* xxi–xxii.

22. Ibid., 5. The next quotation is from p. 82.

23. Ibid., 284.

24. Ibid., 278. See also p. 277. A sympathetic commentary on prejudice in Gadamer is to be found in Richard J. Bernstein, *Beyond Objectivism and Relativism: Science, Hermeneutics and Praxis* (Philadelphia: University of Pennsylvania Press, 1983), 126–31, where the emphasis is placed on Gadamer's argument that we cannot confront the world without prejudices or preconceptions, but that these can then be tested against the prejudices of others, with the result being increased knowledge.

25. Gadamer, *Truth and Method,* 192. For Gadamer's view on romantic hermeneutics, see p. 296.

26. This is a tendency of some postmodernists. As Stanley Rosen puts it for one,

"Derrida's deconstructive hermeneutics is not attentive to the written text in its own terms, largely because he denies the significance of the author's intentions" (*Hermeneutics as Politics* [New York: Oxford University Press, 1987], 57).

27. Erwin Schrödinger, *What Is Life?* (Cambridge: Cambridge University Press, 1992), 162. Dealing with physiology in the nineteenth century, Lorraine Daston and Peter Galison offer a fascinating handling of the subject in "The Image of Objectivity," *Representations* 40 (Fall 1992), 81–128.

28. Stephen Jay Gould, *Time's Arrow, Time's Cycle: Myth and Metaphor in the Discovery of Geological Time* (Cambridge: Harvard University Press, 1987), 149. See also p. 152. I might add that Lyell's model for geology seems to have been history; see his *Principles of Geology,* 1st ed., vol. 1 (Chicago: University of Chicago Press, 1990), 1–4.

29. Raymond Firth, *We, the Tikopia* (Boston: Beacon Press, 1963), 3. Ironically, the preface to this book was written by Malinowski. In it, he speaks of Dr. Firth doing his work "quite independent of anything but the evidence of facts" and lauds cultural anthropology as "the science among social studies," for, after all, "science lives by inductive generalization, and this is only possible either by experiment or comparative study" (viii–xi). The positivist aspiration is strong.

30. Interpretation is a fundamental part of all human knowing because humans became symbol-making animals in the course of evolution. Symbols are different from signs. Signs indicate things; they are signals. A whistle may indicate to a dog that food is on its way. Symbols, in contrast, are abstracted from the sense impressions from which they arise. They can be thought of as substitute signs that have taken on a new dimension: having separated from the original impulse, they can be made into mental images that are timeless and that can be combined and recombined flexibly in the mind. As Susanne K. Langer tells us, "The symbol-making function is one of man's primary activities, like eating, looking, or moving about" (*Philosophy in a New Key* [New York: New American Library, 1948], 32).

Symbol making also gets humans into endless trouble. It can distort reality, create nonexistent beings, and conjure up spectres of the mind—like ideologies—for which people gladly fight and die. It gives rise to dreams and fantasies, anxieties and hopes, that may inspire great works of art or great advances in science but may also lead to psychological breakdown. The wires in the mental switchboard, playing with the ghosts of past sensations, may go haywire.

31. Richard J. Bernstein, *The Restructuring of Social and Political Theory* (Philadelphia: University of Pennsylvania Press, 1976), 145. Cf. Alfred Schutz, *The Phenomenology of the Social World,* trans. George Walsh and Frederick Lehnert (Evanston, Ill.: Northwestern University Press, 1967); and Schutz, *Collected Papers,* vol. 1: *The Problem of Social Reality,* ed. Maurice Natanson (The Hague: Martinus Nijhoff, 1962). Bernstein offers some thoughtful criticism of Schutz (whom he appears to admire), pp. 156–169. I find Schutz the most illuminating and inspiring of the phenomenologists; he goes well beyond Husserl both in clarity and in contributing to our thinking about the social sciences.

32. Clifford Geertz, *The Interpretation of Cultures* (New York: Basic Books, 1973), 5.

33. Geertz himself seems to have swung from a view of culture as "a positive science like any other" (quoted in Christopher Herbert, *Culture and Anomie* [Chicago: University of Chicago, 1991], 18) to a view of culture as singularly interpretive. I am suggesting that we keep our minds and options open.

34. F. A. Hayek, *The Counter-Revolution of Science: Studies on the Abuse of Reason* (Glencoe, Ill.: Free Press of Glencoe, 1955), 20–21. The first quotation from Hayek can remind us of Gyan Prakash's story of the appropriation of modern science into India (see Chapter 1).

35. Habermas criticized Dilthey's embrace of empathy, which should give us pause, but not if we realize that Habermas was attacking empathy as a kind of copy theory of truth rather than judging it as a method or tool for achieving understanding of others, as well as better self-understanding.

36. Hans Blumenberg, *The Genesis of the Copernican World*, trans. Robert M. Wallace (Cambridge, Mass.: MIT Press, 1987; orig. pub. in German, 1975), 60.

37. Shortly before Galileo, Kepler pioneered in the new way of looking at appearances and at the instruments through which they were examined. "His radical answer," says Svetlana Alpers, "was to turn his attention away from the sky and the nature of light rays to the instrument of observation itself: to turn from astronomy to optics. Kepler argued that what was at issue was the precise optics of the images formed behind small apertures in the pinhole camera. . . . To understand our view of sun, or moon, or world, we must understand the instrument with which we view it, an instrument, so Kepler argued, with distortion or errors built in. In this 1604 publication Kepler went on to take the next step of recognizing the necessity of investigating our most fundamental instrument of observation, the eye, which he now in effect described as an optical mechanism supplied with a lens with focusing properties" (*The Art of Describing: Dutch Art in the Seventeenth Century* [Chicago: University of Chicago Press, 1983], 33–34). Two observations are in order. First, we must note that the eye has been objectivized. It is now treated as a mechanical device rather than, say, as a reflection of our soul. Second, the idea of investigating the very lens with which we look at phenomena must be extended now to the human sciences. We must examine the lens—in this case, the theories and concepts of the particular disciplines—with which we observe human phenomena.

38. Quoted in Blumenberg, *Genesis*, 658.

39. Typical was the Royal Society's "Directions for Seamen, bound for far voyages" of 1665–66: "It being the Design of the *Royal Society*, for the better attaining the End of their Institution, to study *Nature* rather than *Books*, and from the Observations, made of the *Phenomena* and Effects she presents, to compose such a History of Her, as may hereafter serve to build a Solid and Useful Philosophy upon" (quoted in Bernard Smith, *European Vision and the South Pacific*, 2d ed. [New Haven: Yale University Press, 1985], 8).

40. Quoted in Blumenberg, *Genesis*, 40.

41. The quotations from Galileo here and below are from *Discoveries and Opin-*

ions of Galileo, trans. Stillman Drake (Garden City, N.Y.: Doubleday, 1957), 182, 179, 186, 187, 194.

42. Blumenberg, *Genesis,* 38.

43. Quoted in Alpers, *Art of Describing,* 17.

44. The seminal essay on linear perspective, especially in regard to painting, is Erwin Panofsky's, first published in 1927: *Perspective as Symbolic Form,* translated from the German by Christopher S. Wood (New York: Zone Books, 1991). Panofsky touches on painterly perspective as an attempt at "distancing," that is, at greater objectification, but qualifies this claim by noting that "the history of perspective may be understood with equal justice as a triumph of the distancing and objectifying sense of the real, *and* as a triumph of the distance-denying human struggle for control; it is as much a consolidation and systematization of the external world, as an extension of the domain of the self" (67–68; my italics). The introduction by the translator is both historical and critical.

Samuel Y. Edgerton, Jr.,'s book *The Renaissance Discovery of Linear Perspective* (New York: Basic Books, 1975) is a beautiful treatment of the subject. His overall thesis is that linear perspective was vital to the scientific revolution that occurred uniquely in the West. The origins of linear perspective are in antiquity, where scenic painting hinted at its possibilities and where Euclid's geometry and then Ptolemy's *Geographia* (which appeared in western Christendom around 1400) supplied the foundations. Key figures in the renaissance of ancient perspective are Brunelleschi, who, possibly inspired by the geographer Toscanelli, conducted the first experiment, using a mirror, to demonstrate linear perspective; and Alberti, who wrote the first treatise on painting showing how to draw in the new mode. Edgerton then describes the development of linear perspective in the service of painting. What was the revolution in perspective? As Edgerton tells us, the medieval painter "saw each element in his composition separately and independently. . . . He was absorbed with the visual world he was representing rather than, as with the perspective painter, standing without it, observing from a single, removed viewpoint." (21) Linear perspective meant "the power to render an abstract image of space in our minds, regulated by an inflexible coordinate framework of horizontals and verticals" (114).

Can we extend Edgerton in one direction and make the case that linear perspective allowed for the same objective repetition of experiences as found in the witnessed experiments of Boyle? For example, as Edgerton tells us, "this new Quattrocento mode of representation was based on the assumption that visual space is ordered a priori by an abstract, uniform system of linear coordinates. The artist need only fix himself in one position for the objective field to relate to this single vantage point. He can then represent the objects in his picture in such a way that the viewer can apprehend the scene *exactly as if he were standing in the same place as the artist*" (7; my italics). Another short quotation supports the point that I am trying to make: "His [Alberti's] perspective construction was intended to copy no specific place. It provided a purely abstract realm which the viewer would discern as a world of order" (30). In sum, the artistic-scientific experiment is one carried out in an abstract, universal world of space and time, whether by seeing exactly as the painter does or doing exactly as the virtuoso does.

Edgerton followed up his first book with another, *The Heritage of Giotto's Geometry: Art and Science on the Eve of the Scientific Revolution* (Ithaca, N.Y.: Cornell University Press, 1991). Here he makes a sustained comparison of science in the West and in China, focusing on the critical presence of linear perspective in one and its absence in the other. As he remarks, after the fourteenth century and only in the West, "space began to be conceived as continuous and uniform in all directions (isotropic)" (6). Linear perspective "demanded that all space, celestial and terrestrial alike, be perceived as having the same physical properties and obeying the same geometrical rules" (17). The consequences of this way of viewing ramified into the development of the telescope and the microscope ("invented according to the same optical principles that underlie Renaissance perspective painting" [20]) and out to the mappings of Toscanelli (with his probable influence on Columbus), Ortelius, and Mercator. Whereas Copernicus's "epochal *De revolutionibus orbium coelestium* of 1543 had only a flat planispheric diagram to illustrate what was to be the most revolutionary stereometric thought in all premodern science" (166), the geographic imagination of the Age of Discovery plotted itself along the meridians and parallels of the celestial globe as if viewed from a detached single viewpoint far out in space.

As can be seen, Edgerton has much to offer. Our particular focus is on the development of a nonprivileged perspective, though one that, once taken, obliges its viewer to see and to know along set lines. As suggested by the comment about Copernicus's flat planispheric representation, even his revolutionary imaginative step out into space, where he saw with God's eye, was of the "two steps forward, one step backward" kind. The triumph of the perspectival look was not achieved in one fell swoop. Nevertheless, it did achieve victory in the West at the time of the Renaissance, and one can link that domination to the question of the Other.

45. We catch the flavor of horror at the thought of the Other in the English essayist Thomas De Quincey. Writing not about primitives but about Chinese, he admits that "I am terrified by the modes of life, by the manners, by the barrier of utter abhorrence placed between myself and *them*, by counter-sympathies deeper than I can analyse. I could sooner live with lunatics, with vermin, with crocodiles or snakes" (quoted by Lewis D. Wurgaft in his review of John Barrell, *The Infection of Thomas De Quincey* [New Haven: Yale University Press, 1991], in the *New York Times Sunday Book Review*, June 2, 1991, p. 27).

46. William Brandon, in *New Worlds for Old* (Athens: Ohio University Press, 1986), examines the way the Western notion of liberty drew sustenance, if not inspiration, from the early explorers' depiction of the New World inhabitants as living in nonhierarchical societies.

47. There is probably a correlation between increased awareness of the Other and

the increased number of roles played by individuals in modern society. As the individual plays a role, he or she must entertain new perspectives, which necessarily includes seeing things from the perspective of other role-players.

48. Michel Foucault, *The Order of Things: An Archaeology of the Human Sciences* (New York: Vintage Books, 1973; orig. pub. in French, 1966). On Foucault, see the Introduction, n. 5. Some scholars, it should be conceded, question Foucault's inclusion as a postmodernist. Foucault sees himself as a kind of anthropologist of modern culture. As he remarked in an interview, his research is "something like an ethnology of the culture to which we belong. In point of fact, I attempt to place myself outside the culture to which we belong" (quoted in Axel Honneth and Hans Joas, *Social Action and Human Nature*, trans. Raymond Meyer [Cambridge: Cambridge University Press, 1988; orig. pub. in German, 1980], 131). Foucault has also described his book *Madness and Civilization* as a "history of the Other."

Another line of thought, to be found among thinkers such as George Herbert Mead and Alfred Schutz, seeks affirmatively to understand the Other as part of an epistemological analysis of intersubjectivity. In their schemes, the Other is essential for the construction of *any* knowledge, whether of a commonsense, daily-understanding kind or of a scientific-theoretical kind. See, for example, Alfred Schutz, *Collected Papers*, vol. 1: *The Problem of Social Reality*, ed. Maurice Natanson (The Hague: Martinus Nijhoff, 1971), 13–26. A fuller development of the history of the idea of the Other would certainly pursue this line of inquiry, involving as it does ideas about sympathy, impartial spectators, and connections.

49. As Derrida is quoted as saying, "Il n'y a pas de hors texte" (There is nothing outside the text). In one of the better books on the subject (from which the Derrida quotation is drawn, p. 157) — Jonathan Culler, *On Deconstruction: Theory and Criticism After Structuralism* (Ithaca, N.Y.: Cornell University Press, 1982) — it seems to be implied at one point that even interpretation is not the aim of postmodern literary theory. We are told that the philosophical writings on which deconstructionists should draw "do not find their justification in the improvement of interpretations" (8). However, this appears to be a momentary confusion, for Stanley Fish is later quoted approvingly to the effect that "everything is constituted by interpretation" (75).

More important for our purposes is Culler's effort to distinguish poststructuralists, who generally can also be called deconstructionists or postmodernists, from structuralists: "Structuralists are convinced that systematic knowledge is possible; post-structuralists claim to know only the impossibility of this knowledge" (22). On a more affirmative note, we are told that "to deconstruct a discourse is to show how it undermines the philosophy it asserts, or the hierarchical oppositions on which it relies, by identifying in the text the rhetorical operations that produce the supposed ground of argument, the key concept or premise" (86). The injunction to carefully tease out the "warring forces of signification within the text" (213) is good hermeneutical advice, reminding careful interpreters of what they have been presumably doing all along but must now do more consciously and systematically. For further reference to postmodernist texts, see the Critical Bibliography.

50. In the words of Nietzsche, the father of most postmodernism, "The most extreme form of nihilism" prevails, for reality is nonexistent except as "a *perspectival appearance* whose origin lies in us" (*The Will to Power,* trans. W. Kaufmann and R. J. Hollingdale [New York: Vintage Books, 1968], 14–15).

51. Louis A. Sass, *Madness and Modernity: Insanity in the Light of Modern Art, Literature, and Thought* (New York: Basic Books, 1992). Although Sass speaks only of modernism, postmodernism is readily included under his rubric.

52. One modern writer, William Burroughs, has sought to hot-wire the mind and to create distortion (he also coined the phrase *heavy metal* to describe a version of rock music).

53. Edward O. Wilson, *On Human Nature* (Cambridge: Harvard University Press, 1978), 58.

54. Sass, *Madness and Modernity,* 14, which also quotes Bleuler.

55. Oliver Sachs, however, has emphasized the way schizophrenics are able to function, often effectively, in society.

56. *Diagnostic and Statistical Manual of Mental Disorders,* 3d ed., rev. (DSM III-R) (Washington, D.C.: American Psychiatric Press, 1987), 404, 395, 398.

57. My esteem for my colleague Frank Sulloway is great, but he is, alas, a typical example of the denigrating tendencies of which I speak. His book *Freud: Biologist of the Mind* (New York: Basic Books, 1979) is a brilliant piece of research, the first on the subject by a trained historian of science. Though critical of the myth surrounding Freud, Sulloway fully recognizes his extraordinary genius and achievement (see, for example, pp. 499–500). More recently, however, he has mounted the Freud-bashing bandwagon, seeing more fraud than Freud from where he now sits. But he has done this without coming to grips with his earlier praise and explaining why it no longer holds. Needless to say, there are fashions in intellectual history, as in other areas, and overestimations of Freud are followed by detestations, and so on, with, one hopes, the indicator eventually coming to rest at a more balanced judgment. For an early article of mine on Freud's claims to offer scientific knowledge of the psyche as against Nietzsche's poetic and philosophical insights into the mind, see "Freud and Nietzsche," *Psychoanalytic Review* 55, no. 3 (1968), 360–75; reprinted in my book *The Leader, the Led, and the Psyche: Essays in Psychohistory* (Hanover, N.H.: Wesleyan University Press, 1990).

58. A classic exposition of Freud's early ideas is his own *Introductory Lectures on Psycho-Analysis (1915–17).* For a sensitive exposition of his basic ideas and methodology, see Jürgen Habermas, *Knowledge and Human Interests* (Boston: Beacon Press, 1971), chaps. 10–12.

59. The two works that I have found especially enlightening are Peter Brooks, *Reading for the Plot: Design and Intention in Narrative* (Cambridge: Harvard Univer-

sity Press, 1984); and Frank Kermode, *The Sense of an Ending* (New York: Oxford University Press, 1967).

60. Quoted in *The Literary Structure of Scientific Argument: Historical Studies*, ed. Peter Dear (Philadelphia: University of Pennsylvania Press, 1991), 165.

61. Jürgen Habermas, *On the Logic of the Social Sciences*, trans. Shierry Weber Nicolsen and Jerry A. Stark (Cambridge, Mass.: MIT Press, 1991), 156.

62. Charles Darwin, *The Voyage of the Beagle* (London: J. M. Dent and Sons, 1960), 229.

63. For a fuller discussion of Freud's use of narrative, see the chapter on Darwin in my *The Leader, the Led, and the Psyche*.

64. Peter Manicas, though with Marx and Max Weber in mind, argues that "in the last analysis, a theoretically informed multi-causal history is the human science which has the most significance for us" (*History and Philosophy of the Social Sciences* [Oxford: Basil Blackwell, 1987], 279).

65. See, for example, Henri Frankfurt et al., *Before Philosophy* (Penguin Books, 1946).

66. Hans Blumenberg, *Work on Myth*, trans. Robert M. Wallace (Cambridge, Mass.: MIT Press, 1990), 3–4. Cf. Robert B. Pippin, "Modern Mythic Meanings: Blumenberg Contra Nietzsche," *History of the Human Sciences* 6 (November 1993), 38–39. The entire issue is devoted to Blumenberg and is well worth consulting.

67. Bronislaw Malinowski, *Argonauts of the Western Pacific* (Prospect Heights, Ill.: Waveland Press, 1984; orig. pub., 1922), 328.

68. For a fundamental analysis of the differences between myth and the novel, see M. M. Bakhtin, *The Dialogic Imagination: Four Essays*, ed. Michael Holquist, trans. Caryl Emerson and Michael Holquist (Austin: University of Texas Press, 1996).

69. Blumenberg, *Work on Myth*, 34.

70. Gananath Obeyesekere, *The Apotheosis of Captain Cook: European Mythmaking in the Pacific* (Princeton, N.J.: Princeton University Press, 1992), 10. In this fascinating book the author exemplifies the comment about mythmaking being prolific in European thought by arguing that the apotheosis of Captain Cook is itself a piece of Western mythmaking; not the Hawaiians but the Europeans deified him. Obeyesekere also points out that "myth is a term from Western thought. It rarely exists as a category or conceptual term in other cultures" (59). I would add that this is probably so because only with the development of the concept of logos can the possibility of conceptualizing mythos arise—which is *not* to say that primitives or non-Westerners are not logical in their thinking processes. Obeyesekere's major opponent in this debate is Marshall Sahlins, who has written a detailed rebuttal, *How 'Natives' Think: About Captain Cook, for Example* (Chicago: University of Chicago Press, 1995). It is also worth noting that myths can be manufactured or contrived with ideological purposes in mind, as with the Nazis. Such myths differ from genuine ones, as Blumenberg argues, in that they are not subject to the "Darwinism of words," the selection that takes place through millennia of storytelling (*Work on Myth*, xxvi).

71. Brooks, *Reading for the Plot*, xii.

72. Tony Tanner, *Jane Austen* (Cambridge: Harvard University Press, 1986), 51.

73. Brooks, *Reading for the Plot*, 27.

74. Sass, *Madness and Modernity*, 156. Peter Brooks gives a luminous analysis of the destabilizing tendencies of modern narrative; see especially *Reading for the Plot*, chap. 9.

75. Quoted in Stephen Greenblatt, *Marvelous Possessions: The Wonder of the New World* (Chicago: University of Chicago Press, 1991), 21. On Diaz, see pp. 128–29.

76. Anthony J. Cascardi, *The Subject of Modernity* (Cambridge: Cambridge University Press, 1992), 81.

77. Daniel Defoe, *Robinson Crusoe* (New York: New American Library, 1961), 7; and Henry Fielding, *Joseph Andrews*, ed. Martin C. Battestin (Middletown, Conn.: Wesleyan University Press, 1967), 10. A splendid treatment of realism in the early novel is Ian Watts, *The Rise of the Novel* (London: Chatto and Windus, 1957).

78. See, for example, the magisterial account by Michael McKeon, *The Origins of the English Novel, 1600–1740* (Baltimore, Md.: Johns Hopkins University Press, 1987).

79. As Peter Manicas puts it, "There is nothing precious in the label 'social science' . . . especially since distinct alternative 'genres,' particularly literature, may quite successfully offer us knowledge of persons and society" (*History and Philosophy of the Social Sciences*, 279).

80. Brooks, *Reading for the Plot*, 219. Balzac's comment comes from his short story "Facino Cane."

81. It is only fitting that this comment on the novel be made by a character in a novel, that is, by a fictitious character who nevertheless speaks truth. The character is Dr. Stephen Maturin, in Patrick O'Brian's novel *The Nutmeg of Consolation* (New York: W. W. Norton, 1991), 253.

82. Sally Beauman, "Encounters with George Eliot," *New Yorker*, April 16, 1994, p. 93.

83. Pierre Bourdieu, *Outlines of a Theory of Practice*, trans. Richard Nice (Cambridge: Cambridge University Press, 1985), 10.

84. There is a major difference between literature and myth, however. It lies in the capacity of the former to move in realistic time and to deal with secular characters. Though borrowing mythical attributes, literature can aim at more mundane truths. Salman Rushdie argues passionately, for example, for the political nature of his work. In regard to contemporary involvements with Nicaragua, Vietnam, and so forth, Rushdie writes, "It seems to me imperative that literature enters such arguments, because what is being disputed is nothing less that *what is the case,* what is truth and what untruth." He goes on: "There is a genuine need for political fiction, for books that draw new and better maps of reality, and make new languages with which we can understand the world" (*Imaginary Homelands* [London: Granta Books, 1992], 100). Rushdie is aware of the difficulties in carrying out this task; he warns against a "highly selective truth, a novelist's truth masquerading as objective

reality" (374; cf. 376). Nevertheless, truth is the grail that the novelist must pursue. Is it strange that a practitioner of magical realism ranges literature among the disciplines that seek to portray objective reality? I think not.

85. Let me give an instance from a realm different from literature. For me, a prime example of a "truthful" experience is the second movement of Beethoven's Piano Trio in D Major, op. 70, no. 1 ("Ghost"). When listening to it, I know that I am in a state of transcendence. I am therefore aware of the reality—the truthfulness—of the *experience* that many people have, which they call transcendent.

86. The term *genetic overshoot* may serve to describe this phenomenon.

87. There has been much argument recently that scientific persuasion shades imperceptibly into rhetorical persuasion. See, for an example, Donald N. McCloskey, *The Rhetoric of Economics* (Madison: University of Wisconsin Press, 1985). As with so much else, we are dealing with a question of degree.

88. A form of posthumous jurisprudence may be history. Indeed, one version of historical studies treats history as a kind of impartial and God-like judgment. A classic expression of this position can be found in Lord Acton. "Historical responsibility," he tells us, "has to make up for the want of legal responsibility. Power tends to corrupt and absolute power corrupts absolutely. Great men are almost always bad men, even when they exercise influence and not authority; still more when you superadd the tendency or the certainty of corruption by authority. . . . The inflexible integrity of the moral code is to me the secret of the authority, the dignity, the utility of history. If we may debase it for the sake of a man's influence, of his religion, of his party, of the good cause which prospers his credit and suffers by his disgrace, then History ceases to be a science, an arbiter of controversy" (quoted in *New York Times*, March 13, 1974). In short, history takes on a long-term and more scientific judicial function, repairing the lapses of contemporary judgments.

Judicial weighing of evidence figured not only in the development of history and hermeneutics but in classical probability theory as well. Most accounts of the development of probability theory trace it to reflections on games of chance. According to Lorraine Daston, in *Classical Probability in the Enlightenment* (Princeton, N.J.: Princeton University Press, 1988), however, there is another source: "The hierarchy of proofs within Roman and canon law led mathematicians to conceive of degrees of probability as degrees of certainty along a graduated spectrum of beliefs, ranging from total ignorance or uncertainty to firm conviction or 'moral' certainty" (14). It appears, then, that testimony can be interpreted either hermeneutically or mathematically, or both.

89. Stanley Rosen, *Hermeneutics as Politics*, 163. The connection of this statement to Foucault's schematics concerning knowledge and power is obvious.

90. Brooks, *Reading for the Plot*, 269, 18.

91. See Pierre Bayle, *Historical and Critical Dictionary* (1697).

92. See the anthology *The Young Hegelians*, ed. Lawrence S. Stepelevich (Cambridge: Cambridge University Press, 1983); and an older but still useful treatment

by Sidney Hook, *From Hegel to Marx: Studies in the Intellectual Development of Karl Marx,* 2d ed. (Ann Arbor: University of Michigan Press, 1962; 1st ed., 1950).

93. Among the "other things" is Marx's claim to scientific certainty. Marxist scholars have argued endlessly about the existence of a young, more hermeneutic (and Hegelian) Marx and an old, more scientific Marx.

94. Martin Heidegger is one source of this concern with representation, with Michel Foucault taking up the torch from him. The topic as a whole appears conceptually muddled. It has, however, stimulated some exceptionally interesting work, like that found in the journal *Representations* and in the book by Timothy Mitchell, *Colonising Egypt* (Cambridge: Cambridge University Press, 1988).

95. My own scurrying in regard to existing MSS on Marx is recorded in my book *The Meaning of Karl Marx* (New York: Oxford University Press, 1984).

96. Sigmund Freud, "Constructions in Analysis," in *The Standard Edition of the Complete Psychological Works of Sigmund Freud,* ed. James Strachey (London: Hogarth Press, 1953–74), vol. 23, p. 262.

97. Brooks, *Reading for the Plot,* 123.

98. See especially Karl Mannheim, *Ideology and Utopia,* trans. Louis Wirth and Edward Shils (New York: Harcourt, Brace, 1952); and Mannheim, *Essays on the Sociology of Knowledge,* ed. Paul Kecskemeti (New York: Oxford University Press, 1952). For an interesting discussion of Mannheim's article "On the Interpretation of *Weltanschauung*" and his move from Dilthey and interpretation to the sociology of knowledge, see Goran Dahl, "Documentary Meaning — Understanding or Critique? Karl Mannheim's Early Sociology of Knowledge," *Philosophy and Social Criticism* 20, no. 1–2 (1994), 103–21.

99. An alternative to the notion of depth (though not a contradiction of it) is presented by the nineteenth-century British philosopher of science William Whewell. He offered a scientific methodology that stands in contrast to the classical method of simplification and experimental replication that he called "consilience (or jumping together) of inductions." Whewell assumed that because sources, or phenomena, were so diverse and numerous, they had to be juxtaposed in such a manner that all of the effects could be brought under one causal explanation. This is the method employed by Darwin in his theory of evolution by natural selection, where innumerable phenomena are made to submit to the one law. Freud followed Darwin in adopting the method of consilience, but in his case it took the shape of the famous, or infamous, concept of overdetermination. With him, numerous phenomena fall into a single causal explanation, say, dream interpretation, where the same word or symbol refers to many elements in the unconscious thought process. When we run the process the other way, a single unconscious drive or pattern of behavior manifests itself in innumerable different outward expressions. The challenge for historians is to use some form of consilience theory along with depth explanations in a convincing and useful way.

100. According to a recent article in the *New York Times,* "Scientists have shown

that people have a covert system in their brains for telling them when decisions are good or bad and that the system, which draws upon emotional memories, is activated long before people are consciously aware that they have decided something. Intuition and gut feelings have a firm biological basis, said Dr. Antonio Damasio, the lead author of a paper describing the experiment in the current issue of the journal Science" (March 4, 1997). As John Keats might ask, Is nothing sacred and immune from science?

Chapter 5. Some Achievements to Date

1. Cf. George W. Stocking, Jr., *Victorian Anthropology* (New York: Free Press, 1987), 59 and passim.

2. See, for example, *Art and Empire: Treasures from Assyria in the British Museum* (An Exhibit at the Metropolitan Museum of Art, May 2–August 13, 1995), ed. J. E. Curtis and J. E. Reade (New York: Metropolitan Museum of Art, 1995; distributed by Harry N. Abrams, N.Y.). Assyria was mentioned in Herodotus and the Bible. Early Arab geographers had somewhat more detailed information based on local tradition. By the twelfth century, a European traveler, the Jewish rabbi Benjamin of Tudela, noted its location, and later travelers commented on its ruins. However, the first scientific examination of Nineveh was not undertaken until 1820–21. It was against this background that Layard, in 1840, started the excavations that moved the field of archaeology to an entirely new level (see *Art and Empire*, 9 and passim).

3. A. J. Ayer, "Man as a Subject for Science," in *Philosophy, Politics and Society, Third Series*, ed. Peter Laslett and W. G. Runciman (New York: Barnes and Noble, 1967), 6.

4. Cf. Claude Blanckaert, "Buffon and the Natural History of Man: Writing History and the 'Foundational Myth' of Anthropology," *History of the Human Sciences* 6(February 1993), especially 36–37.

5. Lambert A. J. Quetelet, *A Treatise on Man and the Development of His Faculties* (Gainesville, Fla.: Scholars' Facsimiles and Reprints, 1969), viii.

6. Peter Manicas puts this nicely when he remarks that "the by now familiar practices and disciplinary divisions in the human sciences were constituted in the 20 or 30 years which span the turn of the nineteenth century, and I argue that this occurred in the United States. Put in other terms, if, as social scientists, we were to imagine ourselves transported to Oxford, the Sorbonne, or Harvard in, say, 1870, we would find almost nothing familiar. There would be no 'departments' of 'sociology' or 'psychology'; the research practices of the faculties and the modes of graduate instruction of those institutions would be for the most part alien. But we would find *very little* which is *not* familiar if we were to make a similar visit to *any* 'department' in *any* American university in 1925" (*A History and Philosophy of the Social Sciences* [Oxford: Basil Blackwell, 1987], 5).

7. On the early history of economic thought, William Letwin, *The Origin of Scientific Economics* (Garden City, N.Y.: Doubleday, 1964); and Joyce Appleby, *Economic Thought and Ideology in Seventeenth-Century England* (Princeton, N.J.: Princeton University Press, 1978), are especially illuminating.

8. The classic work on money as a human invention is Georg Simmel, *The Philosophy of Money*, trans. Tom Bottomore and David Frisby (London: Routledge and Kegan Paul, 1978).

9. An incisive article on the origins of economic theory is Susan Buck-Morss, "Envisioning Capital: Political Economy on Display," *Critical Inquiry* 21 (Winter 1995), 434–67.

10. Shirley P. Burggraf, "The Feminine Economy," in "The Feminine Economy," MS, 6. Burggraf goes on to remark that "until very recent times it [Smith's concept] has really applied only to men because women haven't been very active participants in the market economy." (The published version of Burggraf's MS, *The Feminine Economy and Economic Man* [Reading, Mass.: Addison-Wesley, 1997], 9, gives a slightly reduced version of these passages.)

11. Karl Marx, "Outlines of the Critique of Political Economy," in *Karl Marx, Frederick Engels: Collected Works*, vol. 28 (New York: International Publishers, 1986), 174–75. This is one of Marx's more useful works for our purposes. In spite of his pseudo-Darwinism, Marx grasped perhaps better than any other nineteenth-century thinker, aside from the Darwinians, the centrality of the human species as the subject of investigation in the social sciences. His statement about elephants not producing for tigers does, however, implicitly ignore the Darwinian perception of the ecological interrelatedness of species; the true comparison, of course, would consider whether elephants produce for other elephants. Nevertheless, in terms of conscious exchange, Marx is right.

12. Ibid., 41.

13. Louis Dumont, *Homo Hierarchicus*, trans. Mark Sainsbury, Louis Dumont, and Basia Gulati (Chicago: University of Chicago Press, 1980), 165. Dumont's account should be supplemented with Karl Marx's classic article "On Imperialism in India," conveniently found in *The Marx-Engels Reader*, ed. Robert Tucker (New York: W. W. Norton, 1978).

14. The literature on Smith is extensive. Especially useful in regard to the notion of political economy is the treatment in Donald Winch, *Riches and Poverty: An Intellectual History of Political Economy in Britain, 1750–1834* (Cambridge: Cambridge University Press, 1996). For the manner in which Smith's political economy found its way into the thought of Hegel, see Norbert Waszek, *The Scottish Enlightenment and Hegel's Account of 'Civil Society'* (Dordrecht: Kluwer Academic Publishers, 1988), a book that deserves much more attention than it has generally received.

15. Other pioneers in marginal utility theory were Léon Walras, Carl Menger, Antoine Augustin Cournot, and Alfred Marshall. Marshall marks an important fork where one road led to "pure" economics, marked by mathematicization (although

Marshall warned against its misuse), modeling, and indifference to empiricism and the other led to what can variously be called political economy, institutionalization, or social economy. For an illuminating treatment of this forking, see Neva R. Goodwin, *Social Economics: An Alternative Theory,* vol. 1: *Building Anew on Marshall's Principles* (New York.: St. Martin's Press, 1991).

16. Walter Allan, "The Appeal to Biology," a review of Geoffrey M. Hodgson, *Economics and Evolution: Bringing Life Back into Economics* (Oxford: Polity Press, 1993), in *Times Literary Supplement,* March 4, 1994, p. 22.

17. See a full discussion and useful bibiographical references in Bruce Mazlish, *A New Science: The Breakdown of Connections and the Birth of Sociology* (New York: Oxford University Press, 1989). A complete history of sociology would go back to the work of Scottish Enlightenment thinkers, such as Adam Ferguson and John Millar, and proceed systematically to the present.

18. What was produced were working classes, not a single working class. See E. P. Thompson, *The Making of the English Working Class* (New York: Vintage Books, 1966). For more on this distortion in Marx's perception, see my article "Marx's Historical Understanding of the Proletariat and Class in Nineteenth-Century England," *History of European Ideas* 12, no. 6 (1990), 731-47.

19. The French sociologist Christian Topalov has some interesting things to say about social practice and theory (or its absence). He states: "Modern empirical sociology sprang from social surveys . . . ; social statistics from public health and labor agencies; city planning from local governments and civic groups. All of those sciences developed along a roughly similar intellectual pattern. They constructed objective and often measurable causal sequences which related elements that had been extracted from the reality of workers ways of living. These relationships were constructed for practical purposes of transformation." Topalov thus stresses the practical concerns in the origins of sociology (or a particular form of it). He sees such concerns as having action as an aim and leading to the creation of social institutions (welfare services, workhouses). The institutions are themselves based on an assumption about causal relations — alcoholism is caused by *x* and can be dealt with by doing *y* — and on correct classification of social phenomena. The problem, according to Topalov, however, is that such diagnoses and classifications are often based on incorrect representations. In a specific case, unemployment, he argues that in the nineteenth century many instances of nonemployment "which we would not today name unemployment were thought of under the same category" ("Inventing the Language of Unemployment, 1880-1910: Britain, France and the United States," a paper presented at the seminar "Citizenship and Social Policies," Harvard Center for European Studies, December 16, 1994). The implication is that a hermeneutic approach is required, one going beyond the positivist approach and recognizing the historical ambiguity of current terms and classifications. In short, even positivist-oriented social studies require an interpretive approach.

20. A different sort of book would take up here Weber's reflections on method-

ology in his *Economy and Society* (Berkeley: University of California Press, 1978); and his *Roscher and Knies: The Logical Problems of Historical Economics,* trans. Guy Oakes (New York: Free Press, 1975), as well as the extensive secondary literature on this subject.

21. See further my book *A New Science,* as well as my review of Wolf Lepenies, *Between Literature and Science: The Rise of Sociology,* in *Social Forces* 69 (June 1991), 1249–56.

22. J. W. Burrow, *Evolution and Society: A Study in Victorian Social Theory* (Cambridge: Cambridge University Press, 1966), 83. For the tortuous history of how anthropology emerged in England, from an Ethnological Society and an Aborigines Protection Society, and elsewhere, see Burrow; and Stocking, *Victorian Anthropology,* passim. For an argument that the "science of anthropology" was an invention of the Enlightenment, see Robert Wokler, "Anthropology and Conjectural History," in *Inventing Human Science,* ed. Christopher Fox, Roy Porter, and Robert Wokler (Berkeley: University of California Press, 1995), 31–52; the quotation is from p. 32. Wokler's notes also include useful bibliographical references.

23. See, for example, Margaret Hodgen, *Early Anthropology in the Sixteenth and Seventeenth Centuries* (Philadelphia: University of Pennsylvania Press, 1964); and Anthony Pagden, *The Fall of Natural Man: The American Indian and the Origins of Comparative Ethnology* (Cambridge: Cambridge University Press, 1982).

24. See Hugh West, "The Limits of Enlightenment Anthropology: Georg Forster and the Tahitians," *History of European Ideas* 10, no. 2 (1989), 149.

25. Claude Blanckaert, "Buffon and the Natural History of Man: Writing History and the 'Foundational Myth' of Anthropology," *History of the Human Sciences* 6 (February 1993), 37.

26. Christopher Herbert, *Culture and Anomie: Ethnographic Imagination in the Nineteenth Century* (Chicago: University of Chicago Press, 1991), 112. Herbert, a literary scholar, looks at the origins of anthropology in a most interdisciplinarily rewarding manner.

27. Bernard Smith, *European Vision and the South Pacific,* 2d ed. (New Haven: Yale University Press, 1985), 317.

28. Jonathan Swift, *Gulliver's Travels* (New York: Modern Library, 1950), 166. Swift's great work must be placed in the overall context of travel literature. When Swift has Gulliver write about his first encounter with a Lilliputian, we are immediately put in mind of more authentic travel accounts and their own spiritual diminution of the savage: "I perceived it to be a human creature not six inches high, with a bow and arrow in his hands, and a quiver at his back" (20). Seeing the Brobdingnags as giants perhaps parodies the "tall tales" of Patagonian and other giants (cf. Smith, *European Vision,* 34–37).

29. Stocking, *Victorian Anthropology,* 108.

30. Raymond Williams, *Culture and Society, 1780–1950* (Garden City, N.Y.: Doubleday, 1960), 355.

31. See A. L. Kroeber and Clyde Kluckhohn, *Culture: A Critical Review of Concepts and Definitions* (New York: Vintage Books, 1963).

32. Quoted in Herbert, *Culture and Anomie*, 4.

33. Clifford Geertz, "Ideology as a Cultural System," in *Ideology and Discontent*, ed. David E. Apter (Glencoe, Ill.: Free Press of Glencoe, 1964), 62.

34. Clifford Geertz, *The Interpretation of Cultures* (New York: Basic Books, 1973), 14.

35. Cf. Sherry Ortner, "Theory in Anthropology Since the Sixties," *Comparative Studies in Society and History* 26 (1984), 126–66.

36. Eric R. Wolf, *Europe and the People Without History* (Berkeley: University of California Press, 1982), 18. Wolf is helped to his historical perspective by his beliefs as a Marxist. For an incisive critique of Margaret Mead's book, see Derek Freeman, *Margaret Mead and Samoa: The Making and Unmaking of an Anthropological Myth* (Cambridge: Harvard University Press, 1983).

37. Victor Turner, *Dramas, Fields and Metaphors: Symbolic Action in Human Society* (Ithaca, N.Y.: Cornell University Press, 1974), 36.

38. Terence Turner, "Indigenous Rights, Environmental Protection and the Struggle over Forest Resources in the Amazon: The Case of the Brazilian Kayapo" unpub. MS, January 4, 1995, p. 9.

39. Axel Honneth and Hans Joas, *Social Action and Human Nature*, trans. Raymond Meyer (Cambridge: Cambridge University Press, 1988; orig. pub. in German, 1980), 7. On universals, see Donald E. Brown, *Human Universals* (Philadelphia: Temple University Press, 1991).

40. Bronislaw Malinowski, *Argonauts of the Western Pacific* (Prospect Heights, Ill.: Waveland Press, 1984; orig. pub., 1922), 3. On Malinowski and problems of ethnographic authority, see *The Early Writings of Bronislaw Malinowski*, ed. Robert J. Thornton and Peter Skalnik, trans. Ludwik Krzyzanowski (Cambridge: Cambridge University Press, 1993), notably as it sheds light on Malinowski's youthful philosophical and positivist tendencies; *Man and Culture: An Evaluation of the Work of Bronislaw Malinowski* (London: Routledge and Kegan Paul, 1970; orig. pub., 1957), especially the chapters by Phyllis Kaberry and E. R. Leach; *Observers Observed: Essays on Ethnographic Fieldwork*, ed. George W. Stocking, Jr. (Madison: University of Wisconsin Press, 1983); and James Clifford, *The Predicament of Culture: Twentieth-Century Ethnography, Literature, and Art* (Cambridge: Harvard University Press, 1983), especially chap. 1. Malinowski's own *A Diary in the Strict Sense of the Word* (Stanford, Calif.: Stanford University Press, 1989) is an extraordinary work.

41. Malinowski, *Argonauts*, 3.

42. Bronislaw Malinowski, *The Sexual Life of Savages in North-Western Melanesia* (New York: Harcourt, Brace, 1929), xxv.

43. Malinowski, *Argonauts*, 4.

44. Clifford Geertz, *Works and Lives: The Anthropologist as Author* (Stanford, Calif.: Stanford University Press, 1988), 5.

45. On the internal debate in anthropology, see James A. Boon, *Other Tribes, Other Scribes: Symbolic Anthropology in the Comparative Study of Cultures, Histories, Religions, and Texts* (Cambridge: Cambridge University Press, 1982), 20 and passim; and George Marcus and Michael M. J. Fischer, *Anthropology as Cultural Critique: An Experimental Moment in the Human Sciences* (Chicago: University of Chicago Press, 1986). Wolf Lepenies, *Between Literature and Science: The Rise of Sociology* (Cambridge: Cambridge University Press, 1988; orig. pub. in German, 1985), is on the same subject, though it concerns sociology and literature.

46. For some interesting material on the early evolutionary perspective and the latent shift inherent in it, see Stanley Jeyaraja Tambiah, *Magic, Science, Religion, and the Scope of Rationality* (Cambridge: Cambridge University Press, 1990), especially 46–47. The whole book is germane to our inquiry into the human sciences.

47. Max Harris, "Performing the Other's Text: Bakhtin and the Art of Cross-Cultural Hermeneutics," *Mind and Human Interaction* 4 (April 1993), 94. Robert Burns, with his "O wad some Power the giftie gie us . . ." had already foreshadowed the hermeneutic circle.

48. The emphasis here is on "mere" literary gaze. In fact, great literature can often be described in the same terms used here, that is, as combining both distance and empathy. The similarity is one reason for the increasing rapprochement between anthropology and literature. It also further underlies the view that literature shows scientific aspects and is a part of the human sciences.

49. Robert Wokler, "The Enlightenment Science of Politics," in *Inventing Human Science,* ed. Christopher Fox, Roy Porter, and Robert Wokler (Berkeley: University of California Press, 1995), 323. So, too, John Dunn declares that "in its comprehension of the scope and limits of its own techniques of understanding and its own approach, the academic discipline of political science has been a fairly unmitigated intellectual disaster" ("The Economic Limits to Modern Politics," in *The Economic Limits to Modern Politics,* ed. John Dunn [Cambridge: Cambridge University Press, 1990], 15). Various writers, however, from the seventeenth century on to the present have thought otherwise. To take one example, David Hume wrote an essay, "That Politics May Be Reduced to a Science," in which we find such passages as "consequences almost as general and certain may sometimes be deduced from them [laws and forms of government] as any which the mathematical sciences afford us" and such assertions as that there exist "an universal axiom in politics" (*Hume's Moral and Political Philosophy,* ed. Henry D. Aiken [New York: Hafner Publishing Co., 1948], 296, 298). Hume promises more than he can deliver.

50. Jürgen Habermas, *Theory and Practice,* trans. John Viertel (Boston: Beacon Press, 1973), 42.

51. A "new route" are the words that Machiavelli uses in the introduction to the first book of the *Discourses,* of which *The Prince* was probably intended originally to be a part.

52. For further details, though I would write it up differently now, taking into ac-

count many subsequent writings, see J. Bronowski and Bruce Mazlish, *The Western Intellectual Tradition* (New York: Harper and Row, 1960), chap. 11; as well as Hobbes's *Leviathan* and Locke's *Two Treatises on Government.* In the secondary literature, there are a host of good works, most of which have extensive bibliographies.

53. *The Federalist Papers,* no. 37 (New York: New American Library, 1961), 228–29.

54. Stefan Collini, Donald Winch, and John Burrow, *That Noble Science of Politics: A Study in Nineteenth-Century Intellectual History* (Cambridge: Cambridge University Press, 1983), 16–17. As the authors point out, the "science of politics" in nineteenth-century Britain "embraced much of the territory now assigned to the semi-autonomous dominions of economics and sociology, just as it was itself constituted by unspecified areas of the larger continents of history and philosophy." Their conclusion is that, broadly speaking, their subject "no longer appears on modern maps of knowledge" (3).

55. In writing of Marx, one should also be writing of Friedrich Engels. Their relationship was complicated, however, and for simplicity's sake, I have stayed with the singular, Marx. For one example of their relationship and collaboration, see my article "Marx's Historical Understanding of the Proletariat and Class," which is as much about Engels as about Marx.

56. Thomas Robert Malthus, *Population: The First Essay* (Ann Arbor: University of Michigan Press, 1959), 96–97. Edmund Burke voiced similar sentiments. This view of a political system as an organic whole precludes its treatment as a machine, whose parts can be taken apart and reassembled. The holistic view leads naturally, though not inevitably, to a link with the hermeneutic method.

57. Swift, *Gulliver's Travels,* 279.

58. Charles E. Lindblom and David K. Cohen, *Usable Knowledge. Social Science and Social Problem Solving* (New Haven: Yale University Press, 1979), 23. This wise but unassuming book focuses on the limits of social science. Another version of the authors' position seems to be " 'decisionism,' a term coined by Carl Schmitt [the German political philosopher] . . . now commonly used to refer to an attitude that emphasizes the element of sheer decision in human affairs" (Robert M. Wallace, Translator's Note, in Hans Blumenberg, *Work on Myth,* trans. Wallace [Cambridge, Mass.: MIT Press, 1990], 173).

59. Power, often considered the core idea in political analysis, is believed to resemble value in economics. It is impossible, however, to do much more with power than to notice its manifestations and to praise or condemn its pursuit. An exception is the far-reaching attempt by Michael Mann: *The Sources of Social Power,* vol. 1: *A History of Power from the Beginning to A.D. 1760;* vol. 2: *The Rise of Classes and Nation-States, 1760–1914* (Cambridge: Cambridge University Press, 1989–93).

60. Afsaneh Najmabadi, "States, Politics, and the Radical Contingency of Revolutions: Reflections on Iran's Islamic Revolution," *Research in Political Sociology* 6 (1993), 197–215, offers an excellent analysis of the general question of predictability in politics, especially in regard to revolutions. The author comments that "perhaps a

great deal of our reluctance to accept revolutions as unpredictable has to do with a notion that such acceptance amounts to abdication of theory." Then she adds, "But why should unpredictability imply irrelevance of theory?" I agree with Najmabadi, but believe that the theory to which she is alluding is *historical* theory. As she herself goes on to say, revolutions are made "by unforeseen and unpredictable events. They are highly contingent on a complex sequence that at any point has *many* possibilities, and yet only one possibility eventually consolidates itself" (200). Only after the event can we try to say why what happened, happened. In my view, this is to operate historically, for history of the sort that I am advocating is always freighted with theory. If one wishes to call such an approach a part of political science, I see no reason to call in the disciplinary police.

61. Cf. Chapter 5, n. 72.

62. See "Chinese Historiography in Comparative Perspective," *History and Theory* 35 (December 1996). As a number of the articles in this theme issue suggest, the intentions of Chinese historiography are different from those of the West: instead of being a scientific inquiry into human affairs, it aims at depicting "a metahistorical ideal order which had been realized within the human sphere in the past" (23). It also attempts to explain why humanity has departed from that order and how it might return to it. Thus, the overall point being made in the text seems valid.

63. For an excellent account, see Charles William Fornara, *The Nature of History in Ancient Greece and Rome* (Berkeley: University of California, 1983). On China and India, see Donald E. Brown, *Hierarchy, History and Human Nature: The Social Origins of Historical Consciousness* (Tucson: University of Arizona Press, 1988), which also offers a useful bibliography.

64. K. H. Waters, *Herodotus the Historian: His Problems, Methods and Originality* (Norman: University of Oklahoma Press, 1985), 1, 8. This is one of the most useful books on Herodotus and Greek historiography.

65. Paolo Rossi, *The Dark Abyss of Time: The History of the Earth and the History of Nations from Hooke to Vico*, trans. Lydia G. Cochrane (Chicago: University of Chicago Press, 1984; orig. pub. in Italian, 1979), xiv. This is a magisterial treatment of the subject, with Rossi immersed in original writings of the time, and a fundamentally important work in historiography.

It is also important to note anew that the traffic between geology and history flowed in both directions. Lyell, in the first chapter of his *Principles of Geology* explicitly compares the two inquiries, remarking, "As the present condition of nations is the result of many antecedent changes, some extremely remote, some gradual, others sudden and violent, so the state of the natural world is the result of a long succession of events, and if we would enlarge our experience of the present economy of nature, we must investigate the effects of her operations in former epochs" (*Principles of Geology* [Chicago: University of Chicago Press, 1990], vol. 1, p. 1). Lyell's model for the history of nations was provided by the German historian Berthold Georg Niebuhr in his *Römische Geschichte*, which Lyell quotes (see Martin J. S. Rudwick, "Transposed

Concepts from the Human Sciences in the Early Work of Charles Lyell," in *Images of the Earth*, ed. L. S. Jordaneva and Roy Porter [Chalfont St. Giles, Buckinghamshire: British Society for the History of Science, 1979]). Two observations can be drawn from these facts: (1) a recognition that the geologist is engaged on an essentially interpretive or hermeneutic task, thus underlining the view maintained earlier of a continuity between the natural and human sciences, and (2) a heightened awareness of the way both forms of science, natural and human, fertilize one another (and one could add that the same process takes place within the disciplines of each form of science), with ideas from each inspiring work in the other — what Rudwick has referred to as transposed concepts.

66. Friedrich Meinecke, *Die Entstehung des Historismus*, 2 vols. (Munich, 1936).

67. Cf. Hans Blumenberg's comment, "The Enlightenment saw and evaluated all this [myth and religion] from the perspective of the *terminus ad quem* (the point at which the process terminates); it was incapable of turning its attention to the *terminus a quo* (the point from which the process takes its departure), and it paid for this incapacity with its defeat by historicism" (*Work on Myth*, 19). As the translator explains, Blumenberg uses " 'historicism' to refer to the endeavor . . . to interpret each historical phenomenon as having a unique character that is to be understood as the product of a specific process of historical development. In contrast to this, the Enlightenment, by understanding history as a whole in terms of itself as the 'goal' . . . prevented itself from understanding the unique character of previous epochs in terms of what *they* had overcome (their *terminus a quo* in each case)" (33). We might remember that the rise of historicism correlates with the growing nineteenth-century concern with the spirit of the times and the concept of society as being about unique and changing entities.

68. Language is at the heart of hermeneutics, so anything about linguistics relates to our subject. In the battle between sacred and philological texts in the seventeenth century, for example, much ink was spilt over the question of what the original language of humanity was. If there had been one original, all-pervading universal language, as some theologians held, much of hermeneutics — its translating and transporting functions — would be unnecessary. In such a case, the linguistic turn might simply be a dead end or a dirt road. As is clear, however, there are many languages, and each language undergoes a historical development in which words, ambiguous to start with, change and become historical in their ambiguity as well.

69. Anthony Grafton, "The Footnote from De Thou to Ranke," *History and Theory* 33 (1994), 53, 56–57. This is a very important article in historiography. Grafton's book *The Footnote: A Curious History* (Cambridge: Harvard University Press, 1997) appeared while this book was already in press; therefore, I can only note its existence.

70. Grafton, "Footnote from De Thou to Ranke," 74. It will not have escaped the reader's notice that, by its extensive use of notes, the book that he or she is now reading, *The Uncertain Sciences*, embraces a double narrative, with its inquiring mode. In

addition, the notes are intended to suggest the kind of "archive" necessary and available to the student of the human sciences, for, alas, as noted earlier, in our subject there is no obvious Rankean-like collection of "primary" documents conveniently to be found in one obvious location.

Grafton gives Bayle the credit for both devising and defending this notion of a double narrative. Without disagreeing with this attribution, I would question whether Bayle was countering skepticism, as Grafton argues, or encouraging it, as I would argue, in regard to sacred texts—or, if Bayle is more deeply interpreted, perhaps both. Ernst Cassirer, in fact, emphasizes that Bayle does not start with facts but seeks to discover them. In this process of discovery, however, Bayle generally subverts existing facts rather than proving them, thus contributing to a general air of skepticism. As Cassirer puts it, "One should not imagine that . . . it [the "truth of facts"] can be grasped in immediate sense experience; it can only be the result of an operation no less complex, subtle, and precise than the most difficult mathematical operation. For it is only by the finest sifting, by the most painstaking examination and evaluation of the bits of evidence that the kernel of an historical 'fact' can be isolated." But Cassirer goes on to say that "Bayle's genius, paradoxically enough, does not lie in the discovery of the true but the discovery of the false" (*The Philosophy of the Enlightenment,* trans. Fritz C. A. Koelln and James P. Pettegrove [Princeton, N.J.: Princeton University Press, 1951; orig. pub. in German, 1932], 205–6). Most historians after Bayle have settled for a degree of certainty about most of their facts rather than focusing on the uncertainty involved.

71. Quoted in Ken Auletta, "What Wouldn't They Do," *New Yorker,* May 17, 1993, p. 48.

72. Two outstanding historians who appear to embrace in some of their work the views cited here are Simon Schama and Jonathan Spence. For comments on the former, see Cushing Strout, "Border Crossings: History, Fiction, and *Dead Certainties,*" and, on the latter, my article "The Question of *The Question of Hu,*" both in *History and Theory* 31, no. 2 (1992), on pp. 153–62 and 143–52, respectively. For the view of history as mere art, see Hayden White's frequently cited book *Metahistory: The Historical Imagination in Nineteenth-Century Europe* (Baltimore, Md.: Johns Hopkins University Press, 1973). On an earlier work of White's, see my review-essay "From History to Sociology," *History and Theory* 1, no. 2 (1961), 219–27, which deals with his translation of Carlo Antoni's book of the same name (for which White also wrote a tendentious introduction). On a somewhat different tack, and with more sense, William H. McNeill, in *Mythistory and Other Essays* (Chicago: University of Chicago Press, 1986), argues that history and myth are closely aligned, with history being in some ways another form of myth.

73. In discussing ancient history, M. I. Finley has some wise things to say: "If the historian does nothing else, he arranges events in a temporal sequence. . . . When Thucydides selected the incidents at Corcyra and Potidaea for a detailed narrative . . . his choice was dictated by a decision about causes. . . . The choice of events which

are to be arranged in a temporal sequence, which are to be interrelated, necessarily rests on a judgment of an inherent connection among them" (*The Use and Abuse of History,* rev. ed. [Penguin Books, 1975], 67). In making a judgment the historian also has to decide how much weight to give to immediate and to remote causes and how to judge the difference between causes and pretexts. Given the multiplicity of causes and the confusion about their importance, the narrator is necessarily impelled to impose a simplified order upon them and to single out particular connections. The novelist Patrick O'Brian gets at the problem when he writes, "In a venereal engagement between a man and a woman the events occur in turn, in a sequence of time; each can be described as it arises. Whereas in a martial contest so many things happen at once, that even the ablest hand must despair of drawing the appearance of a serial thread from the confusion." "For example," O'Brian has one of his characters say, "I have never yet heard two accounts of the battle of Trafalgar that consist with one another in their details" (*The Fortune of War* [New York: W. W. Norton, 1991], 54). None of this stops O'Brian from giving splendid accounts of various naval engagements. Perhaps the same could be said of a more classical author, Leo Tolstoy, in his novel *War and Peace.*

74. Philosophers of history, like Hegel, do try to find a definitive perspective, a final viewpoint. For Hegel, it was for Reason, or the Idea, to know from the beginning, as potential, what had to be at the end, as actuality. The resemblance of his Reason to God is evident. Hegel's desires to a large extent became Marx's, and Marx, in his more pretentious, rigid moments, also thought that he knew the inevitable outcome of history, a point from which all else could be judged. Earlier, Kant had also wrestled with this problem of history and the need for a final perspective. Blumenberg puts the matter well when he writes of Kant as seeking to compare human history "to the apparent motion of the planets, the irregularity of which could finally be ended by choosing a point of view that was different from the terrestrial one. The question then is whether there is a comparable special point of view—optically unattainable, but accessible to reason—for the ups and downs of history as well" (*Genesis of the Copernican World,* 604).

The difference between philosophy of history, with its scientific pretensions, and what I have been calling scientific history is also well illustrated by Blumenberg in his reflections on the history of science. He remarks, "The retrospective angle of vision of the history of science does not perceive the byways, detours, and blind alleys because it acknowledges as an 'event' only what led further; in this way the course of this history gets the appearance of a logical sequence, but each of its components loses the quality of action. But historiography has to understand events as actions, and it does so by trying to imagine the distribution of possibilities and probabilities in a situation at the time" (667). Thus, historians are constantly in a dilemma. They must try to construct a causal, connected explanation of the past, often through ruthless selectivity, which reveals one sort of truth. And they must try to reconstruct the actuality and the actions of what happened in the confused, complicated way it took place. In fact, they must seek to do both things at once. I should add that the speculative philosophy of history, practiced by Kant, Hegel, Marx and others, is not without value. For some of its contribution to history and the social sciences, see my book *The Riddle of History: The Great Speculators from Vico to Freud* (New York: Harper and Row, 1966).

75. For one example among many, see the discussion of sixteenth-century Catholics and Reformers in Stephen Greenblatt, *Renaissance Self-Fashioning: From More to Shakespeare* (Chicago: University of Chicago Press, 1980), 109–14.

76. In talking of archives, we should also note the problems with their use. For example, Ranke, using the Venetian *relazioni* — the ambassadorial reports — did not subject them to interpretive scrutiny: Were they reporting what they had actually seen or what they needed to write in order to convince their superiors? Further, his archives were political and diplomatic, slanting his history in that direction rather than, say, toward cultural or economic history. Even more serious, and applicable to all archival research, is the problem of the "locked trunk in the attic." Historians are initially at the mercy of what documents are accidentally left behind (many a graduate student's career has been made on the exploitation of one archival source). Increasingly, however, historians aware of this problem have been giving up the passive position — using the *Sitzfleisch* famously required in the archives — and actively seeking and, in that sense, constructing the needed archival materials.

77. Cf. Mark Bevir, "Objectivity in History," *History and Theory* 33, no. 3 (1994), 339.

78. A. R. Louch, *Explanation and Human Action* (Berkeley: University of California Press, 1969), viii.

79. Peter Winch, *The Idea of a Social Science and Its Relation to Philosophy* (London: Routledge and Kegan Paul, 1958).

80. Peter T. Manicas, *A History and Philosophy of the Social Sciences* (Oxford: Basil Blackwell, 1987), 4. This is a most important and, in many ways, successful book on its subject. For a critical review, see Brian Fay, *History and Theory* 27, no. 3 (1988), 287–96. For a review of Manicas's book in the context of four others on more or less the same subject, see David A. Hollinger, "Giving at the Office in the Age of Power/Knowledge," *Michigan Quarterly Review* 29 (Winter 1990), 123–32.

81. The project of which I speak has been going forward under the inspiration of Tosun Aricanli and various colleagues.

Chapter 6. The Uncertain Sciences

1. Charles E. Lindblom and David K. Cohen, *Usable Knowledge: Social Science and Social Problem Solving* (New Haven: Yale University Press, 1979), 47. One problem is that two values cannot be maximized on the same scale at once. Although economists try to deal with the problem by drawing two curves and seeing where

they intersect, their science basically impels them in the direction of maximizing one value at a time. Another problem, indicated in Chapter 5, n. 19, concerns the interpretive challenge presented by the very terms *employment* and *unemployment*.

2. Adam Smith, *The Wealth of Nations,* ed. Bruce Mazlish (Indianapolis: Bobbs-Merrill, 1961), 149.

3. Perry Anderson, *Lineages of the Absolutist State* (London NLB: Humanities Press, 1974), 426. The next quotation from Anderson is from p. 57. Marx was a brilliant occasional commentator on cultural matters, but he left essentially no room for the cultural as an autonomous or semiautonomous causal factor in his overall scheme.

4. Norbert Elias, *State Formation and Civilization,* vol. 2: *The Civilizing Process,* trans. Edmund Jephcott (Oxford: Basil Blackwell, 1982), 4. The initial volume is *The History of Manners,* trans. Edmund Jephcott (New York: Pantheon Books, 1978). The central role of social stratification (touched upon conceptually in Chapter 3) in the development of manners and the absolutist state is obvious.

5. Reinhart Koselleck, *Critique and Crisis: Enlightenment and the Pathogenesis of Modern Society* (Cambridge, Mass.: MIT Press, 1988), 183–84. On some of what follows, see p. 64.

6. Hans Blumenberg, *The Genesis of the Copernican World,* trans. Robert M. Wallace (Cambridge, Mass.: MIT Press, 1987), 163.

7. Nicholas Rupke, *The Great Chain of History* (Oxford: Clarendon Press, 1983); and Martin J. S. Rudwick, *The Great Devonian Controversy: The Shaping of Scientific Knowledge Among Gentlemanly Specialists* (Chicago: University of Chicago Press, 1985).

8. Elias, *State Formation and Civilization,* 160.

9. Hugh Stretton, *The Political Sciences: General Principles of Selection in Social Science and History* (New York: Basic Books, 1969), raises such questions as these in a useful fashion. See especially pp. 93 and 231.

10. Thomas McCarthy, "Introduction," in Jürgen Habermas, *On the Logic of the Social Sciences,* viii.

11. George Herbert Mead, *Mind, Self, and Society,* ed. Charles W. Morris (Chicago: University of Chicago Press, 1934), 329. See Ernst Mayr, *The Growth of Biological Thought* (Cambridge: Harvard University Press, 1982). Cf. Donald R. Griffin, *The Question of Animal Awareness,* rev. ed. (New York: Rockefeller University Press, 1981), 30.

12. Rudwick, *Great Devonian Controversy,* 4.

13. Denis Wood, *The Power of Maps* (New York: Guilford Press, 1992), 43.

14. Keith Michael Baker, *Condorcet* (Chicago: University of Chicago Press, 1975), 260–61.

15. *Karl Marx, Frederick Engels Collected Works* (New York: International Publishers, 1986), vol. 28, p. 42. Whatever reservations one might have as to details of this statement, it points us in the right direction.

16. Michel Foucault, *Discipline and Punish: The Birth of the Prison,* trans. Alan Sheridan (New York: Vintage Books, 1979), 304–5. As a history of the prison, this is an idiosyncratic work. Cf. Michael Ignatieff, *A Just Measure of Pain: The Penitentiary in the Industrial Revolution, 1750–1850* (New York: Columbia University Press, 1978); and John Bender, *Imagining the Penitentiary: Fiction and the Architecture of Mind in Eighteenth-Century England* (Chicago: University of Chicago Press, 1987), whose author, though much influenced by Foucault, makes his own independent argument.

17. On consciousness, see Daniel C. Dennett, *Consciousness Explained* (Boston: Little, Brown, 1991). On Descartes and the human-machine question, see my book *The Fourth Discontinuity: The Co-Evolution of Humans and Machines* (New Haven: Yale University Press, 1993), 14–26.

18. The quotation from Wedgwood is taken from *The Rise of Capitalism,* ed. David S. Landes (New York: Macmillan, 1966), 67; and that from Carlyle is from his "Signs of the Times," *Thomas Carlyle: Selected Works,* ed. Alan Shelston (Penguin Books, 1980), 67. There is a nice irony in the fact that Wedgwood was one of Charles Darwin's grandparents. Although a deterministic mechanism might lurk in Darwin's work — he was a philosophical materialist — the basic thrust of his theories was toward uncertainty and emergence. Huxley, it is true, tended to carry Darwin's theories in a mechanical direction. See my *Fourth Discontinuity,* pp. 97–99, 141–46, for this point, as well as for a fuller discussion, pp. 175, 220, of what follows concerning man as "prosthetic god."

19. It is true, as noted before, that we are dependent on our senses for our knowledge of the external world; and if our senses were changed, our knowledge would presumably be different. The extension of our sight by telescopes and X-rays does open a world of sensory knowledge denied to our forebears. In that sense, the human is a changing animal. As suggested earlier, the subject studied by the human sciences is best conceived of as a problematic species.

20. I have suggested the name *combot* for such computer-robots (although second thought suggests *compubot* as a better coinage, avoiding the agonistic associations with the term *combat*). For further discussion, see my *Fourth Discontinuity,* chap. 11.

21. See my book *A New Science* (New York: Oxford University Press, 1989), 114–25.

22. Isaiah Berlin, "Does Political Theory Still Exist?" *Philosophy, Politics and Society (Second Series),* ed. Peter Laslett and W. G. Runciman (Oxford: Basil Blackwell, 1962), 8. Cf. my earlier remarks in this chapter under the subheading "Human as Machine."

23. Steven Shapin, *A Social History of Truth: Civility and Science in Seventeenth-Century England* (Chicago: University of Chicago, 1994), xxvi. In this book, Shapin is expanding on the notion of witnesses first broached in his book *Leviathan and the Air-Pump: Hobbes, Boyle, and the Experimental Life,* with Simon Schaffer (Princeton, N.J.: Princeton University Press, 1985). The general subject of witnesses in early modern history of science has become a special area of research. Mario Biagioli, in *Galileo*

Courtier (Chicago: University of Chicago Press, 1993), has explored the principles of legitimation in regard to the Italian academies of science and the princely courts. His long essay "Etiquette, Interdependence, and Sociability in Seventeenth-Century Science" (*Critical Inquiry* 22 [Winter 1996], 193–238), is a fascinating comparative study. Where Shapin points to gentlemen, Biagioli argues that in the Italian academies it was generally the prince whose status conveyed legitimacy to the results of science.

24. Rudwick, *Great Devonian Controversy,* 425.

25. *The Correspondence of Charles Darwin,* vol. 1: *1821–1836* (Cambridge: Cambridge University Press, 1985), 513–14.

26. Rudwick, *Great Devonian Controversy,* 169.

27. John Weightman, "How Wise Was Montaigne?" *New York Review of Books,* November 5, 1992, p. 32.

28. John Locke, *An Essay Concerning Human Understanding* (any edition), bk. I, chap. 4, par. 23.

29. Michael Welbourne, *The Community of Knowledge,* Scots Philosophical Monograph no. 9 (Aberdeen: Aberdeen University Press, 1986), 32. This book is one of the most incisive on the topic. It is eminently suggestive in extending the Habermasian community to the scientific community. Welbourne wishes to defend the role of authority from the withering and unrealistic attack on it by Descartes and his followers. Welbourne correctly accepts that misinformation forms part of the communication of knowledge and that its presence must not undermine all belief in authority. As he points out, the requirement that knowledge be certain is "the road to pessimism about knowledge" (56). He concludes: "Our basic notion of enquiry is what I have called market-place enquiry" (81).

Also very helpful is Helen E. Longino, *Science as Social Knowledge: Values and Objectivity in Scientific Inquiry* (Princeton, N.J.: Princeton University Press, 1990). As Longino explains her purpose, inspired or provoked by feminist critiques of science, "I abandoned a negative goal—rejecting the idea of value-free science—for a positive one—developing an analysis of scientific knowledge that reconciles the objectivity of science with its social and cultural construction" (ix).

30. F. A. Hayek, *The Counter-Revolution of Science: Studies on the Abuse of Reason* (Glencoe, Ill.: Free Press of Glencoe, 1955), 59.

31. Peter Novick, in his long and exhaustively researched book *That Noble Dream: The "Objectivity Question" and the American Historical Profession* (Cambridge: Cambridge University Press, 1988), recapitulates the principal elements of the idea of objectivity as follows: "The assumptions on which it rests include a commitment to the reality of the past, and to truth as correspondence to that reality; a sharp separation between knower and known, between fact and value, and above all between history and fiction. Historical facts are seen as prior to and independent of interpretation: the value of an interpretation is judged by how well it accounts for the facts; if contradicted by the facts it must be abandoned. Truth is one, not perspectival. Whatever patterns exist in history are 'found,' not 'made.' Though successive

generations of historians might, as their perspectives shifted, attribute different significance to events in the past, the meaning of those events was unchanging" (1–2). This seems a very old-fashioned definition of objectivity. Many of the key questions implicitly go begging. If the past is not real, what is it? *Merely* a human construction, spun out of thin air? Does the dismissal of a "real" past entail as well the dismissal of a "real" natural world (even though we can experience that world only in terms of our senses and thoughts, as with the historical world)? Ought an interpretation not be held accountable in regard to facts?

Many of the other assumptions have little or nothing to do with the ideal and practice of objectivity as I have sought to deal with it. The correspondence theory is hardly necessary for claims to objectivity (see, for example, William Dray, *Laws and Explanations in History* [London: Oxford University Press, 1957]). Perspective must be correlated with objectivity and the reach for truth, not dismissed from discussion. Patterns are, of course, human constructions, whether we claim to find them or to make them. No reputable "scientific" historian other than one made of straw believes that the meaning of events is final and unchanging (perhaps a religious historian might take this position). In sum, whatever the iniquities of past, positivist-minded historians, the discussion of the subject of objectivity needs to proceed in more realistic terms than those found in Novick's definition.

The whole subject is an enormous one, going well beyond the remarks of either Novick or myself. In regard to natural science, I have found highly instructive the article by Lorraine Daston and Peter Galison, "The Image of Objectivity," *Representations* 40 (Fall 1992). Other works to be consulted are the various chapters in *Rethinking Objectivity*, ed. Allan Megill (Durham, N.C.: Duke University Press, 1994), especially the chapter by Johannes Fabian, which picks up again some of the arguments in our treatment of anthropology and also offers a useful bibliography on the subject of objectivity; and the papers for "Social History of Objectivity," a symposium, collected in *Social Studies of Science* 22, no. 4 (1992) — see especially the contributions by Lorraine Daston and by Peter Dear.

32. Quoted in Stephen Greenblatt, *Marvelous Possessions: The Wonder of the New World* (Chicago: University of Chicago, 1991), 34.

33. Cf. Blumenberg, *Genesis*, 655.

34. See Tian Yu Cao, "The Kuhnian Revolution and the Postmodernist Turn in the History of Science," *Physes* 30 (December 1993), 6. Cf. Michael Polanyi, *Personal Knowledge* (Chicago: University of Chicago, 1958). In a personal communication, however, Dmitri Shalin argues that "it was not Polanyi who pioneered the term; [Charles] Peirce talks about 'the community of inquirers' and Dewey builds on this notion by pointing out at some length that democracy and science share this emphasis on community-building and inquiry" (letter of November 2, 1995). Thus, we must also reckon with a pragmatist notion of scientific community.

35. Stephan Jay Gould, *Time's Arrow, Time's Cycle: Myth and Metaphor in the Discovery of Geological Time* (Cambridge: Harvard University Press, 1987), 126.

36. *A Vital Rationalist: Selected Writings from Georges Canguilhem*, trans. Arthur Goldhammer (New York: Zone Books, 1994), 263.

37. Gaston Bachelard, *The New Scientific Spirit*, trans. Arthur Goldhammer (Boston: Beacon Press, 1984), 136.

38. Alfred Schutz, *Collected Papers*, vol. 1, ed. Maurice Natanson (The Hague: Martinus Nijhoff, 1962), 48, 49.

39. Philip Mirowski, *More Heat Than Light: Economics as Social Physics, Physics as Nature's Economics* (Cambridge: Cambridge University Press, 1989), details the exchanges involved between physics and economics.

40. Anthony Giddens, "Jürgen Habermas," in *The Return of Grand Theory in the Human Sciences*, ed. Quentin Skinner (Cambridge: Cambridge University Press, 1985), 130. Cf. Karen Shabetai, "Facts Are Stubborn Things," *Critical Review* 3, no. 1 (1989), 110. For the ideas of Habermas himself, see his *On the Logic of the Social Sciences*, trans. Shierry Weber Nicolsen and Jerry A. Stark (Cambridge, Mass.: MIT Press, 1991).

41. Carl Schmitt, quoted in Jean L. Cohen and Andrew Arato, *Civil Society and Political Theory* (Cambridge, Mass.: MIT Press, 1992), 201–2. This 771-page book by Cohen and Arato is a compendium on its subject, approaching it as a part of political philosophy.

42. Quoted in Bernard Bailyn, "Jefferson and the Ambiguities of Freedom," *Proceedings of the American Philosophical Society* 137, no. 4 (1993), 499.

43. Robert K. Merton, *Social Theory and Social Structure*, enl. ed. (New York: Free Press, 1968), 607. Cf. the discussion in Piotr Sztompka, *Robert K. Merton: An Intellectual Profile* (London: Macmillan, 1986), 53.

44. *The Autobiography of Charles Darwin and Selected Letters*, ed. Francis Darwin (New York: Dover, 1958), 25. In this account, Darwin claims that he "worked on true Baconian principles, and without any theory collected facts on a wholesale scale" (42). As a glance at Darwin's work shows, this is patently false; a stream of theorizing always surrounded even the most trivial of his observations. It becomes clear, therefore, that Darwin's Baconian statement is a bow to what he took to be the prevailing orthodoxy: Look, it is the facts that speak for themselves. Darwin's "grouping"—a form of consilience—hardly springs, unmediated by mind, from the "facts."

45. Richard Hooke, *Micrographia* (New York: J. Cramer-Weinheim and Hafner Publishing Co., 1961), n.p.

46. Ibid., 28.

47. Nietzsche seems to have intuited this need when he lamented the fact that "the task of *incorporating* knowledge and making it instinctive is only beginning to dawn on the human eye and is not yet clearly discernible"—from his *Gay Science*, quoted in Alexander Nehamas, *Nietzsche* (Cambridge: Harvard University Press, 1985), 26. Nehamas's book is generally pertinent for a number of our concerns about the human sciences.

48. The latest version of this ancient cavil is reported in the *New York Times*,

about a blind theologian, Sheik Abdel-Aziz Ibn Baaz, who recently issued a fatwa condemning those who say the earth is not flat (February 12, 1995).

49. Cohen and Arato, _Civil Society and Political Theory_, 7. The next quotation is from p. ix.

50. Krishan Kumar, "Civil Society: An Inquiry into the Usefulness of an Historical Term," _British Journal of Sociology_ 44, no. 3 (1993), 376–77. This is an especially useful and thoughtful treatment of its subject.

51. Anthony J. La Vopa, "Conceiving a Public: Ideas and Society in Eighteenth-Century Europe," _Journal of Modern History_ 64 (March 1992), 79. This incisive review article about Koselleck's _Critique and Crisis_ and Habermas's _Structural Transformation of the Public Sphere_ ends up as an important commentary on the concept of civil society itself. All sorts of things turned out to play a role in the construction of civil society. As William M. Reddy informs us, "The appearance of handbooks and reference works on many subjects in this period was an important facet of the emergence of a public sphere in Western Europe." (His specific concern is with Savary des Bruslons's _Dictionnaire universel du commerce_, first published between 1723 and 1730 and reissued at least six more times up to 1784.) See Reddy's "Structure of a Cultural Crisis: Thinking About Cloth in France Before and After the Revolution," in _The Social Life of Things_, ed. Arjun Appadurai (Cambridge: Cambridge University Press, 1986), 265.

52. Lorraine Daston, _Classical Probability in the Enlightenment_ (Princeton, N.J.: Princeton University Press, 1988), starts with the question "What does it mean to be rational?" and explores the answer given by such classical probabilists as Pascal, Bernouilli, and Laplace, who applied their talents to setting up a "model of rational decision, action, and belief under conditions of uncertainty" (xi). They assumed that mathematical probabilities measured both "objective frequencies of events" and "subjective degrees of belief," which they held to be equivalent (xiii). Thus, they could apply probability theory to assessments of the credibility of witnesses and opinion in general.

53. The reality was rather different. For an excellent account of the Freemasons, for example, see Margaret C. Jacob, _Living the Enlightenment: Freemasonry and Politics in Eighteenth-Century England_ (New York: Oxford University Press, 1991).

54. Cohen and Arato, _Civil Society and Political Theory_, 91.

55. _Hegel's Philosophy of Right_, trans. T. M. Knox (Oxford: Clarendon Press, 1942), 266 (Add. 116, Paragraph 182). For more on Hegel's relation to Adam Smith and economics, treated as a realm of civil society, see Lawrence Dickey, _Hegel: Religions, Economics, and the Politics of Spirit, 1770–1807_ (Cambridge: Cambridge University Press, 1987).

56. Cohen and Arato, in _Civil Society and Political Theory_, for example, trace a line from Carl Schmitt to Jürgen Habermas. They also discuss critics of civil society, such as Hannah Arendt and Michel Foucault.

Foucault is especially important in regard to civil society, as well as to our more general concerns. As we have noted, Foucault was adamant that knowledge was power, meaning that the effort at scientific knowledge was not a disinterested inquiry about truth but rather an exercise of dominion. Both the natural and the human sciences come under Foucault's censure, but he is especially severe about the human sciences. Discussing the bourgeoisie's juridical pretensions, Foucault declared that "the general juridical form that guaranteed a system of rights that were egalitarian in principle was supported by these tiny, everyday, physical mechanisms, by all those systems of micropower that are essentially nonegalitarian, and asymmetric that we call the disciplines" (Michel Foucault, *Discipline and Punish: The Birth of the Prison,* trans. Alan Sheridan [New York: Vintage Books, 1979], 222). Studying asylums, penal systems, and sexual classifications, Foucault sought to show how, as Arato and Cohen summarize it, "the use of medical, psychological, sociological expertise, of statistical data, in short of empirical information and nonlegal languages within legal discourse to make one's case, is proof that the disciplines have penetrated the juridical structures, and rendered them positive, empirical, functional, and quasi-disciplinary themselves" (264; for another excellent summary, see p. 667 n. 59). Thus, for Foucault, the disciplines of the human sciences, positive and empirical as they aspire to be, mainly serve disciplinary, i.e., punitive, functions. It would seem, therefore, that Foucault believes that he has implicitly undercut the claims of a Habermasian civil society. As I would see it, he has taken Gramsci's notion of hegemony and given it "specialized" features. Nevertheless, however one-sided his presentation, Foucault offers many suggestive insights and a particular critical perspective on the human sciences. Acknowledging the large grains of truth in Foucault's criticisms, both of civil society and its disciplinary institutions, Cohen and Arato also elaborately illustrate the weaknesses in his position; and the reader is referred to their discussion.

57. *Globalization* refers specifically to a process taking place historically. *Globalism* is often used as a synonym, but some critics object to the term because it appears to project a deterministic triumph of capitalism everywhere. When the term *globalism* is used here, however, it is simply as a synonym for *globalization,* a process contingent and uncertain in its effects.

58. This list of factors was first offered in my "Introduction," in *Conceptualizing Global History,* ed. B. Mazlish and R. Buultjens (Boulder, Colo.: Westview Press, 1993), 1–2. Differing in significant ways from what is generally called world history, global history is an initiative dating from about 1989, taken in order to gain some historical perspective on globalization while it is happening. Carol Gluck and Raymond Grew are my coeditors of the Global History series published by Westview Press, of which the volume cited above is the first. An especially useful article on the topic is Michael Geyer and Charles Bright, "World History in a Global Age," *American History Review* 100 (October 1995), 1034–60. Sociologists have been before

historians in recognizing the importance of globalization; see, for example, Roland Robertson, *Globalization: Social Theory and Global Culture* (London: Sage, 1992). An extensive literature on the subject is gradually coming into existence.

59. Locke, *Essay Concerning Human Understanding,* bk. IV, chap. 15, par. 5. Another story bearing on the same point relates to the Hindi text *Bhugolsār* (1841), written by an Indian astronomer, Omkar Bhatt, to defend the teaching of Western science in his country against the criticisms of the Brahmins. In the form of a debate, Bhatt has a guru explain to his disciple that, although Hindu knowledge is ancient, "the geography of the *Bhagwāt* is mere description, and not all geographical knowledge has been produced by the Hindus. The siddhants do not even describe travels south of the equator. . . . The Westernizers have seen the entire globe" (Gyan Prakash, "Science as a Sign of Modernity in Colonial India," unpub. MS, March 1994, p. 9).

60. As noted in Chapter 1, it used to be thought that there was a struggle to the death between science and the Christian religion (or at least theology). A. D. White's classic *A History of the Warfare of Science with Theology in Christendom* (1896) and the address of John Draper to the British Association for the Advancement of Science, later expanded into a book entitled *History of the Conflict Between Religion and Science* (1875), appear to be forceful statements of this view. As a matter of history, however, the warfare thesis, whatever it may be philosophically, is now seen as a grossly simple view. See, for example, John Hedley Brooke, *Science and Religion: Some Historical Perspectives* (Cambridge: Cambridge University Press, 1991), which offers a thoughtful, modern historical treatment of the relations of science and religion from the seventeenth century to the present. Brooke points out that the relation is a complex one. "There is no such thing," he argues, "as *the* relationship between science and religion. It is what different individuals and communities have made of it in a plethora of different contexts" (321). Brooke also offers a substantial bibliographic essay.

61. A comment by Nietzsche is pertinent here. He remarks in *Beyond Good and Evil* (in *Basic Writings of Nietzsche,* trans. and ed. Walter Kaufmann [New York: Modern Library, 1992], 259), "Perhaps the day will come when the most solemn concepts which have caused the most fights and suffering, the concepts 'God' and 'sin,' will seem not more important to us than a child's toy and a child's pain seem to the old — and perhaps the 'old' will then be in need of another toy and another pain — still children enough, eternal children."

62. The survey on science, conducted by Louis Harris and Associates, was reported in the *New York Times,* April 21, 1994; that on religion, a USA/CNN/Gallup Poll, in *USA Today,* April 2, 1994. A different sort of trend does seem to be showing itself in Europe. In the Netherlands, the national Bureau of Standards reports that 18 percent of the Dutch said they belonged to no church or religion in 1960, but the number reached 40 percent in 1995 (*New York Times,* March 10, 1997). What should be made of all this would require a separate essay.

63. See Hans Blumenberg, *Work on Myth*, trans. Robert M. Wallace (Cambridge, Mass.: MIT Press, 1990), especially xi–xiv, chap. 1.

Chapter 7. "Da Capo," or Back to the Beginning

1. Stephen Gleenblatt, "Introduction," *Representations* 49 (Winter 1995), 7. For Habermas's response to criticism of his work, see, for example, his book *The Philosophical Discourse of Modernity: Twelve Lectures*, trans. Frederick G. Lawrence (Cambridge, Mass.: MIT Press, 1987). Also to be consulted is Seyla Benhabib, *Situating the Self: Gender, Community, and Postmodernism in Contemporary Ethics* (New York: Routledge, 1992).

2. Quoted in Frank E. Manuel, *A Portrait of Newton* (Cambridge: Harvard University Press, 1968), 143.

3. Newton's self-image was belied in his actual life, for he was often overcome by passions when unconstrained by scientific method, and sometimes led astray even when so bounded.

4. Mario Biagioli, *Galileo, Courtier: The Practice of Science in the Age of Absolutism* (Chicago: University of Chicago Press, 1993), 241–42. Later, Biagioli seems to undercut his own observation by admitting a "progressive shift from a scientific discourse framed by considerations of 'honor' to one centered around the notion of 'scientific credibility' " (354). The latter discourse would appear to be on the way to creating a kind of Habermasian truth community, or what I refer to as the practice of scientific method.

5. See Chapter 6, n. 34.

6. Piotr Sztompka, *Robert K. Merton: An Intellectual Profile* (Houndmills, England: Macmillan, 1986), 85–86.

7. Reason is not the only guide to human action. It is always accompanied by emotion, especially in regard to human relations. The nature of emotions is therefore of the utmost importance. A recommended focus for future work would be on the ways in which reason can achieve knowledge about the emotions and about how emotions can guide reason.

8. Some thinkers believe otherwise and speak of the end of science or the limits of science. Oswald Spengler was one of the early twentieth-century proponents of such a view. Often, of course, the debate on the end of science forms part of the debate on the idea of progress. For one recent brief survey of such thinking, see John Horgan, "The Twilight of Science," *Technology Review*, July 1996, pp. 50–61; as well as his book *The End of Science: Facing the Limits of Knowledge in the Twilight of the Scientific Age* (Reading, Mass.: Helix Books/Addison-Wesley, 1996).

9. Fernand Braudel, quoted in Christopher Hill, review of Braudel's *Civilisation matérielle et capitalisme*, vol. 1 (1967), in *History and Theory* 8, no. 2 (1969), 302.

10. Michael Dertouzos, quoted in the *MIT Tech Talk,* April 3, 1996, p. 4.

11. Quoted in Colin Jones, "The Great Chain of Buying: Medical Advertisement, the Bourgeois Public Sphere, and the Origin of the French Revolution," *American Historical Review* 101, no. 1 (February 1996), 27. Cf. Robert Mauzi, *L'idée du bonheur dans la littérateur française au XVIIIe siècle* (Paris, 1960), especially 300–314.

12. Meredith Small, "Rethinking Human Nature (Again)," *Natural History* 9 (1995), 8.

13. Georges Gusdorf, *Les sciences humaines et la pensée occidentale,* vol. 1: *De l'histoire des sciences à l'histoire de la pensée* (Paris: Payot, 1966), 195. The French word *conscience* has a double meaning: on one side, "conscience," or a moral sense, and, on the other side, "consciousness." In the phrase cited, Gusdorf has in mind the second meaning. See further D. R. Kelley, "Gusdorfiad," *History of the Human Sciences* 3 (1990), 123–40.

14. Norbert Elias, *The Symbol Theory,* ed. Richard Kilminster (London: Sage Publications, 1991), 50.

15. Other important forerunners of the idea being presented here were George Herbert Mead — see his *Mind, Self and Society: From the Standpoint of a Social Behaviorist* (Chicago: University of Chicago Press, 1934), 329 — and Emile Durkheim.

16. Daniel C. Dennett, quoted in the *New York Times Book Review,* November 10, 1991, p. 59.

17. Daniel C. Dennett, *Consciousness Explained* (Boston: Little, Brown, 1991), 21. This is an impressive argument (although it is always dangerous to talk about the last surviving mystery), which must be read in its entirety. Needless to say, there is much controversy both about consciousness and about Dennett's book on the subject. In this light, however, it is well to ponder Stephen Jay Gould's comment that "perhaps that greatest and most effective of all evolutionary inventions, the origin of human consciousness, required little more than an increase of brain power to a level where internal connections became rich and varied enough to force this seminal transition" (*Natural History* 10 [1985], 25).

18. Charles Darwin, "Old and Useless Notes," in *Darwin on Man: A Psychological Study of Scientific Creativity,* by Howard E. Gruber (New York: E. P. Dutton, 1974), 385, 386.

19. Ibid., 233.

20. As Dennett, in *Consciousness Explained,* puts it, "You can open up subjects' skulls (surgically or by brain-scanning devices) to see what is going on in their *brains,* but you must not make any *assumptions* about what is going on in their *minds,* for that is something you can't get any data about while using the intersubjectively verifiable methods of physical science" (70). He goes on to compare consciousness to something like a narrative stream, distributed around the brain but without a single location (135). For more on consciousness in animals besides the human, see Douglas R. Griffin, *The Question of Animal Awareness,* rev. ed. (New York: Rockefeller University Press, 1981).

21. See Julian Jaynes, *The Origin of Consciousness in the Breakdown of the Bicameral Mind* (Boston: Houghton Mifflin, 1976). For his summary of the factors "at work in the great transilience from the bicameral mind to consciousness" (grounded in both the right and left hemispheres of the brain), consult p. 221. A central aspect of human consciousness, it should be noted, is the ability to distinguish self and nonself. How this happens is a continuing subject of scientific research.

22. G. F. W. Hegel, *The Philosophy of History*, trans. J. Sibree (New York: Colonial Press), 17. Hegel thought of his search in regard to consciousness as a form of science. As he declared in another work, "The systematic development of truth in scientific form can alone be the true shape in which truth exists. To help to bring philosophy nearer to the form of science—that goal where it can lay aside the name of *love* of knowledge and be actual *knowledge*—that is what I have set before me" (*The Phenomenology of Mind*, trans. J. B. Baillie [New York: Harper Torchbooks, 1967], 70). The further question is what Hegel meant by "science."

23. Although I have singled out Hegel, many other philosophers, ancient and modern, loom large in the history of thought about consciousness. Of those in modern times, John Locke is especially important. Indeed, as Christopher Fox points out, "outside of several minor uses of the word itself, the earliest written use of the term *consciousness* in the [English] language is by John Locke" (*Locke and the Scriblerians: Identity and Consciousness in Early Eighteenth-Century Britain* [Berkeley: University of California Press, 1988], 12). Fox's book is an interesting study of the debate linking consciousness to self and to a sense of individuality persisting over time, especially by Locke and his contemporaries. (On the origins of the discipline of psychology as a context for this debate, see also Fox's "Defining Eighteenth-Century Psychology: Some Problems and Perspectives," in *Psychology and Literature in the Eighteenth Century*, ed. Christopher Fox [New York: AMS Studies in the Eighteenth Century, 1987].)

Locke was the great analyst and proponent of sense impressions. Hegel sought to go beyond him by looking more closely at the actual phenomenon, i.e., phenomenology, of what is involved in being conscious of a sense impression. According to Hegel, "Consciousness is, on the one hand, consciousness of the object, on the other consciousness of itself; consciousness of what to it is true, and consciousness of its knowledge of that truth" (*Phenomenology of Mind*, 141). He continues, "In sense-experience pure being at once breaks up into the two 'thises' . . . one this as I, and one as object" (150). He then traces what he refers to as the truth of self-certainty through its realization of itself in relation to nature, to other selves similar to the self, and to the Ultimate Being of the world. It is in this context that he writes of Lordship and Bondage, as a moment in the process of the self coming to consciousness. These reflections on consciousness go in a different direction from those of Locke and his contemporaries and lead Hegel to pursue the notion that consciousness became historical consciousness.

24. Along with Freud, William James must be listed as a major inquirer into

consciousness. As he insisted in his article "Does Consciousness Exist?" (1904), the word *consciousness* does not stand for an entity, but represents an "aboriginal staff or quality of being. . . . [It is] a function in experience which thoughts perform" (quoted in Dmitri N. Shalin, "Modernity, Postmodernism, and Pragmatist Inquiry: An Introduction," *Symbolic Interaction* 16, no. 4 [1993], 314). See also, of course, James's *Principles of Psychology* (1890). Although our triumvirate's sense of certainty in regard to their own work, including that on consciousness, is different from the notion of the uncertain sciences being advanced here, their ideas (along with those of many others) are fundamental to the latter conception.

25. Hans Blumenberg, *The Genesis of the Copernican World*, trans. Robert M. Wallace (Cambridge, Mass.: MIT Press, 1987), 486. The very title of Blumenberg's book should be noted: he is offering a modern version of genesis, displacing the biblical account.

26. Ibid., 312–13, also 629.

27. Cf. Alfred Schutz, *Collected Papers*, vol. 1,: *The Problem of Social Reality*, ed. Maurice Natanson (The Hague: Martinus Nijhoff, 1971), 5–6.

28. Cf. Antonio Pérez-Ramos, *Francis Bacon's Idea of Science and Maker's Knowledge Tradition* (Oxford: Clarendon Press, 1988), 135, 114.

29. Lee Cronbach, quoted in Robert McC. Adams, "Rationales and Strategies for Social Science Research," *Bulletin: The American Academy of Arts and Sciences* 39 (April 1981), 31.

30. Such changes are generally given institutional form, which in turn redounds upon consciousness. My emphasis here is upon the latter, but, as I insisted earlier, consciousness must and does manifest itself in actions and subsequent institutionalization.

31. In Chapter 7, n. 23, I mentioned Hegel's treatment of Lordship and Bondage as moments in the development of consciousness. In this development, the notion of the Other also figures. As Hegel remarks, "Self-consciousness has before it another self-consciousness; it has come outside itself. This has a double significance. First it has lost its own self, since it finds itself as an *other* being; secondly, it has thereby sublated that other, for it does not regard the other as essentially real, but sees its self in the other." He continues, "It must cancel this its other. To do so is the sublation of that first double meaning, and is therefore a second double meaning. First, it must set itself to sublate the other independent being, in order thereby to become certain of itself as true being, secondly, it thereupon proceeds to sublate its own self, for this other is itself " (*Phenomenology of Mind*, 229). The key word for Hegel is *recognition*. We achieve consciousness of ourselves when we are recognized by an other and, in turn, recognize ourselves in that other.

32. Empirical investigations on why some people believe in magic have been undertaken, showing them, for example, "poor . . . at assessing probabilities" (*Economist*, October 30, 1993, p. 99). Without passing judgment on the validity of that work, we can recognize that such work suggests the need for additional empirical research situated in a firm theoretical context.

33. There is opposition to the claim of a common historical consciousness. James Baldwin, the black American writer, tells us that he "brought to Shakespeare, Bach, Rembrandt, to the stones of Paris, to the cathedral at Chartres, and to the Empire State building, a special attitude. These were not really my creations, they did not contain my history; I might search them in vain forever for any reflection of myself. I was an interloper; this was not my heritage" (quoted in Gerald Early, "Three Notes on the Meaning of Diversity in America," paper prepared for "At the End of the Century: Looking Back to the Future," a conference at the Library of Congress, November 3–5, 1994). While appreciating Baldwin's sense of alienation, I think that we hear in his words the voice of a disappointed lover. Besides, his statements are not true. Willy-nilly, his history does overlap with the names and stones he cites; even rejected, they are part of his heritage. Among other things, Baldwin is writing in English, the language of Shakespeare.

Contrast Baldwin with Salman Rushdie, another "outsider." "We [Indian writers in England] can quite legitimately claim as our ancestors the Huguenots, the Irish, the Jews; the past to which we belong is an English past, the history of immigrant Britain. Swift, Conrad, Marx are as much our literary forebears as Tagore or Ram Mohan Roy" (*Imaginary Homelands* [London: Granta Books, 1991], 20).

A charged atmosphere surrounds a discussion of this sort, especially in the closing decades of the twentieth century. Humanity is not one simple construct; and historical consciousness is not a monolithic entity, uniformly shared around the world. The consciousness spoken of here is heterogeneous, protean in nature and constantly in process. Nevertheless, because thoughts and their material embodiments can be acquired by all peoples as a cultural addition—has this not been the history of the human species?—increasingly the findings of the natural and human sciences have been infiltrating into what I have been calling our common historical consciousness. It is a prerequisite of future work in this area that the human sciences take into account the experiences, thoughts, and actions of others than Westerners. A reexamination of the social sciences in these terms is a pressing task for the coming century.

34. Jacob Bronowski has captured exquisitely what is involved. "The most remarkable discovery made by scientists is science itself. The discovery must be compared in importance with the invention of cave-painting and of writing. Like these earlier human creations, science is an attempt to control our surroundings by entering into them and understanding them from inside. And like them, science has surely made a critical step in human development which cannot be reversed" ("The Creative Process," *Scientific American* 199 [September 1958], 5).

35. Rushdie, *Imaginary Homelands*, 377. Not only people living in modern times enjoy self-consciousness. Japanese poets of the eight century, for example, show exquisite sensitivity and self-awareness. What we are witnessing today is an extension of self-awareness and an increased coupling of it to historical consciousness.

36. Historical consciousness has many other aspects aside from those supplied by the human sciences, music, visual art, and literature. One can think immediately of

ethnic memories and religious beliefs. How these elements and those who hold them interact with the truth claims of scientific communities is a subject of great importance. Indeed, the human sciences must take such elements as part of their subject of study. Our main concern here, however, is with historical consciousness as it provides an alternative to, or facsimile of, a scientific truth community that comprises all of humanity.

37. Quoted in the preface to Mary Shelley, *Frankenstein*, ed. James Rieger (Chicago: University of Chicago Press, 1982), xxxii.

38. Friedrich Nietzsche, *The Will to Power*, trans. W. Kaufmann and R. J. Hollingdale (New York: Vintage Books, 1968), 14–15.

39. Anthony Cascardi, *The Subject of Modernity* (Cambridge: Cambridge University Press, 1992), 302.

40. Theodore Ziolkowski, *German Romanticism and Its Institutions* (Princeton, N.J.: Princeton University Press, 1990), contains, in its second chapter, entitled "The Mine: Image of the Soul," many affirmative allusions to mines. In contrast, a highly negative reaction to mining can be found in Lewis Mumford, who speaks of the mine, in *Technics and Civilization* (New York: Harcourt, Brace, 1934), as "the first completely inorganic environment to be created and lived in by man . . . the environment alone of ores, minerals, metals . . . if the miner sees shapes on the walls of his cavern, as the candle flickers, they are only the monstrous distortions of his pick or his arm: shapes of fear" (69–70). There is no hint of the sublime, which, as Ziolkowski points out, had its attractions for the German romantics, in Mumford's treatment (ironically, as Diane Greco reminds me, Mumford's reaction is itself of a romantic nature).

Rosalind Williams, in *Notes on the Underground* (Cambridge, Mass.: MIT Press, 1990), takes the position that "the opposition of surface and depth may well be rooted in the structure of the human brain. . . . In any case, the metaphor of depth is a primary category of human thought" (7–8). She instances stories of descent into the underworld, produced even in such environments as the flat Kalahari desert. My emphasis, however, is specifically on the nineteenth-century romance with depth, with the metaphor itself being "deepened." Dostoevsky's *Notes from the Underground* is an obvious case in point. See also my article "A Triptych: Freud's *The Interpretation of Dreams*, Rider Haggard's *She*, and Bulwer-Lytton's *The Coming Race*," in *Comparative Studies in Society and History* (October 1993), 726–45, especially 738–44.

41. Christopher Herbert, *Culture and Anomie: Ethnographic Imagination in the Nineteenth Century* (Chicago: University of Chicago Press, 1991), 327 n. 1. Herbert views the exploitation of the idea of "deep" in the nineteenth century as indebted to "Wesley and to the Christian tradition of original-sin theology that his evangelism helped inject into modern awareness" (254). See also Jules David Law, *The Rhetoric of Empiricism: Language and Perception from Locke to I. A. Richards* (Ithaca, N.Y.: Cornell University Press, 1993). Given our concern with depth, Law's chapters on Burke, where the sublime is emphasized, and on Hazlitt are the most pertinent, although the entire book relates to the topic.

42. Neo-Darwinian theory does emphasize mutation as a means of rapid biological change, along with the slower processes of natural selection.

43. This is so largely because, as Stanley Tambiah puts it, "although one must take important recognition of the nature of scientific revolutions and the shifts in paradigms, and although one must accept the provisional nature of extant scientific knowledge, yet I think it is sensible to hold that in principle the laws of physics or chemistry have to be the same everywhere in this world" (*Magic, Science, Religion, and the Scope of Rationality* [Cambridge: Cambridge University Press, 1990], 114). The problem is whether there are laws of that kind concerning human nature.

44. One may argue, as William MacKinley Runyan points out, that humans can be affected by the human sciences even when unaware of them. For example, Watsonian-behavioral or Spock-Freudian approaches to child rearing can affect both infants and their parents. Likewise, the social policies or actions of leaders under the influence of social scientific knowledge can affect others deprived of such knowledge. Runyan is right. But such observations in no way weaken my theses about self-understanding or historical consciousness. They simply mean that the details involved are varied and complex and must themselves be studied carefully from within specific human sciences (such as psychology and sociology). Climates of opinion, spirits of the age, leadership roles, pedagogy, the transmission of ideas — these are some of the areas studied so far. In all such cases, what I have been calling historical consciousness is the whole against which particular influences operate. See W. T. Jones, *The Romantic Syndrome* (The Hague: Martinus Nijhoff, 1961), for an interesting analysis of this problem.

45. See *Charles Bubbage and His Calculating Engines*, ed. Philip Morrison and Emily Morrison (New York: Dover, 1961), xx.

46. This task becomes especially poignant on the brink of the millennium. Human weakness and stupidity seem to manifest themselves everywhere. Given current ethnic hatreds; increased belief in the Devil, UFOs, astrology, and miracles; overt distrust of science and the scientific mode of thinking, and so forth, there hardly appears to be ground for optimism. At such moments, one must turn to the consolations of history and take a long-range view of the human species.

Critical Bibliography

This bibliography is highly selective. It is not meant to be comprehensive but to stimulate further inquiry. It makes no pretense to be exhaustive on any of its subjects (moreover, with a few exceptions, most of the books noted are in English). Because there is no set archive on the challenges and condition of the human sciences, this is an effort to help create one.

The notes to this book should be treated as part of the bibliography; in a few cases, works mentioned there will be commented on again here. Almost all of the books listed contain their own bibliographies. For more in-depth treatments of certain topics, such internal bibliographies must be relied upon.

The books and articles are listed under headings, but many of the items could go under more than one heading. No attempt is made to list the books that would figure under the rubrics of the specific human sciences, including economics, anthropology, sociology, political science, and history. Interested readers can find these for themselves, beginning perhaps with the notes in Chapter 5. Nor is there any attempt to cite the literature on, for example, pragmatists involved with our subject, such as William James, Charles Peirce, George Herbert Mead, and John Dewey, or, with a few critical exceptions, important contemporary thinkers on the philosophy of the social sciences, such as Richard Rorty, Hilary Putnam, and Charles Taylor. Others will surely attend to them. The selection of single figures listed here may illustrate the direction in which future bibliographies might proceed.

In the compilation of this bibliography, Jonathan Eastwood, who brought unusual dedication to the task, served as a valued research assistant and is primarily responsible for a number of the entries.

Organizations and Periodicals

In the Introduction, I pointed out the lack of institutional supports for the study of human science(s). Here, let me note those that do exist. Cheiron, an organization devoted to the history and philosophy of the human sciences, meets once a year. Its primary focus is psychology; indeed, a peripheral branch of the discipline of psychology has been at the forefront of interest in the human sciences (even though its focus is far from that taken in this book). John I. Brooks III kindly introduced me to another group, the Forum for History of Human Science (FHHS), founded as an interest group within the History of Science Society (HSS), which publishes a newsletter.

The journal *History of the Human Sciences* started publication in 1985–86. Other journals that frequently have articles related to our general subject are *Journal of the History of Behavioral Science, Philosophy of the Social Sciences, History and Theory, Theory Culture and Society, Representations,* and *Social Research.* There are many others.

Study groups in various places are concerned with the subject in one way or another, including Cornell Medical/Payne Whitney and the Committee on Social Thought at the University of Chicago.

History and Theory

Two of the most compendious and helpful treatments of much of what falls within the human sciences are Peter T. Manicas, *A History and Philosophy of the Social Sciences* (Oxford: Basil Blackwell, 1987); and Scott Gordon, *The History and Philosophy of Social Science* (London: Routledge, 1991). They are, however, largely restricted to the social sciences (although Manicas accepts other genres, such as literature, as offering "knowledge of persons and society" [279]). Manicas suggests further readings on the subjects that he covers; Gordon, in his 690-page book, does not.

Manicas focuses on the structure as well as the content of the social sciences, but he makes little effort to carry his story beyond World War I. He argues against empiricism, by which he means positivism, as not merely "an untenable philosophy of the human sciences, but as a philosophy of *any* science" (5). In contrast, he favors "realist human science," which draws upon the philosophy and program of Marx and Engels, however flawed they are. He defines a realist as one who "holds that a valid scientific explanation can appeal to the in principle non-observable" (10).

Gordon's book alternates chapters on the history of political theory, classical and neoclassical economics, the methodology of history, and the development of sociological theory with chapters on the philosophical issues involved in social and natural laws, modeling, the idea of harmonious order, utilitarianism, positivism, Marxism, and the biological basis of the social sciences — and much more. His own predilections are for neoclassical economics and utilitarianism, and one result is that all kinds of other approaches and topics are missing from his book. For a full critique, see my review in the *American Historical Review* (October 1992), 1176–77.

A pioneering work that is rather more restrictive in reach is Hugh Stretton, *The Political Sciences* (New York: Basic Books, 1969). Stretton argues against the view of social science as an objective, value-free form of knowledge; instead, he says that all such activity has a committed and persuasive intent. He offers much detail from sociology, history, and political science to support his argument. Although some of the argument seems outdated, partly as a result of his own and others' work, the book is rich in suggestiveness and insight. See also my review in *The Nation*, March 9, 1970, pp. 277–79.

Another book that deserves mention here is Richard J. Bernstein, *The Restructuring of Social and Political Theory* (Philadelphia: University of Pennsylvania Press, 1976). Bernstein has an extraordinary ability to summarize and clarify often-convoluted philosophical positions and to offer fresh interpretations of them. See also all of his subsequent work, including *Beyond Objectivism and Relativism: Science, Hermeneutics, and Praxis* (Philadelphia: University of Pennsylvania Press, 1983) and *The New Constellation: The Ethical-Political Horizons of Modernity/Postmodernity* (Cambridge: Polity Press, 1991).

James Bohman, in *New Philosophy of Social Sciences: Problems of Indeterminacy* (Cambridge: Polity Press, 1991), seeks to avoid both the old logic of social science, with its reductionist models of behavior, and postmodernism, with its deconstructionist relativism. His new logic claims that explanation is possible along Habermasian interpretive lines. A more basic introduction to the subject is Vernon Pratt, *The Philosophy of the Social Sciences* (London: Methuen, 1978). Now somewhat outdated—devoid of any mention of hermeneutics, for example—this modest book is a straightforward exposition of some of the main issues and topics in the field of the social sciences. It starts from such questions as What is the nature of the human being—machine or animal? and What the nature of society? then makes its way through problems concerning meaning, experience, explanation, objectivity, functionalism, and more. For someone totally ignorant of the subject, this might be a good book with which to start.

Geoffrey Hawthorn, *Plausible Worlds: Possibility and Understanding in History and the Human Sciences* (Cambridge: Cambridge University Press, 1991), is an examination of the worth of counterfactuals—hypothetical "other worlds"—and their usefulness for historians and social scientists. The author allows for some valid counterfactuals and offers four case studies to illustrate his position. Although Hawthorn is an admirable scholar, this particular book seems convoluted; see the review by Charles Maier, in *London Review of Books*, February 13, 1992, pp. 11–12.

The Return of Grand Theory in the Human Sciences, ed. Quentin Skinner (Cambridge: Cambridge University Press, 1985), has become a classic among books examining and generally criticizing the positivist view of science and society, a view that rejects grand philosophical systems in favor of empirical theories of social behavior and development. Skinner seeks to recover Grand Theory—the word was used first by C. Wright Mills as a term of abuse—in an anti-positivist, postmodern form (surely a paradox) in the shape of a series of short essays written by specialists; the essays on Gadamer, Derrida, Foucault, Kuhn, and Habermas are of special interest.

Alasdair MacIntyre, *After Virtue*, 2d ed. (Notre Dame, Ind.: University of Notre Dame Press, 1984). Although only one chapter (8) is directly concerned with our inquiry, the whole book can be so viewed. MacIntyre's aim is to vindicate a premodern — not a postmodern — view, rooted in Aristotle's account of the virtues, against the modern view, characterized by more than two centuries of error. For him, the Enlightenment Project is a fall into emotivism, positivism, and fragmentation, with few redeeming features. The book is vigorously and often engagingly written, with an exposition of what virtues look like on an Aristotelian basis in chapters 14 and 15; even those who do not share his sweeping dislike of modernity may find much with which to agree.

One of the great virtues of Ronald Inden's *Imagining India* (Oxford: Basil Blackwell, 1990) is that the author approaches the human sciences from an Indian perspective. Inspired, however, by R. G. Collingwood and, to a lesser extent, Foucault and Gramsci, he attacks Western essentialism, in the form of Indology, i.e., orientalism in regard to India. Emphasizing agency, he joins the attack on positivist social science, focusing especially on discourse about caste and Divine Kingship. The end result is an important treatment of both Indian themes and the human sciences. Inden's book takes up the theme of Edward Said's *Orientalism* (New York: Pantheon Books, 1978) — although the latter is innocent of overt reflection on the human sciences — and, more distantly, the theme of Benedict Anderson's *Imagined Communities: Reflections on the Origin and Spread of Nationalism* (London: Verso, 1983), though without making reference to it or its notions.

Paul Diesing, *Patterns of Discovery in the Social Sciences* (Chicago: Aldine, 1971), offers a typology identifying four ways philosophers talk about social science. For a succinct account, see pp. 4–5 in Richard H. Brown, *A Poetic for Sociology: Toward a Logic of Discovery for the Human Sciences* (Cambridge: Cambridge University Press, 1977). Brown embraces a cognitive aesthetic that sees sociology as neither a natural science nor a fine art but a form of poetics, or semiotics; thus, knowledge is seen as existing as knowledge "only in terms of some universe of discourse, some system of meaning" (8). This approach, which is no longer novel, connects with the social construction of science. It is exemplified by Brown in his summaries of the principal schools of sociological thought, where he seeks to show how each grasps a part of the truth (which it constructs). There are interesting chapters on point of view, metaphor, and irony. In spirit, this book is close to mine in the expressed belief in the connectedness of the natural and human sciences, although it concentrates on the human science of sociology.

Woodruff D. Smith, *Politics and the Sciences of Culture in Germany, 1840–1920* (New York: Oxford University Press, 1991), explores why the concept of culture became the central feature of nineteenth-century German thought and how it functioned as an ideology. Smith sees the turn to the culture concept as fostered by the crisis in liberalism. Dealing with such figures as W. H. Riehl, Rudolf Virchow, Wilhelm Wundt, and Friedrich Ratzel and the development of geography, ethnology, historical economics, and and other fields as cultural sciences, the author seems to offer the fascinating thesis or implication that the effort to establish the cultural sciences proceeded mainly under the banner of nomothetic, lawlike science, not hermeneutics. This is an important book both in its detail and in its overall theses as to the political and social construction of cultural sciences.

Anne Harrington, *Reenchanted Science: Holism in German Culture from Wilhelm II to Hitler* (Princeton, N.J.: Princeton University Press, 1996), takes up a different piece of the story. The author treats of the reaction to Machine science (a movement spearheaded by Rudolf Virchow, mentioned by Smith) in the name of a "holistic" science of life. This view manifested itself mainly in the biological and psychological sciences with a definite political accent, and Harrington singles out four somewhat obscure figures (at least today) to illustrate her thesis.

The *"Science of Man" in the Scottish Enlightenment: Hume, Reid and Their Contemporaries*, ed. Peter Jones (Edinburgh: Edinburgh University Press, 1989), is a collection of essays devoted mainly to Hume. The essays by Harvey Chisick and Donald Livingston are of most interest in the light of our particular concerns.

More specific in its focus but impressive in its analytic power is Norbert Waszek, *The Scottish Enlightenment and Hegel's Account of "Civil Society"* (Dordrecht: Kluwer Academic Publishers, 1988). Waszek sees Hegel as responding to despair about the course of history, despair mainly manifest in the spheres of religion and politics, by appealing to reason. In doing so, he turned to the thinkers of the Scottish Enlightenment and their attempts to understand reality, especially in its scientific and economic forms. As Hegel lectured his students in 1818, "At first I can only ask you to trust in science, to believe in reason, and to have confidence and faith in yourself. The courage of truth, faith in the power of 'Geist' is the foremost condition in the study of philosophy; the human being ought to honour and consider itself worthy of the highest" (4).

Inventing Human Science: Eighteenth-Century Domains, ed. Christopher Fox, Roy Porter, and Robert Wokler (Berkeley: University of California Press, 1995) is a splendid collection devoted to the origins of the human sciences in the Age of Reason and the inquiry into the science of human nature before that inquiry became broken into professional disciplines. The domains treated, to give them their modern names, are anthropology, psychology, sociology, political economy, and political science. In addition, there are essays on sex and gender and on nonnormal science, e.g., animal magnetism and phrenology. See my review of the book in the *American Historical Review* (April 1997), 444–45.

Appearing as this bibliography was in press, Roger Smith, *The Human Sciences*, The Norton History of Science (New York: W. W. Norton, 1997), treats of the origin, growth, and consolidation of various areas of study that have come to be known today as the human sciences. Though the book is over one thousand pages long, the author skillfully shapes his account around psychology, while also dealing with sociology, linguistics, economics, and anthropology. His is a major accomplishment in delineating large parts of the history of the human sciences.

William M. Reddy, *The Rise of Market Culture: The Textile Trade and French*

Society, 1750–1900 (Cambridge: Cambridge University Press, 1984), is a meticulous attempt to demonstrate that the classical economists' concept of labor, entrepreneurship, and the market did not correspond with the reality of the textile industry in France in the first half of the nineteenth century. Reddy claims that "political economy, which set out merely to describe, was now becoming so thoroughly entrenched that it had become a model to prescribe" (70). In this quotation can be seen the overlap with our own interests.

Whereas *The Rise of Market Culture* seeks to do for early industrial France what E. P. Thompson's book *The Making of the English Working Class* (London: V. Gollancz, 1963) did for early industrial England, Reddy's next book, *Money and Liberty in Modern Europe: A Critique of Historical Understanding* (Cambridge: Cambridge University Press, 1987), is a much more ambitious work theoretically but is a less pathbreaking work in regard to the concerns of the human sciences.

The Sign of Three: Dupin, Holmes, Peirce, ed. Umberto Eco and Thomas A. Sebeok (Bloomington: Indiana University Press, 1983), is an enormously suggestive and important book on the scientific method to be used in the human sciences. In this collection of essays, many of the authors relate the work of the philosopher Charles Peirce to that of detective story writers, such as Edgar Allan Poe and Arthur Conan Doyle. Peirce's idea about abduction — i.e.,the making of retrospective predictions, the reasoning from consequent to antecedent, which is separate from deduction and induction — resembles subtly the method of consilience (see the entry on William Whewell). It is the variant of scientific method used in detective stories and in medicine, as well as in psychoanalysis and other branches of the human sciences, and must be distinguished from the Galilean approach, which ignores the qualitative. Carlo Ginzburg's contribution, "Morelli, Freud, and Sherlock Holmes: Clues and Scientific Method," is an absolute gem. It and the chapter by Thomas A. Sebeok and Jean Umiker-Sebeok, " 'You Know My Method': A Juxtaposition of Charles S. Peirce and Sherlock Holmes," are required reading.

The Sign of Three draws heavily on Régis Messac's *Le 'Détective Novel' et l'influence de la pensée scientifique* (Geneva: Slatkine Reprints, 1975; orig. pub., 1929), a wonderfully erudite book on its subject that merits reading in its own right. Messac, however, describes Sherlock Holmes's method as mainly deductive-inductive and does not identify the abductive method as such.

Modernist Impulses in the Human Sciences, 1870–1930, ed. Dorothy Ross (Baltimore, Md.: Johns Hopkins University Press, 1994), is a collection of essays of varied worth. Those that seem to bear most on our subject are Dorothy Ross's introduction and the essays by David Joravsky and Theodore M. Porter. For those especially interested in American social science, see Ross's book *The Origins of American Social Science* (Cambridge: Cambridge University Press, 1991).

Postmodernism and Social Theory: The Debate over General Theory, ed. Steven and David G. Wagner (Oxford: Blackwell, 1992), has a number of interesting essays on its topic. Those by Jonathan Turner, David G. Wagner, Richard Harvey Brown, and

Craig Calhoun seem especially pertinent to our concerns. Pauline Marie Rosenau, *Post-Modernism and the Social Sciences: Insights, Inroads, and Intrusions* (Princeton, N.J.: Princeton University Press, 1992), is an attempt to deal with the postmodernist challenge by a social scientist who, while seeking to give a neutral account of the conflict, does not seem particularly enamored with the new ideas. Nevertheless, the author does give a useful description of postmodern notions.

The numerous works of Anthony Giddens are all of interest. Especially pertinent is *The Consequences of Modernity* (Stanford, Calif.: Stanford University Press, 1990), with its attention to the problematic of the social sciences and to the subject of globalization.

Two books, already cited in the notes, are of unusual value: Helen E. Longino, *Science as Social Knowledge: Value and Objectivity in Scientific Inquiry* (Princeton, N.J.: Princeton University Press, 1990); and Michael Welbourne, *The Community of Knowledge*, Scots Philosophical Monographs no. 9 (Aberdeen University Press, 1986). Both are the work of philosophers but speak incisively to the concerns of human scientists. See also Alexander Rosenberg, *Philosophy of Social Science* (Boulder, Colo.: Westview Press, 1980); and Paul A. Roth, *Meaning and Method in the Social Sciences* (Ithaca, N.Y.: Cornell University Press, 1987).

Open the Social Sciences: Report of the Gulbenkian Commission on the Restructuring of the Social Sciences (Stanford, Calif.: Stanford University Press, 1996) is the work of Immanuel Wallerstein and a number of colleagues. It is mainly concerned with the structure of the disciplines as they are practiced in universities today and has little or no interest in concepts or ideas as such. An older, still valuable book that is concerned with ideas is *Rationality*, ed. Bryan R. Wilson (New York: Harper and Row, 1970). Wilson starts from the view that Weber "came closest to recognizing the limitations of rational procedures of enquiry and understanding. But he did not raise doubts about the criteria of rationality itself" (vii). What follows are important reprintings of significant essays by Peter Winch (who gives a short version of *The Idea of a Social Science*), Ernest Gellner (who offers a moderate defense of functionalism), I. C. Jarvie, Alasdair MacIntyre (who critiques Winch), and others. Peter Winch's *The Idea of a Social Science* (London: Routledge and Kegan Paul, 1958) has become a standard item in discussions about the social sciences. A follower of Collingwood, Winch opposes functionalism and the view that social interaction is modeled on the interaction of forces in a physical system.

Ralf Dahrendorf, *Essays in the Theory of Society* (Stanford, Calif.: Stanford University Press, 1968), is another older book, but it is still of great value for its urbane and thoughtful handling of important topics. A major part of the book is a translation of Dahrendorf's *Home Sociologicus*, with its emphasis on internalization of roles: "In its most frightening aspect the world of *homo sociologicus* is a 'brave new world' or a '1984,' a society in which all human behavior has become calculable, predictable, and subject to permanent control" (58). Thus, the sociologist's dilemma is that he or she must study humans as *homo sociologicus* while helping to liberate them, for a human being has an "intelligible moral character" (86) as well as roles to play. Attention should also be paid to Dahrendorf's later writings.

Other classics to consider are Robert K. Merton, *Social Theory and Social Structure: Toward the Codification of Theory and Research* (Glencoe, Ill.: Free Press of Glencoe 1949), a collection of chapters advocating theories of the middle range as well as a number of chapters on the sociology of knowledge; and Robert S. Lynd, *Knowledge for What? The Place of Social Science in American Culture* (Princeton, N.J.: Princeton University Press, 1967), which, first published in 1939, appeared in its ninth hardcover edition in 1970, a history that attests to its influence. Lynd claims a crisis in the culture and social science of America—his sole focus—and then seeks to offer solutions. Methodologically, he stresses holism, but he recognizes that individuals make up institutions and that personality and culture studies are therefore important. He assumes that predictability is possible and that we can also learn "new modes of behavior" (206).

Outstanding in its sensibility is Charles E. Lindblom and David K. Cohen, *Usable Knowledge: Social Science and Social Problem Solving* (New Haven: Yale University Press, 1979). The authors stress political interaction over analysis, partly because they do not believe that the latter can give verifiable knowledge. The orientation of the book is toward social action and social science, rather than the human sciences more generally and the problem of knowledge therein.

Positivism

General Works

Christopher G. A. Bryant, *Positivism in Social Theory and Research* (London: MacMillan, 1985), is an excellent historical account of the debate over positive methodology in social science in France, Germany, and the United States. It includes an interesting discussion of logical positivism and the critique of the Frankfurt School.

Leszek Kolakowski, *The Alienation of Reason: A History of Positivist Thought* (Garden City, N.Y.: Anchor Books, 1969), treats positivism as merely a philosophical doctrine. Kolakowski finds its intellectual antecedents in medieval scholarship and treats Hume, rather than Comte, as its father.

Positivism and Sociology, ed. Anthony Giddens (London: Heinemann Educational Books, 1974), complements Kolakowski's work by dealing with positivism almost exclusively in terms of how the doctrine has been formulated and understood by sociologists. It contains excellent selected readings from Weber, Schutz, Gellner, Habermas, and Herbert Marcuse, among others.

The Positivist Dispute in German Sociology, Theodor Adorno et al., trans. Glyn Adey and David Frisby (New York: Harper Torchbooks, 1976), develops the debate between Adorno and Popper, which began with Popper's twenty-seven theses on the logic of the social sciences.

Johan Heilbron, *The Rise of Social Theory* (Cambridge: Polity Press, 1995), which is mainly about sociology, is an attempt to intellectually rehabilitate Comte. See the review in *History of the Human Sciences* 9 (February 1996), 113–21, and the reply by Heilbron in the next issue (May 1996).

Georges Canguilhem, *Etudes d'histoire et de philosophie des sciences* (Paris: Librairie Philosophique J. Vrin, 1983), is an interesting work, especially the chapters on Auguste Comte and Charles Darwin.

W. M. Simon, *European Positivism in the Nineteenth Century* (Ithaca, N.Y.: Cornell University Press, 1963), contains a brief history of Comte and the institutionalization of positivism. The main part of the book is devoted to a descriptive and explanatory account of "the diffusion of positivism" within certain disciplines in nineteenth-century France, England, and Germany.

Francis Bacon

Chapter 2 and its notes have a fairly extensive consideration of the literature on Bacon. Here I note two additional items. Lisa Jardine, *Francis Bacon: Discovery and the Art of Discourse* (Cambridge: Cambridge University Press, 1974), looks at Bacon against the background of sixteenth-century dialectic handbooks, where method is conceived as both a means of discovery and an art of discourse or presentation. The argument is that Bacon, whom the author considers a "well-educated gentleman with a good (but not scholarly) grounding in the curriculum subjects" (7), sought to make a distinction between these two activities as crucial for the progress of knowledge. This is a specialist work. Though not on Bacon as such, Anthony Grafton, *Defenders of the Text: The Traditions of Scholarship in an Age of Science, 1450–1800* (Cambridge: Harvard University Press, 1991), has some interesting chapters defending the humanists denigrated by Bacon and Descartes and placing figures like Kepler more favorably in the classical tradition.

Auguste Comte

There is a large literature on the founder of modern positivism, some cited in the notes to Chapter 2. In addition, a recent contribution is Mary Pickering, *Auguste Comte: An Intellectual Biography*, vol. 1 (Cambridge: Cambridge University Press, 1993. This 776-page opus is an exhaustively researched biography that also traces the nature of Comte's thought and its development. Though not profound in its philosophical treatment, Pickering's book does catch the essential elements of Comte's pioneering effort at founding a science of sociology, especially his flexibility concerning the methods appropriate to the new discipline. In the end, as Pickering shows, Comte's own sociology is less positive than many histories of positivism would have it. Kenneth Thompson, *Auguste Comte: The Foundation of Sociology* (New York: John Wiley and Sons, 1975), is more restricted in its interest and takes a somewhat uncritical attitude toward its subject. *Comte and Positivism: The Essential*

Writings, ed. Gertrud Lenzer (Chicago: University of Chicago Press, 1975), is a broad survey of Comte's thought in the form of well-chosen excerpts from his voluminous writings. An analytical introduction by Lenzer is an attempt to reevaluate Comte's significance, particularly with reference to modern positivism.

The Vienna Circle and Logical Positivism

Gordon Baker, *Wittgenstein, Frege, and the Vienna Circle* (New York: Basil Blackwell, 1988) is, in its first half, a fairly detailed descriptive analysis of the logic of Frege and Wittgenstein and, in its second, largely a survey of varieties of conventionalism, with much detail given to the interchange of ideas between Wittgenstein and members of the Vienna Circle. That section is useful for those who wish to achieve an intellectual-historical understanding of logical positivism.

Allan Janik and Stephen Toulmin, *Wittgenstein's Vienna* (New York: Simon and Schuster, 1973), is an interesting re-creation of the context in which the ideas of the Vienna Circle took root.

Viktor Kraft, *The Vienna Circle: The Origin of Neo-Positivism—A Chapter in the History of Recent Philosophy* (New York: Greenwood Press, 1953), is a brief and readable piece on the intellectual history of the Vienna Circle up to 1938 written from the perspective of a participant. Though essentially a survey, it is clear and accurate.

The Legacy of Logical Positivism, ed. Peter Achinstein and Stephen F. Barker (Baltimore, Md.: Johns Hopkins University Press, 1969), is a series of philosophical essays on the import and status of logical positivism after the demise of the Vienna Circle. See especially the contributions of Herbert Feigl, a member of the group, and Stephen Toulmin. Essays by Carl Hempel, Michael Scriven, and Hilary Putnam deal with logical positivism and the human sciences.

Rediscovering the Forgotten Vienna Circle: Austrian Studies on Otto Neurath and the Vienna Circle, ed. Thomas E. Eubel (Boston: Kluwer Academic Publishers, 1991), is a revaluation of Otto Neurath's role in the Vienna Circle and of his importance for contemporary analytic philosophy. Friedrich Stadler's essay "The Social Background and Position of the Vienna Circle at the University of Vienna" constitutes a much needed piece of historical scholarship. A brief historical paper on Austrian philosophy by Rudolf Haller is also included in the book. See also Haller's essay "The First Vienna Circle."

Edmund Runggaldier, *Carnap's Early Conventionalism: An Inquiry into the Historical Background of the Vienna Circle* (Amsterdam: Rodopi, 1984), is a relatively clear introduction to the early thought of one of the Vienna Circle's central figures. For further critical discussion of Carnap, see *Rudolf Carnap, Logical Empiricist*, ed. Jaakko Hintikka (Dordrecht: D. Reidel Publishing Co., 1975); and *The Philosophy of Rudolf Carnap*, ed. Paul Arthur Schlipp (La Salle, Ill.: Open Court, 1963).

Science and Philosophy in the Twentieth Century: Basic Works of Logical Empiricism, 6 vols., ed. Sahotra Sarkar (New York: Garland Publishing, 1996), is an overwhelmingly comprehensive collection of primary source documents of logical empiricism, with much attention given to members of the Vienna Circle. There is a series introduction, as well as historical introductions to each volume. Excellent bibliographies serve as a guide to further research. Volume 6 consists of modern appraisals of the Vienna Circle's legacy.

On Science

Reason, Experiment, and Mysticism in the Scientific Revolution, ed. M. L. Righini and William R. Shea (New York: Science History Publishers, 1975), has a number of interesting articles on its topic, especially the one by Paolo Rossi. Everything written by Paolo Rossi is of great interest. Besides his book *The Dark Abyss of Time: The History of the Earth and the History of Nations from Hooke to Vico,* trans. Lydia G. Cochrane (Chicago: University of Chicago Press, 1984; orig. pub. in Italian, 1979), cited in the notes, see his *Philosophy, Technology, and the Arts in the Early Modern Era,* trans. Salvator Attanasio (New York: Harper and Row, 1970).

Stanley Jeyaraja Tambiah, *Magic, Science, Religion, and the Scope of Rationality* (Cambridge: Cambridge University Press, 1990), is especially strong in its treatment of religion, which it places in the context of science and culture. This is a rich book of great pertinence to the human sciences at large.

Catherine Wilson, *The Invisible World: Early Modern Philosophy and the Invention of the Microscope* (Princeton, N.J.: Princeton University Press, 1995), is a splendid treatment of its announced subject, the microscope and its effect on philosophy (which also supplied the context for the instrument's development). The author also inquires into the general problem of truths and appearances and has wise things to say about the relation of scientific perceptions to reality.

Gerald Holton, *Thematic Origins of Scientific Thought,* rev. ed. (Cambridge: Harvard University Press, 1988; orig. pub., 1973), elaborates a concept of themata, defined as "those fundamental preconceptions of a stable and widely diffused kind that are not resolvable into or derivable from observation and analytic ratiocination" (13–14), which is then developed and illustrated with case studies ranging from Kepler to Einstein. The book is an in-depth look at how natural science is carried out, with implications for the human sciences; Bohr's formulation of complementarity appears especially relevant.

On Mapping

The literature on mapping is huge. Although the subject is pertinent for us, it is also somewhat tangential. See, for illustrative purposes, P. J. Marshall and Glyndwr Williams, *The Great Map of Mankind: British Perceptions of the World in the Age of Enlightenment* (London: J. M. Dent and Sons, 1962). The authors offer a dense net of quotations and citations but little in the way of larger views; still, the material is of much interest.

Hermeneutics

Gerald R. Bruns, *Hermeneutics Ancient and Modern* (New Haven: Yale University Press, 1992), has a useful introduction and chapters on what the author calls the inventory of hermeneutics. Of most interest is chapter 7, on Luther, and chapter 8, on Wordsworth. Notes 1 and 2 of the introduction offer an excellent brief bibliography on hermeneutics.

The Literary Structure of Scientific Argument: Historical Studies, ed. Peter Dear (Philadelphia: University of Pennsylvania Press, 1991), has a number of interesting essays relating natural science to literature. On the same topic, though convinced that "it is simply no good pretending that science and literature represent complementary and mutually sustaining endeavors to reach a common goal" (15), see the article by the Nobel Prize winner P. B. Medawar: "Science and Literature," *Encounter* (January 1969), 15–23. Lionel Gossman, *Between History and Literature* (Cambridge: Harvard University Press, 1990), is a collection of nine essays dealing with literature and the (other) human sciences both theoretically and historically.

For work on the linguistic turn in history, see Dominic LaCapra, *Rethinking Intellectual History: Texts, Contexts, Language* (Ithaca, N.Y.: Cornell University Press, 1983); and for a particular application of interest, see Naomi Tadmor, " 'Family' and 'Friend' in *Pamela*: A Case-Study in the History of the Family in Eighteenth-Century England," *Social History* 14 (October 1989), 289–306. For additional consideration of the relation of literature to human science, in this case, to sociology, see Wolf Lepenies, *Between Literature and Sociology: The Rise of Sociology* (Cambridge: Cambridge University Press, 1988).

The Hermeneutic Tradition: From Ast to Ricoeur, ed. Gayle L. Ormiston and Alan D. Schrift (Albany: State University of New York Press, 1990), is a collection of essays on Gadamer, Betti, Habermas, and Ricoeur, among others. It includes an excerpt from Schleiermacher's 1819 Lectures on Hermeneutics, an introduction by the editors, and a select bibliography on the essays published and more general reading. The same editors have produced another book to be consulted: *Transforming the Hermeneutic Context: From Nietzsche to Nancy* (Albany: State University of New York Press, 1990).

The Interpretative Turn: Philosophy, Science, Culture, ed. David R. Hiley, James F. Bohman, and Richard Shusterman (Ithaca, N.Y.: Cornell University Press, 1991), contains many of the papers presented at a conference held at the University of California, Santa Cruz, on the theme "Interpretation and the Human Sciences." The contributors are in philosophy, literature, law, and political science (none in history), but the overall perspective is philosophical, with much internal debate invoking the names of Charles Taylor, Thomas Kuhn, Richard Rorty, all of whom are contributors to the volume.

Michael Serres, Hermes: Literature, Science, Philosophy, ed. Josue V. Harari and David F. Bell (Baltimore, Md.: Johns Hopkins University Press, 1982), is a selection from Serres's work that was published in five volumes under the title *Hermes* between 1968 and 1980); there is a useful introduction. Serres, who also published a number of other works, crosses and recrosses the domains of myth, science, and literature (for example, writing on La Fontaine, Molière, Zola, and Verne) in an original, if eccentric, fashion. In his view, there is always local, not universal truth, and existing conceptual and institutional categories serve mainly to block passage between the natural and the human sciences. He also appears to believe that science is a culture that perhaps has reached its limit and that classical science was an expression as well as an instrument of discipline and death (Serres was a pupil of Foucault's, but Serres then went off on his own). It is in these terms that he employs his hermeneutical skills, which are real.

Stanley Rosen, *Hermeneutics as Politics* (New York: Oxford University Press, 1987), is a highly learned and thoughtful treatment of its subject. Among its theses are that hermeneutics has an intrinsically political nature and that postmodernism, whose characteristic obsession is hermeneutics, is, in fact, though conceived by its proponents as an attack on the Enlightenment, a continuation of it. Kant plays an important role in Rosen's analysis of the problem, as does Alexandre Kojève. Generally hostile to postmodernism, the author has written a difficult and subtle work well worth close reading.

Richard E. Palmer, in *Hermeneutics: Interpretation Theory in Schleiermacher, Dilthey, Heidegger, and Gadamer* (Evanston, Ill.: Northwestern University Press, 1969), laments that "one of the greatest impediments to the historical development of hermeneutics has been that it has no home in an established discipline" (69). If one wished to remedy this situation, one could do worse than start with Palmer's book, which, although three decades have passed since its publication, is a neutral, comprehensive, and informed work on the subject.

Almost all of Paul Ricoeur's texts are germane to the subject of hermeneutics, and readers are referred to other sources for their listing. One, *Hermeneutics and the Human Sciences: Essays on Language, Action and Interpretation,* trans. John B. Thompson (Cambridge: Cambridge University Press, 1981), offers historical vignettes of some of the key figures in hermeneutical thought and gives some idea of Ricoeur's own views.

A short paper, "Science, the Constitution and the Judicial System," *Proceedings of the American Philosophical Society* 132, no. 3 (1988), 247–51, by a Nobel Prize winner in medicine, Baruch S. Blumberg, has a number of wise things to say about certainty in science and the jurisprudence system, which can be treated as a part of hermeneutics.

Peter L. Berger and Thomas Luckmann, *The Social Construction of Reality: A Treatise in the Sociology of Knowledge* (Garden City: N.Y.: Doubleday, 1966), is one of the early arguments in English that "reality is socially constructed and that the sociology of knowledge must analyze the processes in which this occurs" (1). The book is still valuable for its treatment of a number of European sociologists.

David J. Levy, *Political Order: Philosophic Anthropology, Modernity, and the Challenge of Ideology* (Baton Rouge: Louisiana State University Press, 1987), focuses on ontology rather than epistemology and seeks to understand nature and human nature in terms of philosophical anthropology rather than in sociobiological terms. Levy's heroes are Max Scheler, Nicolai Hartmann, Mircea Eliade, Martin Heidegger, Eric Vogelin, Arnold Gehlen, Helmuth Plessner, Hans Joas, and Hans-Georg Gadamer, about all of whom he has informed remarks to make. Levy is opposed to Jürgen Habermas' position. In general, Levy is critical of modernity and wishes a return through philosophical anthropology to the secure position of humans in nature provided earlier by myth. Levy's book is an extremely valuable, demanding, and suggestive work for anyone interested in the human sciences.

Also focusing on work in philosophical anthropology is Axel Honneth and Hans Joas, *Social Action and Human Nature,* trans. Raymond Meyer (Cambridge: Cambridge University Press, 1988; orig. pub. in German, 1980). The authors treat not only of the usual philosophical anthropologists, such as Gehlen, but also of George Herbert Mead, Michel Foucault, and Norbert Elias. In connection with Mead especially, but also with some of the other concerns taken up in the Honneth and Joas's book, see Dmitri N. Shalin's long and thoughtful essay "Modernity, Postmodernism, and Pragmatist Inquiry: An Introduction," *Symbolic Interaction* 16, no. 4 (1993), 303–31.

Ernst Cassirer, *An Essay on Man: An Introduction to a Philosophy of Human Culture* (New Haven: Yale University Press, 1944), is an admirable example of philosophical anthropology. It offers an extended treatment of humans as the symbolic creature par excellence and of how symbolic activity plays itself out in myth and religion, language, art, history, and science. As Cassirer concludes, in these areas "man can do no more than to build up his own universe — a symbolic universe that enables him to understand and interpret, to articulate and organize, to synthesize and universalize his human experience" (278).

Having read this book many years ago, I was startled to discover on rereading it while my own book was in press what an inspiration it had obviously been for me, especially in regard to the notion of historical consciousness. To explore what is similar in and what is different between Cassirer's *Essay* and my book would be illuminating.

In addition to the *Essay,* one should consult Cassirer's four volumes on the problem of knowledge, his four volumes on the philosophy of symbolic forms, and many of his shorter books as well.

The Human Species

Work in biology, anthropology, and related areas is changing so rapidly — new pale-ontological discoveries crop up weekly, changing our view as to the time of the species' emergence and development — that journal and newspaper accounts may be as useful as serious and systematic books. Rather than seeking to deal with any of the latter in bibliographic form, let me just say that magazines like *Natural History, Science,* and *Scientific American* often carry stories relevant to the study of the human species as I have been defining it.

Though published more than a quarter of a century ago, *Man the Hunter,* ed. Richard B. Lee and Irven DeVore (Chicago: Aldine Publishing Co., 1968), is still a very suggestive collection. So, too, is another book with the same editors, *Kalahari Hunter-Gatherers: Studies of the !Kung San and Their Neighbors* (Cambridge: Harvard University Press, 1976).

Along very different lines is Henrika Kuklick, *The Savage Within: The Social History of British Anthropology, 1885–1945* (Cambridge: Cambridge University Press, 1991), which is a sober, detailed treatment of "the elements of culture, power, class, and institution in the development of a specialized intellectual enterprise" (5) — in this case, the discipline of anthropology. Kuklick's work falls into the category of books on the history of science that treat of the "production of social knowledge" (26). On its specific topic, compare Adam Kuper, *Anthropology and Anthropologists: The Modern British School,* 3d rev. and enl. ed. (London: Routledge, 1996), and Jack Goody, *The Rise of Social Anthropology in Britain and Africa, 1918–1970* (New York: Cambridge University Press, 1995).

Among the useful works on emergence are the next ones listed in this section. An important volume is *Emergence or Reduction? Essays on the Prospects of Nonreductive Physicalism,* ed. Ansgar Beckermann, Hans Flohr, Jaegwon Kim (New York: Walter de Gruyter, 1992). Especially pertinent are Achim Stephan's historical paper on emergence, Jaegwon Kim's essay on emergence, and Paul Teller's evaluation of emergence in contemporary discourse.

Ernst Mayr, *The Development of Biological Thought* (Cambridge, Mass.: Harvard University Press, 1982), is a very thorough history of biology written by an eminent contemporary practitioner. Although some of the material may be becoming a bit dated in this fast-changing field, the book offers a clear exposition of the concept of emergence, along with discussion of some of the applications of this concept to biology. Extremely useful is the single note on this topic, which takes the form of a guide to further research on the development of the concept of emergence.

Philip McShane, *Randomness, Statistics, and Emergence* (Notre Dame, Ind.: University of Notre Dame Press, 1970), is a fairly technical philosophical work focusing in part on "emergent laws," which the author takes to be "irreducible higher systematizations of aggregates of random but determinate lower events." The technicality of the book, in conjunction with the specificity of its philosophical perspective, may keep it of somewhat limited use to nonphilosophers. It is nevertheless a very interesting defense of the emergentist view.

Michael Polanyi, *The Tacit Dimension* (Garden City, N.Y.: Anchor Books, 1967), consists of three lectures, first delivered in 1962, centering on tacit knowledge. The second, on emergence, is arguably the most coherent philosophical formulation and defense of the concept available.

Major Thinkers

William Whewell

One of the most interesting figures in the mid-nineteenth century, often overlooked, whose thinking is pregnant for our concerns, though he does not write on the human sciences as such, is William Whewell. A good introduction to his original writings is *Theory of Scientific Method*, ed. Robert E. Butts (Indianapolis: Hackett Publishing Co., 1989), especially "Novum Organon Renovatum," which forms part of Whewell's *Philosophy of the Inductive Sciences Founded upon Their History* (1840; 2d ed., 1847). Here, readers encounter at first hand Whewell's ideas about "colligation" and "consilience" and his theory of induction, later called abduction by Charles Pierce (see *The Sign of Three*, listed under History and Theory, above). *William Whewell: Selected Essays on the History of Science*, ed. Yehuda Elkana (Chicago: University of Chicago Press, 1984), serves the same purpose for Whewell's history of the inductive sciences as does Butts, *Theory of Scientific Method*, for the philosophy.

Menachem Fisch, *William Whewell, Philosopher of Science* (Oxford: Clarendon Press, 1991), is an attempt to understand Whewell's philosophy by emphasizing the process by which his views took shape, showing how Whewell started as a Baconian, came under the influence of Kantian idealism, and then emerged with his own subtle, complicated synthesis in which the boundary between fact and theory is made contiguous rather than set rigidly apart. Fisch's book is a serious philosophical treatment of Whewell's theories and their development.

Richard Yeo, *Defining Science: William Whewell, Natural Knowledge, and Public Debate in Early Victorian Britain* (Cambridge: Cambridge University Press, 1993), focuses not only on Whewell's ideas but also on Whewell's vocation as a "metascientist" who served as a critic of science just as Coleridge or others served as critics of culture. In fact, Yeo argues that Whewell was instrumental in moving people from thinking of science in culture to thinking of a scientific culture, coining the term *scientist*, for example, along the way. Yeo seeks to reconstruct Whewell's complicated movement of thoughts about the relation of science to religion and the way this affected his views on empiricism, idealism, and scientific method. One result was that, in principle, Whewell considered that morality was as open to scientific thinking as the physical sciences. Yeo's work has an extensive bibliography.

Wilhelm Dilthey

Charles R. Bambach, *Heidegger, Dilthey, and the Crisis of Historicism* (Ithaca, N.Y.: Cornell University Press, 1995), is mainly concerned with the relation between historicism and modernity. The book includes, however, a fine philosophical-historical chapter on Dilthey, situating him in the context of nineteenth-century German neo-Kantian philosophy.

H. A. Hodges, *The Philosophy of Wilhelm Dilthey* (Westport, Conn.: Greenwood Press, 1974), is a technical discussion of Dilthey's philosophical work, with an emphasis upon his relation to neo-Kantianism. Though originally published in 1952, it is still one of the most comprehensive philosophical approaches to Dilthey available in English.

Rudolf A. Makkreel, *Dilthey: Philosopher of the Human Studies* (Princeton, N.J.: Princeton University Press, 1975), is an important descriptive analysis of Dilthey's work. It includes a valuable discussion of the theoretical background to Dilthey's writings, as well as an exhaustive bibliography of Dilthey's work with critical commentary on it, limited only by the 1975 publication date.

Theodore Plantiga, *Historical Understanding in the Thought of Wilhelm Dilthey* (Lewiston, N.Y.: Edwin Mellen Press, 1992), is a reinterpretation of Dilthey's thought, concentrating largely on his conception of the nature of understanding. The book is not merely a study of this conception, however, and might serve well as a general introduction to Dilthey's philosophy of the human sciences. See also Jacob Owensby, *Dilthey and the Narrative of History* (Ithaca, N.Y.: Cornell University Press, 1994).

H. P. Rickman, *Wilhelm Dilthey: Pioneer of the Human Studies* (Berkeley: University of California Press, 1979), is a brief and somewhat thin introduction to Dilthey's thought. The author's strategy is to introduce Dilthey into contemporary discourse by focusing on where his thinking might be relevant to that discourse. See the chapters on Dilthey and the human sciences and on understanding. For more detail and breadth, see the author's *Dilthey Today: A Critical Appraisal of the Contemporary Relevance of His Work* (New York: Greenwood Press, 1988).

Ilse Nina Bulhof, *Wilhelm Dilthey: A Hermeneutic Approach to the Study of History and Culture* (The Hague: Martinus Nijhoff, 1988), is an excellent discussion of Dilthey's approach to the human sciences devoted largely but not exclusively to his hermeneutics.

Dilthey and Phenomenology, ed. Rudolf Makkreel and John Scanlon (Lanham, Md.: University Press of America, 1987), is a series of essays collected for the purpose of elevating the perception of Dilthey's significance in the history of philosophy. See especially Michael Ermarth's and Frithjof Rodi's chapters.

Michael Ermarth, *Wilhelm Dilthey: The Critique of Historical Reason* (Chicago: University of Chicago Press, 1978), arguably the best critical commentary on Dilthey in scope and analytical rigor, includes a fine biographical component, with a supplementary discussion on the context of German Idealism, from within which he was working. There is also excellent discussion of Dilthey's hermeneutics and his philosophy of the human sciences.

Hans Blumenberg

Hans Blumenberg, *The Legitimacy of the Modern Age*, trans. Robert M. Wallace (Cambridge, Mass.: MIT Press, 1983), is a novel approach to the long-standing debate between proponents of romanticism and champions of Enlightenment rationality, in the form of a critique of theories of secularization. Blumenberg attempts to salvage what he takes to be many "legitimate" concepts and attitudes of modernity by offering an original explanatory scheme that does not make reference to secularized eschatology. An informative and detailed translator's introduction is clear and accessible, situating Blumenberg's thought in its historical and intellectual context.

Hans Blumenberg, *The Genesis of the Copernican World*, trans. Robert M. Wallace (Cambridge, Mass.: MIT Press,1987), is, in part, a monumental historical account of the Copernican revolution, differing from other approaches in the extent to which it takes into account systematic extra-scientific factors. Most of the work is devoted to Copernicus's influence and to the impact of his ideas on human consciousness. The translator's introduction is clear and insightful.

Hans Blumenberg, *Work on Myth*, trans. Robert M. Wallace (Cambridge, Mass.: MIT Press, 1985), is a systematic approach to myth that attempts to travel a middle road between Enlightenment conceptions of historical progress from myth to logos. and romantic conceptions of myth, which see it as either superior to or isomorphic to rationality. Robert Wallace has again written a fine translator's introduction, this one illuminating the relations between this and Blumenberg's other works.

Hans Blumenberg, "An Anthropological Approach to the Contemporary Significance of Rhetoric," in *After Philosophy: End or Transformation?* ed. Kenneth Baynes et al. (Cambridge, Mass.: MIT Press, 1987), is an excellent, short introduction to Blumenberg's thought. The author once again charts a reasonable middle course between what he seems to take to be antiquated, overambitious Enlightenment thought and relativistic romanticism. In this particular case, he situates himself in the age-old debate between those who believe that philosophy has access to absolute Truth or the Good and those who believe that philosophy is just rhetoric and therefore has no such access. Blumenberg concedes that philosophy is rhetoric, but attempts to reconstruct our conception of rhetoric to avoid entirely relativistic consequences.

History of the Human Sciences 6 (November 1993) is a special issue devoted entirely to Blumenberg's work. See especially David Ingram's essay contrasting Blumenberg's approach to philosophical issues in the history of science with the approach of Kuhn. See also Charles Turner's interpretive essay on Blumenberg and Weber.

Robert B. Pippin, "Blumenberg and the Modernity Problem," *Review of Metaphysics* 40 (1987), 535–57, is an excellent critical study of Blumenberg's approach to modernity. Pippin doubts the extent to which the questions that Blumenberg believes modernity answers are central to modernity itself. Pippin also argues that Blumenberg, by his own logic, demonstrates the legitimacy of the premodern age, as well as the modern. For more critical comment, see D. Ingram, "Blumenberg and the Philosophical Grounds of Historiography," *History and Theory* 29, no. 1 (1990), 1–15.

Robert M. Wallace, "Introduction to Blumenberg," *New German Critique*, no. 32 (1984), 93–108, focuses mostly on Blumenberg's *Work on Myth*. Wallace is the English translator of that work; this essay overlaps to a fair degree with his translator's introduction.

Robert M. Wallace, "Progress, Secularization and Modernity: The Löwith-Blumenberg Debate," *New German Critique*, no. 22 (1981), 63–79, sets forth a contrast between Blumenberg's explanation for the rise of Whig philosophies of history and Karl Löwith's "theory of progress as secularized eschatology." Wallace offers a modest vindication of Blumenberg's view. This essay overlaps considerably with Wallace's introduction to Blumenberg's *Legitimacy of the Modern Age*. It is, however, a very clear and accessible piece.

Jürgen Habermas

A bibliography of Habermas's works, with translations and reviews, no longer up-to-date, is given in Thomas McCarthy, *The Critical Theory of Jürgen Habermas* (Cambridge, Mass.: MIT Press, 1981), which also provides an excellent summary of Habermas's ideas and writings. For our purposes, the following works by Habermas seem most pertinent: *The Structural Transformation of the Public Sphere* (Cambridge, Mass.: MIT Press, 1962); *On the Logic of the Social Sciences* (Cambridge, Mass.: MIT Press, 1967); *Knowledge and Human Interests* (Boston: Beacon Press, 1968); and *The Theory of Communicative Action*, 2 vols. (Boston: Beacon Press, 1984). Habermas has responded to criticisms of his work in *The Philosophical Discourse of Modernity*, trans. Frederick G. Lawrence (Cambridge, Mass.: MIT Press, 1987), as well as in other writings. In this connection, see also Seyla Benhabib, *Situating the Self: Gender, Community, and Postmodernism in Contemporary Ethics* (New York: Routledge, 1992).

On Habermas in the context of the Frankfurt School, i.e., for the critical, Marxist background, see Rolf Wiggershaus, *The Frankfurt School: Its History, Theories, and Political Significance*, trans. Michael Robertson (Cambridge, Mass.: MIT Press, 1994), 537–63 and passim. This is an exceptionally rich book on its overall subject. A useful group of essays on Habermas can be found in *Habermas: Critical Debates*, ed. John B. Thompson and David Held (Cambridge, Mass.: MIT Press, 1982). Since its publication, there has been a ceaseless flow of other works on the German thinker. Jon D. Wisman, "The Scope and Goals of Economic Science: A Habermasian Perspective," in *Economics and Hermeneutics*, ed. Don Lavoie (London: Routledge, 1990), offers a useful, brief summary of Habermas's positions.

Habermas's *Structural Transformation of the Public Sphere*, one of his more accessible writings, has had an unusually strong effect on historians, especially those studying civil society. Given our own interest, let me add a number of citations to those given in the notes on that topic. In general, see *Habermas and the Public Sphere*, ed. Craig Calhoun (Cambridge, Mass.: MIT Press, 1992), especially the essays by Geoff Eley and by Keith Michael Baker; Dena Goodman, "Public Sphere and Private Life: Toward a Synthesis of Current Historiographical Approaches to the Old Regime," *History and Theory* 31 (1992), 1–20; and Margaret C. Jacob, "The Mental Landscape of the Public Sphere: A European Perspective," *Eighteenth-Century Studies* 28 (Fall 1994), 95–113. On the gendering of public space, see Joan Landes, *Women and the Public Sphere in the Age of the French Revolution* (Ithaca, N.Y.: Cornell University Press, 1988); and Sarah Maza, *Private Lives and Public Affairs: The Causes Célèbres of Prerevolutionary France* (Berkeley: University of California Press, 1993). A book that might be overlooked in this connection, one that also goes beyond the immediate subject of the public sphere to the whole question of women's role in the science of the time, is Geoffrey V. Sutton, *Science for a Polite Society: Gender, Culture, and the Demonstration of Enlightenment* (Boulder, Colo.: Westview Press, 1995).

Norbert Elias

Norbert Elias, *Reflections on a Life* (Cambridge: Polity Press, 1944), consists of a biographical interview and an autobiographical essay, which serve as a useful context for Elias's own work. His book *The Symbol Theory*, ed. Richard Kilminster (Newbury Park, Calif.: Sage, 1991), was originally written for the journal *Theory, Culture, and Society* (Sage) in the later years of Elias's life, when he was almost blind. It is occasionally repetitious, but overall it is a wonderful and suggestive book ranging beyond conventional disciplinary lines. *The Society of Individuals*, trans. Edmund Jephcott (Oxford: Basil Blackwell, 1991), on which Elias was also working at the end of his life, is a series of essays examining the concepts of individual and society. He is most famous for *The Civilizing Process*, 2 vols. (New York: Pantheon Books, 1978 and 1982; orig. pub. in German, 1939). In volume 1, *The History of Manners* (1978), trans. Edmund Jephcott, Elias (who, incidentally, was an assistant to Karl Mannheim), mainly studied the aristocratic strata, rather than the bourgeoisie, as the principal agents in developing rationalization. An interesting piece that traces Elias's trajectory is Wolf Lepenies, "Norbert Elias: An Outsider Full of Unprejudiced Insight," *New German Critique*, no. 15 (Fall 1978), 57–64. Lepenies makes the point that Freud's notion that "external compulsions were increasingly internalized" is "in agreement with certain basic assumptions of Elias' theory" on the development of manners (61).

Michel Foucault

An outstanding intellectual biography, though anathema to some of Foucault's followers, is James Miller, *The Passion of Michel Foucault* (New York: Simon and

Schuster, 1993). Miller treats Foucault's life and ideas as inextricably intertwined, and does so with great knowledge of both. It should be read, however, in conjunction with Didier Eribon, *Michel Foucault*, trans. Betsy Wing (Cambridge: Harvard University Press, 1991). Eribon was a friend of Foucault's and ran in the same circle; thus he has much inside information concerning the context of Foucault's intellectual ideas, as well as his life.

Michael Clark, *Michel Foucault, an Annotated Bibliography: Tool Kit for a New Age* (New York: Garland Publishing, 1983), an extremely well-organized and exhaustive documentation of Foucault's writings and secondary sources about him, is an excellent starting point for research, though somewhat dated, for discussion of Foucault now constitutes an industry that has continued in full swing since the publication of the book. Joan Nordquist, *Michel Foucault (II): A Bibliography* (Santa Cruz, Calif.: Reference and Research Services, 1992), is not as useful as Clark's annotated bibliography but is valuable for its list of sources published between 1983 and 1992. Secondary sources are found below the primary texts to which they principally relate.

Critical Essays on Michel Foucault, ed. Peter Burke (Brookfield, Vt.: Ashgate Publishing Co., 1992), is a diverse set of critical essays on (1) Foucault's historical work on madness, (2) his formulations in the "archaeology of knowledge," (3) his *Discipline and Punish*, and (4) his work on the history of sexuality. Most of the essays are short—in fact, many are book reviews—but they are substantive. See Geertz's commentary in particular.

The Foucault Reader, ed. Paul Rabinow (New York: Pantheon Books, 1984), has a number of interesting selections, but see especially "What Is Enlightenment?" See also *Foucault: A Critical Reader*, ed. David Couzens Hoy (New York: Basil Blackwell, 1986).

Gary Gutting, *Michel Foucault's Archaeology of Scientific Reason* (New York: Cambridge University Press, 1989), is an excellent discussion of Foucault's earlier work, with attention paid to his intellectual precursors Bachelard and Canguilhem. It includes detailed analyses and evaluations of Foucault's work up to 1969. This book is arguably the most clearly and succinctly written discussion of Foucault's work in this period.

Lois McNay, *Foucault: A Critical Introduction* (Cambridge, England: Polity Press, 1994), is a detailed discussion of Foucault's later work, focusing on his approaches to the self and to government, the latter an aspect of his work that has received relatively little critical attention. This book complements Gutting's treatment of the early Foucault very well.

A short, popular introduction is David R. Shumway's *Michel Foucault* (Boston: Twayne, 1989). Note should also be made of *Michel Foucault: Beyond Structuralism and Hermeneutics*, ed. Hubert Dreyfus and Paul Rabinow (Chicago: University of Chicago Press, 1983).

Alan Sheridan, *Michel Foucault: The Will to Truth* (New York: Tavistock Publications, 1980), is an insightful, descriptive account of Foucault's work that succeeds in making clear much of what seems obscure about this illusive thinker.

Among journal issues devoted to the French thinker, see *History of European Ideas* 14, no. 3 (1992), a special issue edited by Desmond Bell; and *History of the Human Sciences* 3 (February 1990).

Thomas Kuhn

Thomas Kuhn, *The Structure of Scientific Revolutions*, 2d ed. (Chicago: University of Chicago Press, 1970), the classic formulation of Kuhn's "historical philosophy of science," is an attempt to explain large-scale conceptual change in science. The second edition includes a detailed response to critics in the form of a postscript.

Criticism and the Growth of Knowledge, ed. Alan Musgrave and Imre Lakatos (Cambridge: Cambridge University Press, 1970), is an indispensable volume devoted to criticisms of Kuhn's work, with a brief response to each. Two essays are especially notable: Margaret Masterman claims to have discerned twenty-one uses of the paradigm concept in *The Structure of Scientific Revolutions;* and Imre Lakatos defends his modified, quasi-sociological, Popperian approach as an alternative to Kuhn's.

Thomas Kuhn, *The Essential Tension* (Chicago: University of Chicago Press, 1977), is a diverse series of historical and philosophical essays written both before and after *The Structure of Scientific Revolutions*. The later pieces are more important insofar as they exhibit refinement in response to criticism.

Larry Laudan, *Progress and Its Problems* (Berkeley: University of California Press, 1977), is, in part, a response to what some have characterized as Kuhn's "sociological" approach. Laudan attempts to salvage the notion of progressivity and rationality in science but without reference to increasing verisimilitude being required.

W. H. Newton-Smith, *The Rationality of Science* (New York: Routledge, 1994), includes a chapter on Kuhn's thought that is perhaps the most clearly organized summary available. It makes clearer than Kuhn himself the definition and scope of the paradigm concept.

See also *Paradigms and Revolutions: Appraisals and Applications of Thomas Kuhn's Philosophy of Science*, ed. Gary Gutting (Notre Dame, Ind.: University of Notre Dame Press, 1980).

Index

Abductive method, 8
Accumulation: of interpretations, 127; problem of, 22–23, 24, 246*n*23
Achievements in the human sciences, 130–34, 175, 215, 234
Acton, Lord, 267*n*88
Adorno, Theodor W., 62, 248*n*1
Age of Discovery, 6, 27–31, 136, 148
Aims of the human sciences, 15–16; interpretation, 23–25; power, 20–22; prediction, 17–18, 20; prescription, 18–20, 187–89; and problem of accumulation, 22–23, 24, 246*n*23; understanding, 16–17
Air pump dispute, 45–48
Alienation, 102, 145, 293*n*33
Alpers, Svetlana, 250*n*34, 260*n*37
Anatomy, 131–32, 151
Anaximander, 114
Anderson, Perry, 177–79
Anthropology, 7, 11, 13, 244*n*2; comparative, 76; Darwin's influence on, 35; development of, as a science, 147–58; and European exploration, 104; evolutionary theory in, 35, 150–51, 156–57; hermeneutical, 155; holism in, 87, 105, 153; and imperialism, 76, 153, 255*n*18; instruments in, 79; interpretation in, 96; origin, 31, 132; and the Other, 28, 29–30, 104–5, 153, 154; philosophical, 10, 93, 243*n*1; unintended consequences in, 179–80

Anthropomorphism, 42, 98, 132, 148
Arato, Andrew, 203, 205, 286*n*56
Archaeology, 131, 133, 199
Aristotle, 41, 42, 135, 158, 203
Arnold, Matthew, 230
Art, 94, 216
Artificial intelligence, 185, 254*n*14
Asimov, Isaac, 17, 18
Astrology, 99
Astronomy, 77, 78, 99, 142, 179
Auden, W. H., 98
Australopithecus afarensis, 68, 81
Authorship, 95, 258*n*26
Ayer, A. J., 131–32

Babbage, Charles, 233
Bachelard, Gaston, 58, 196
Bacon, Francis, 4, 133; and idols, 40, 248*n*9; and invention, 228; and modernism, 75; and positivism, 37–44, 84, 233, 247*n*1, 248*n*2
Baker, Keith, 183
Bakhtin, Mikhail, 157
Baldwin, James, 293*n*33
Barbarossa, Frederick, 199–200
Barth, John, 1
Bauer, Bruno, 123
Bayle, Pierre, 101, 123, 166, 278*n*70
Beethoven, Ludwig van, 267*n*85
Behaviorism, 15, 60

Printed in the United States
by Baker & Taylor Publisher Services